Dynamical Systems and Their Applications (DSTA) — In Memory of Prof. Dr. José A. Tenreiro Machado

Dynamical Systems and Their Applications (DSTA) — In Memory of Prof. Dr. José A. Tenreiro Machado

Alexandra M.S.F. Galhano
António Lopes
Carlo Cattani

MDPI

Basel • Beijing • Wuhan • Barcelona • Belgrade • Novi Sad • Cluj • Manchester

Editors

Alexandra M.S.F. Galhano
Faculdade de Ciências
Naturais, Engenharias
e Tecnologias
Universidade Lusófona
Porto
Portugal

António Lopes
Mechanical Engineering
University of Porto
Porto
Portugal

Carlo Cattani
Engineering School (DEIM)
University of Tuscia
Viterbo
Italy

Editorial Office
MDPI AG
Grosspeteranlage 5
4052 Basel, Switzerland

This is a reprint of articles from the Special Issue published online in the open access journal *Mathematics* (ISSN 2227-7390) (available at: www.mdpi.com/journal/mathematics/special_issues/Dynamical_Systems_Applications).

For citation purposes, cite each article independently as indicated on the article page online and as indicated below:

Lastname, A.A.; Lastname, B.B. Article Title. *Journal Name* **Year**, *Volume Number*, Page Range.

ISBN 978-3-7258-2434-2 (Hbk)
ISBN 978-3-7258-2433-5 (PDF)
doi.org/10.3390/books978-3-7258-2433-5

Contents

About the Editors vii

Preface ix

Duarte Valério, Manuel D. Ortigueira and António M. Lopes
How Many Fractional Derivatives Are There?
Reprinted from: *Mathematics* **2022**, *10*, 737, doi:10.3390/math10050737 1

Maryam Al-Kandari, Latif A-M. Hanna and Yuri Luchko
Operational Calculus for the General Fractional Derivatives of Arbitrary Order
Reprinted from: *Mathematics* **2022**, *10*, 1590, doi:10.3390/math10091590 19

Pshtiwan Othman Mohammed, Hari Mohan Srivastava, Dumitru Baleanu, Rashid Jan and Khadijah M. Abualnaja
Monotonicity Results for Nabla Riemann–Liouville Fractional Differences
Reprinted from: *Mathematics* **2022**, *10*, 2433, doi:10.3390/math10142433 36

Barakah Almarri, Ali Hasan Ali, António M. Lopes and Omar Bazighifan
Nonlinear Differential Equations with Distributed Delay: Some New Oscillatory Solutions
Reprinted from: *Mathematics* **2022**, *10*, 995, doi:10.3390/math10060995 50

Rui Liu, Michal Fečkan, Donal O'Regan and Jinrong Wang
Controllability Results for First Order Linear Fuzzy Differential Systems
Reprinted from: *Mathematics* **2022**, *10*, 1193, doi:10.3390/math10071193 60

Touria Karite, Adil Khazari and Delfim F. M. Torres
Regional Controllability and Minimum Energy Control of Delayed Caputo Fractional-Order Linear Systems
Reprinted from: *Mathematics* **2022**, *10*, 4813, doi:10.3390/math10244813 71

Xuan Wang, Didi Hu and Chenghua Gao
Asymptotic Regularity and Existence of Time-Dependent Attractors for Second-Order Undamped Evolution Equations with Memory
Reprinted from: *Mathematics* **2022**, *10*, 2198, doi:10.3390/math10132198 87

Nicolae Herisanu and Vasile Marinca
A Solution Procedure Combining Analytical and Numerical Approaches to Investigate a Two-Degree-of-Freedom Vibro-Impact Oscillator
Reprinted from: *Mathematics* **2021**, *9*, 1374, doi:10.3390/math9121374 108

Sorin Dumitru, Andra Constantin, Cristian Copilusi and Nicolae Dumitru
Impact Dynamics Analysis of Mobile Mechanical Systems
Reprinted from: *Mathematics* **2021**, *9*, 1776, doi:10.3390/math9151776 125

Ping-Kun Tsai, Cheng-Han Li, Chia-Chun Lai, Ko-Jung Huang and Ching-Wei Cheng
Approximation Solution for the Zener Impact Theory
Reprinted from: *Mathematics* **2021**, *9*, 2222, doi:10.3390/math9182222 151

Zai-Yin He, Abderrahmane Abbes, Hadi Jahanshahi, Naif D. Alotaibi and Ye Wang
Fractional-Order Discrete-Time SIR Epidemic Model with Vaccination: Chaos and Complexity
Reprinted from: *Mathematics* **2022**, *10*, 165, doi:10.3390/math10020165 163

Jocelyn Sabatier
Probabilistic Interpretations of Fractional Operators and Fractional Behaviours: Extensions, Applications and Tribute to Prof. José Tenreiro Machado's Ideas
Reprinted from: *Mathematics* **2022**, *10*, 4184, doi:10.3390/math10224184 . . . . **181**

About the Editors

Alexandra M.S.F. Galhano

Alexandra Galhano is Professor at the Faculty of Natural Sciences, Engineering, and Technology of Lusofona University in Porto, Portugal, since 2021. She was previously professor at the Faculty of Engineering of the University of Porto (1992–1998) and at the Porto Polytechnic School (1998–2021). Her main scientific research interests are systems modeling, dynamics, multidimensional scaling, and fractional calculus. She has been married to José A. Tenreiro Machado since 1982.

António Lopes

António M. Lopes has a degree in mechanical engineering (FEUP), a Ph.D. in mechanical engineering (FEUP), a Master's degree in electrical engineering (FEUP), and a degree in electrical engineering (FCT-UC). He is an integrated researcher at the Associated Laboratory for Energy, Transport, and Aeronautics (LAETA), Institute of Science and Innovation in Mechanical Engineering and Industrial Engineering (INEGI). He is the coordinator of the research group Intelligent Systems & Control (ISC)/LAETA-INEGI.

His areas of research are automation, robotics, complex systems modeling, and fractional order systems. He is the author of more than 230 papers in *SCI* journals and of 4 books.

He is a member of the Editorial Board of the *Journal of Applied Nonlinear Dynamics* (L&H Sci. Pub.), *International Journal of Dynamics and Control* (Springer), *International Journal of Advanced Robotic Systems* (SAGE), *Advances in Mechanical Engineering* (SAGE), *Mathematical Problems in Engineering* (Hindawi), *Entropy* (MDPI), *Mathematics* (MDPI), *Open Engineering* (DeGruyter), *PlosOne* (Plos), *Discover Mechanical Engineering* (Springer), *Mechanics Based Design of Structures and Machines* (Taylor & Francis), *Foundations* (MDPI), *Journal on Teaching Egineering* (U.Porto), and *International Journal of Mechanical Engineering Education* (Sage). He is also an editor-in-chief of the *Journal of Environmental Accounting and Management* (L&H Sci. Pub.) (SCOPUS and JCR), *Journal of Machine Design and Automation Intelligence* (DeGruyter), and *Fractal and Fractional-Section Complexity* (SCOPUS and JCR). He is the co-editor of the *Handbook of Fractional Calculus with Applications* (DeGruyter). He was identified as one of the 2% most influential researchers in the world in "Industrial Engineering and Automation" (Stanford University study, 2020, 2021, 2022, and 2023).

Carlo Cattani

Carlo Cattani has been a full professor of mathematical physics and applied mathematics at the Department of Economics, Engineering, Society, and Enterprise (DEIM) of Tuscia University (VT), Italy, since 2015. He has been previously appointed as a professor/research fellow at the Dept. of Mathematics, University of Rome "La Sapienza" (1980–2004) and Dept. of Mathematics, University of Salerno (2004–2015). He was a research fellow of the Italian Council of Research, CNR (1978–1980) and a visiting research fellow at the Physics Institute of the Stockholm University (1987–1988) and UPV-Valencia (Spain) in 2000. His main scientific research interests include numerical and computational methods, mathematical models and methods, fractals, fractional calculus, data analysis, computer methods and simulations, artificial intelligence, and computational linguistics. He is the author of more than 350 scientific papers in international journals; he has made significant contributions to fundamental topics such as numerical methods, dynamical systems, fractional calculus, wavelets, nonlinear waves, data analysis, neural networks, and artificial intelligence. He is the founder and editor-in-chief of *Fractal and Fractional*; he is the author and editor of several books on wavelets, fractals, nonlinear waves, computational methods, and AI.

He has been awarded as an honorary distinguished professor at the Harran University (2022), an honorary professor at the Azerbaijan University (2019), and at the BSP University, Ufa-Russia (2009) for "his contribution in research and international cooperation". During 2018–2021, he was appointed as an adjunct professor at the Ton Duc Thang University (HCMC Vietnam). Since 2015, he has been a corresponding member of the Italian "Accademia Peloritana dei Pericolanti."

He has been listed as the World's Top 2% Scientist (Stanford) since 2019 in the fields of applied mathematics, numerical and computational mathematics, and artificial intelligence.

Preface

The theory of dynamical systems has evolved from linear to nonlinear and then to complex systems. New modeling and control techniques have been developed and applied in various fields, such as physics, mechanics, electronics, economy, finance, geophysics, and biology, to mention a few. Nonlinear and complex dynamical systems, as well as their related concepts of chaos, bifurcations, criticality, symmetry, memory, scale invariance, fractality, fractionality, and other rich features, have attracted researchers from many areas of science and technology that are involved in system modeling and control, with applications in real-world problems. However, at present, there are still many unsolved problems, and new theoretical developments and applications are needed in order to describe and control more accurately dynamical systems with linear, nonlinear, and complex behavior.

This volume focuses on the modeling and control of dynamic systems. It gathers linear and nonlinear dynamics, systems modeling, advanced control theory, complex systems, fractional calculus, fractals, entropy, information theory, chaos, self-organization, and criticality contributions. This volume presents the manuscripts of the Special Issue by *Mathematics* as a tribute to Professor J. A. Tenreiro Machado, who unexpectedly passed away on October 6, 2021. Studies regarding fractional calculus represent a significant part of this volume, with the modeling and control of dynamic systems being the most represented. Moreover, there is one article that studies the dynamics of a fractional-order discrete-time susceptible–infected–recovered epidemic model and another one that presents an analysis of one of Professor J. A. Tenreiro Machado's works serving as a basis for a deeper reflection on the origins of fractional behaviors and their modeling.

Alexandra M.S.F. Galhano, António Lopes, and Carlo Cattani
Editors

Σ mathematics

MDPI

Article

How Many Fractional Derivatives Are There?

Duarte Valério [1], Manuel D. Ortigueira [2] and António M. Lopes [3,*]

[1] IDMEC, Instituto Superior Técnico, Universidade de Lisboa, 1049-001 Lisboa, Portugal; duarte.valerio@tecnico.ulisboa.pt

[2] Centre of Technology and Systems–UNINOVA and Department of Electrical Engineering, NOVA School of Science and Technology of NOVA University of Lisbon, Quinta da Torre, 2829-516 Caparica, Portugal; mdo@fct.unl.pt

[3] LAETA/INEGI, Faculty of Engineering, University of Porto, 4200-465 Porto, Portugal

* Correspondence: aml@fe.up.pt

Abstract: In this paper, we introduce a *unified fractional derivative*, defined by two parameters (order and asymmetry). From this, all the interesting derivatives can be obtained. We study the one-sided derivatives and show that most known derivatives are particular cases. We consider also some myths of Fractional Calculus and false fractional derivatives. The results are expected to contribute to limit the appearance of derivatives that differ from existing ones just because they are defined on distinct domains, and to prevent the ambiguous use of the concept of fractional derivative.

Keywords: fractional calculus; fractional derivative; signals and systems

MSC: 26A33

check for updates

Citation: Valério , D.; Ortigueira, M.D.; Lopes, A.M. How Many Fractional Derivatives Are There? *Mathematics* **2022**, *10*, 737. https://doi.org/10.3390/math10050737

Academic Editor: Viktor N. Orlov

Received: 24 January 2022
Accepted: 23 February 2022
Published: 25 February 2022

1. Introduction

Four centuries after the first reference to the possibility of non integer order derivatives, the presently termed Fractional Calculus (FC) has reached a crossroads where multiple definitions are mixed, causing a huge confusion that makes life very difficult for those who only intend to make applications in Science and Engineering. In fact, the first reference was found in a letter from Leibniz to J. Bernoulli [1]. Although Euler (1730), Fourier (1822), and Abel (1823) touched on the problem, the true father of FC was Liouville (1832) [2,3], in spite of their many difficulties to impose their vision, due to a main obstacle: At that time, the inverse Laplace integral was unknown. Therefore, Liouville could not find a simple way of expressing a function in terms of exponentials that were the basis for his findings. Anyway, the main definitions we find today are based on the formulæ presented by Liouville, mainly the Riemann–Liouville [4], (Dzherbashian)-Caputo [5,6], and Grünwald-Letnikov [4] definitions. However, and based on these derivatives, new ones have been proposed alongside these, such as Hadamard's [6] or Marchaud's [4]. Consequently, the number of currently existing fractional derivatives (FDs) is so high, which became the biggest obstacle to the diffusion of FC in Science and Engineering.

If we also consider the pseudo-derivatives and the disguised integer order derivatives, we conclude that the situation is really confused and confusing. Trying to introduce some order in the field, Oliveira and Machado, first, and Teodoro et al., more recently [7,8], listed such derivatives and introduced a classification according to some specified criteria. However, these papers included some operators that can hardly be classified as FDs. On the other hand, in recent years a great discussion took place in forums, conferences and articles, where there is a great confusion between the concepts of system and derivative. In a sequence of papers, Ortigueira and Machado tried to clarify the situation by proposing a coherent definition of FD [9], a description of FD suitable for applications in Science and Engineering [10], and introducing the FD in the context of fractional linear systems [11].

In this text, we make another step to clarify the situation, by introducing a step-down procedure. We recover the *unified fractional derivative* (UFD) obtained in [11] after a four step unification, and proceed as if this UFD were a *mother derivative*, defined by two parameters (order and dissymmetry), from which all the derivatives listed in [7,8] emerge as particular cases [12]. We will work in the context of Laplace and Fourier transforms. This includes most of the interesting functions and distributions used in practical applications.

We could go further by considering the tempered FDs [13,14], but we will not do so, in order to keep ourselves in the context of the derivatives introduced in [7,8]. On the other hand, we directed our attention to what we can call *shift invariant derivatives*, without considering the scale invariant ones such as the Hadamard [6] and the quantum [15] derivatives, as well as the discrete-time derivatives [16–18]. We will also not consider variable order derivatives [17]. Some operators, sometimes called FDs, are analyzed and put in the correct framework.

The paper is outlined as follows. In Section 2, we define *fractional derivative*. The UFD and its main properties is introduced in Section 3. The derivatives to referred in [7,8] are then obtained sequentially as particular cases through suitable choice of the parameters and working domain (Section 4). The other operators that, according to our framework, cannot be considered as FD will be treated in Section 5. Section 6 concludes the paper with a reflection on which definitions ought to be chosen.

Remark 1. *We adopt here the following assumptions:*

- *We work on* $\mathbb{R}$.
- *We use the two-sided Laplace transform (LT):*

$$F(s) = \mathcal{L}[f(t)] = \int_{\mathbb{R}} f(t)e^{-st}\,\mathrm{d}t, \tag{1}$$

  where $f(t)$ is any function defined on $\mathbb{R}$ and $F(s)$ is its transform, provided that it has a non empty region of convergence (ROC).
- *The Fourier transform (FT), $\mathcal{F}[f(t)]$, is obtained from the LT through the substitution $s = i\kappa$, with $\kappa \in \mathbb{R}$.*

2. What Is a Fractional Derivative?

In Signal Processing, independently of the applications to Electrical, Mechanical, Biomedical or any other Engineering field, there is a very simple way of defining a FD: *a FD is a linear operator described by Bode diagrams that are straight lines* [11]. In terms of the FT we can write

$$\mathcal{F}\left[D_\theta^\gamma f(t)\right] = |\omega|^\gamma e^{i\theta\frac{\pi}{2}\mathrm{sgn}(\omega)}\mathcal{F}[f(t)] \qquad t,\omega \in \mathbb{R}, \tag{2}$$

where $\gamma, \theta \in \mathbb{R}$, and D_θ^γ represents the derivative. The operator

$$\Psi_\theta^\gamma(\omega) = |\omega|^\gamma e^{i\theta\frac{\pi}{2}\mathrm{sgn}(\omega)} \qquad \omega \in \mathbb{R} \tag{3}$$

is the *frequency response* of the derivative. Functions $A(\omega)$ (amplitude) and $\phi(\omega)$ (phase), when represented in a logscale for $\omega > 0$, are expressed by straight lines. These can be used to define a FD.

However, we need a criterion independent of any transform. As in [9], we define as FD an operator that verifies the following (wide sense) criterion.

Definition 1. *An operator is considered a FD in wide sense if it enjoys properties* **P** *defined as:*

P1 *Linearity*
The operator is linear.

P2 *Identity*
The zero order derivative of a function returns the function itself.

P3 *Backward compatibility*
When the order is integer, FD gives the same result as the ordinary derivative.

P4 *The index law*

$$D^{\alpha}D^{\beta}f(t) = D^{\alpha+\beta}f(t) \tag{4}$$

holds for $\alpha < 0$ and $\beta < 0$.

P5 *The generalized Leibniz rule*

$$D^{\alpha}[f(t)g(t)] = \sum_{i=0}^{\infty}\binom{\alpha}{i}D^{i}f(t)D^{\alpha-i}g(t) \tag{5}$$

holds. As is clear, when $\alpha = N \in \mathbb{Z}^{+}$, we obtain the classical Leibniz rule.

The index law property can be modified to include positive orders. This leads to the *strict sense criterion*. This criterion has the same five conditions, but P4 is modified to:

P'4 The generalized index law

$$D^{\alpha}D^{\beta}f(t) = D^{\alpha+\beta}f(t) \tag{6}$$

holds for any α and β.

This is very important because it allows the existence of the inverse derivative: the *anti-derivative*. It is convenient to state the differences between anti-derivative and "primitive":

- The anti-derivative is unique.
- The anti-derivative is a left and right inverse, while any primitive is only right inverse.

These criteria allow us to clarify the situation of some "disguised" order one derivatives and put out some frauds.

3. Unified Fractional Derivative

In [11], a UFD incorporating most of the useful derivatives was presented and its properties studied. It is described as follows.

Definition 2. *Let $f(t)$ be a function defined on $\mathbb{R}$ ($\mathbb{C}$), and let $\alpha > -1$ if $\theta \neq \pm\alpha$, or $\alpha \in \mathbb{R}$ if $\theta = \pm\alpha$. We define a UFD of GL type by*

$$D_{\theta}^{\alpha}f(t) := \lim_{h\to 0^{+}} h^{-\alpha}\sum_{n=-\infty}^{+\infty}(-1)^{n}\frac{\Gamma(\alpha+1)}{\Gamma\left(\frac{\alpha+\theta}{2}-n+1\right)\Gamma\left(\frac{\alpha-\theta}{2}+n+1\right)}f(t-nh), \tag{7}$$

where α is the derivative order and θ the asymmetry parameter. We define also a general integral formulation for the unified anti-derivative through

$$D_{\theta}^{-\alpha}f(t) = \frac{1}{\sin(\alpha\pi)\Gamma(\alpha)}\int_{\mathbb{R}} f(t-\tau)\sin\left[(\alpha+\theta\cdot\mathrm{sgn}(\tau))\frac{\pi}{2}\right]|\tau|^{\alpha-1}\,d\tau, \tag{8}$$

where sgn(.) *denotes the signum function.*

The integral in (8) can be regularized in order to become valid for positive orders [12]. This is done with the substitution

$$f(t-\tau) \quad\to\quad f(t-\tau) - \sum_{0}^{N-1}\frac{(-1)^{m}f^{(m)}(t)}{m!}\tau^{m}, \tag{9}$$

with $N = \lfloor\gamma\rfloor + 1$.

Some known properties of this derivative can be drawn [12,15]. The main ones are:

1. Fourier transformation
It was introduced above in (2). It permits obtaining (8) from (7), using the convolution theorem. It has another consequence:

$$D^{\alpha}_{\theta}f(t) = \cos\left(\theta\frac{\pi}{2}\right)D^{\alpha}_{0}f(t) + \sin\left(\theta\frac{\pi}{2}\right)D^{\alpha}_{1}f(t). \tag{10}$$

2. Eigenfunctions
Let $f(x) = e^{i\kappa x}$, $\kappa, x \in \mathbb{R}$. Then,

$$D^{\beta}_{\theta}e^{i\kappa x} = |\kappa|^{\beta}e^{i\frac{\pi}{2}\theta\cdot\mathrm{sgn}(\kappa)}e^{i\kappa x}, \tag{11}$$

meaning that the sisoids are the eigenfunctions of the UFD with eigenvalue $\Psi^{\beta}_{\theta}(\kappa) = |\kappa|^{\beta}e^{i\frac{\pi}{2}\theta\cdot\mathrm{sgn}(\kappa)}$.
3. Periodicity in θ
The UFD is periodic in θ with period 4

$$D^{\beta}_{\theta}f(x) = (-1)^{n}D^{\beta}_{\theta+2n}f(x), \ \ n \in \mathbb{Z}, \tag{12}$$

as we observe from (2).
4. Additivity and commutativity of the orders

$$D^{\beta_1}_{\theta_1}D^{\beta_2}_{\theta_2}f(x) = D^{\beta_1+\beta_2}_{\theta_1+\theta_2}f(x). \tag{13}$$

In particular, conjugating (13) with (12),

$$D^{\beta_1}_{0}D^{\beta_2}_{0}f(x) = D^{\beta_1+\beta_2}_{0}f(x), \tag{14}$$

$$D^{\beta_1}_{1}D^{\beta_2}_{1}f(x) = -D^{\beta_1+\beta_2}_{0}f(x), \tag{15}$$

$$D^{\beta_1}_{0}D^{\beta_2}_{1}f(x) = D^{\beta_1+\beta_2}_{1}f(x), \tag{16}$$

$$D^{\beta_1}_{\beta_1}D^{\beta_2}_{\beta_2}f(x) = D^{\beta_1+\beta_2}_{\beta_1+\beta_2}f(x), \tag{17}$$

$$D^{\frac{\beta}{2}}_{\theta}D^{\frac{\beta}{2}}_{-\theta}f(x) = D^{\beta}_{0}f(x), \tag{18}$$

$$D^{\frac{\beta}{2}}_{\theta}D^{\frac{\beta}{2}}_{1-\theta}f(x) = D^{\beta}_{1}f(x). \tag{19}$$

5. Existence of inverse derivative
From (13), the anti-derivative exists when $\beta_2 = -\beta_1$ and $\theta_1 = -\theta_2$. Therefore,

$$D^{\beta}_{\theta}D^{-\beta}_{-\theta}f(x) = D^{-\beta}_{-\theta}D^{\beta}_{\theta}f(x) = f(x). \tag{20}$$

6. Identity operator
According to (13) and (20), the identity operator is defined by

$$D^{0}_{0}f(x) = f(x). \tag{21}$$

Suitable choices of these parameters allow us to recover the causal, anti-causal, and bilateral (acausal) derivatives. The particular, most interesting, cases are obtained from (7) and (8). In terms of the frequency response, we have

- forward derivative, $\theta = \alpha$

$$\Psi^{\alpha}_{\alpha}(\omega) = |\omega|^{\alpha}e^{i\alpha\frac{\pi}{2}\mathrm{sgn}(\omega)} = (i\omega)^{\alpha} \tag{22}$$

- backward derivative, $\theta = -\alpha$

$$\Psi^{\alpha}_{-\alpha}(\omega) = |\omega|^{\alpha}e^{-i\alpha\frac{\pi}{2}\mathrm{sgn}(\omega)} = (-i\omega)^{\alpha} \tag{23}$$

- Riesz derivative and potential, $\theta = 0$

$$\Psi_0^\alpha(\omega) = |\omega|^\alpha \tag{24}$$

- Feller derivative and potential, $\theta = 1$

$$\Psi_1^\alpha(\omega) = i|\omega|^\alpha \text{sgn}(\omega) \tag{25}$$

- Hilbert transform, $\alpha = 0, \theta = 1$

$$\Psi_1^0(\omega) = e^{i\frac{\pi}{2}\text{sgn}(\omega)} = i\text{sgn}(\omega). \tag{26}$$

From these expressions and using (7) it is possible to devise numerous derivatives by:

1. choosing particular values of the parameters α and θ,
2. restricting the domain of the function at hand.

4. One-Sided Derivatives

4.1. The GL Derivatives

The great importance of one-sided derivatives in applications leads us to study them in detail. Taking (7) with $\gamma = \alpha$ and $\theta = \pm\alpha$ we obtain the forward (left) and backward (right) GL derivatives that with some manipulation can be written as

$$D_\pm^\alpha f(t) := \lim_{h\to 0^+} h^{-\alpha} \sum_{n=0}^{+\infty} \frac{(-\alpha)_n}{n!} f(t \mp nh). \tag{27}$$

As it is easy to verify, the forward derivative $+$ is causal, while the backward $-$ is anti-causal. This kind of derivatives was proposed first by Liouville [2].
For functions with LT (or FT), we can write

$$\mathcal{L}\left[D_\theta^\alpha f(t)\right] = (\pm s)^\alpha F(s), \qquad \pm\Re(s) \geq 0, \tag{28}$$

where $F(s) = \mathcal{L}[f(t)]$. As is known from the study of the LT, the LT of a right (left) function, $f(t) = 0,\ t < a(t > a),\ a \in \mathbb{R}$ has a ROC defined by $\Re(s) > 0$ ($\Re(s) < 0$). If $f(t)$ is absolutely of square integrable, then it has FT and we can write

$$\Psi_\pm^\alpha(s) = (\pm s)^\alpha, \qquad \pm\Re(s) \geq 0. \tag{29}$$

Function $\Psi_\pm^\alpha(s) = (\pm s)^\alpha$ with suitable ROC is the transfer function (TF) of the derivative (also called differintegrator).

4.2. The Impulse Response

The relations (28) and (29) suggest the existence of two operators that, convolved with suitable functions, give their FDs. For $\alpha > 0$ the LT inverse of s^α does not exist as regular function. However, it has a generalized inverse represented by the pseudo-function

$$\psi_\pm^\alpha(t) = \pm\frac{t^{-\alpha-1}}{\Gamma(-\alpha)} u(\pm t), \tag{30}$$

where $+$ $(-)$ corresponds to the causal (anti-causal) case. The pseudo-function (30) is called impulse response (IR) of the differintegrator. The use of the convolution resulting from (28) would allow us to obtain the derivative $D_\pm^\alpha f(t)$. For $\alpha < 0$, there is no particular difficulty and we obtain from (8)

$$D_\pm^{-\alpha} f(t) = \frac{1}{\Gamma(\alpha)} \int_0^\infty f(t \mp \tau)\tau^{\alpha-1}\, d\tau. \tag{31}$$

The expression corresponding to the anti-causal case was proposed almost with the above form by Liouville [3]. The causal case was deduced from the anti-causal by Serret [1,19]. Liouville's formula included a factor $(\pm 1)^{-\alpha}$ to ensure that

$$\mathcal{L}D_{\pm}^{-\alpha}f(t) = s^{-\alpha}F(s) \tag{32}$$

for both expressions, although the ROC will be defined by $\pm R_e(s) > 0$. In most texts, such factor is removed. In time problems it must be kept, but, in space applications, it plays no relevant role. Therefore, we will omit it in the following. The finite domain versions of (31) are called RL integrals:

$$D_{a+}^{-\alpha}f(t) = \frac{1}{\Gamma(\alpha)}\int_{a+}^{t} f(t-\tau)\tau^{\alpha-1}\,\mathrm{d}\tau \tag{33}$$

and

$$D_{b-}^{-\alpha}f(t) = \frac{1}{\Gamma(\alpha)}\int_{t}^{b-} f(t-\tau)\tau^{\alpha-1}\,\mathrm{d}\tau. \tag{34}$$

We will assume these forms in the following.

4.3. Liouville's Derivatives

Liouville [2] noted that (31) becomes singular when the order is positive (i.e., when it corresponds to a derivative). This problem can be solved with the regularization [15,17]

$$D_{f}^{\alpha}f(t) = \frac{1}{\Gamma(-\alpha)}\int\limits_{0}^{\infty}\tau^{-\alpha-1}\left[f(t-\tau) - \sum_{0}^{N-1}\frac{(-)^m f^{(m)}(t)}{m!}\tau^m\right]\mathrm{d}\tau, \tag{35}$$

that we will call regularized Liouville (L) derivative. The first regularization of the Liouville integral (31) was done by Marchaud [4]. However, their regularization only verifies the derivative property of the LT if the order is less than 1.

Instead of a regularization, Liouville devised a trick for solving the singularity problem that we can describe as

$$s^{\alpha} = s^{N}s^{\alpha-N} = s^{\alpha-N}s^{N}, \tag{36}$$

where $N > 0$ verifies $N > \alpha$. The most usual choice is $N = \lceil\alpha\rceil$. Basically, it consists of transferring the singular behavior to an integer order derivative. The first approach leads to the so-called Liouville derivatives [4],

$$D_{+}^{\alpha}f(t) = \frac{1}{\Gamma(N-\alpha)}\frac{\mathrm{d}^N}{\mathrm{d}t^N}\int_{0}^{\infty} f(t-\tau)\tau^{N-\alpha-1}\,\mathrm{d}\tau \tag{37}$$

and

$$D_{-}^{\alpha}f(t) = \frac{(-1)^N}{\Gamma(N-\alpha)}\frac{\mathrm{d}^N}{\mathrm{d}t^N}\int_{0}^{\infty} f(t+\tau)\tau^{N-\alpha-1}\,\mathrm{d}\tau, \tag{38}$$

that can be rewritten as

$$D_{+}^{\alpha}f(t) = \frac{1}{\Gamma(N-\alpha)}\frac{\mathrm{d}^N}{\mathrm{d}t^N}\int_{-\infty}^{t} f(\tau)(t-\tau)^{N-\alpha-1}\,\mathrm{d}\tau \tag{39}$$

and

$$D_{-}^{\alpha}f(t) = \frac{(-1)^N}{\Gamma(N-\alpha)}\frac{\mathrm{d}^N}{\mathrm{d}t^N}\int_{t}^{\infty} f(\tau)(t-\tau)^{N-\alpha-1}\,\mathrm{d}\tau. \tag{40}$$

This one is sometimes called "Weyl derivative" [8].

Liouville's second procedure leads to what can be called Liouville–Caputo (LC) derivatives [20]:

$$D_{+}^{\alpha}f(t) = \frac{1}{\Gamma(N-\alpha)}\int_{-\infty}^{t} f^{(N)}(\tau)(t-\tau)^{N-\alpha-1}\,\mathrm{d}\tau \tag{41}$$

and

$$D^{\alpha}_{-}f(t) = \frac{(-1)^N}{\Gamma(N-\alpha)} \int_t^{\infty} f^{(N)}(\tau)(t-\tau)^{N-\alpha-1}\,\mathrm{d}\tau. \tag{42}$$

Remark 2. *We must remark that:*

1. *These derivative definitions are equivalent for functions with LT or FT.*
2. *For some particular classes of functions this may not be correct, e.g., the L derivatives are better than the LC. Possible causes for this are:*
    - *The convolution makes the functions smoother;*
    - *The derivative may introduce roughness or spikes.*
3. *These 2 + 2 formulations lead to most derivatives described in [7,8]. We must reinforce something very important: all the above defined derivatives are valid for functions defined in any interval of $\mathbb{R}$. This means that we do not need to change the definitions to accommodate them to the domain of the function at hand. One thing is the definition, another one is the computation of the derivative. It is a situation similar to the one we find in the LT or FT. We do not need to change the definitions to agree with the domain of the function. Therefore, most derivative definitions in [7,8] have no reason to be considered as autonomous derivatives.*
4. *These derivatives do not introduce any initial conditions.*
5. *For a given derivative, there is always an anti-derivative.*

These derivative formulations using (28) and the convolution are related as Figure 1 illustrates, and collected in Table 1, where

$$\binom{a}{b} = \begin{cases} \dfrac{\Gamma(a+1)}{\Gamma(b+1)\Gamma(a-b+1)}, & \text{if } a, b, a-b \notin \mathbb{Z}^- \\[2ex] \dfrac{(-1)^b\Gamma(b-a)}{\Gamma(b+1)\Gamma(-a)}, & \text{if } a \in \mathbb{Z}^- \wedge b \in \mathbb{Z}_0^+ \\[2ex] 0, & \text{if } [(b \in \mathbb{Z}^- \vee b-a \in \mathbb{N}) \wedge a \notin \mathbb{Z}^-] \vee (a,b \in \mathbb{Z}^- \wedge |a| > |b|) \end{cases} \tag{43}$$

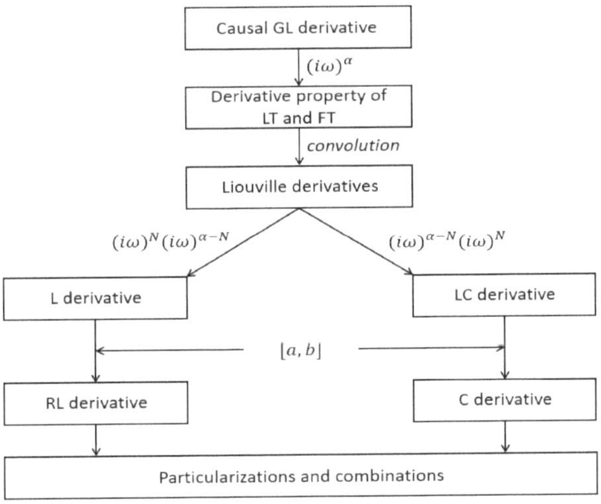

Figure 1. Hierarchical relation of FDs.

Table 1. Derivatives defined on $\mathbb{R}$.

Name	Definition	Parameters		Domain
L	$D^{\alpha}_{-\infty}f(x)=\frac{1}{\Gamma(N-\alpha)}\frac{\mathrm{d}^N}{\mathrm{d}x^N}\int_{-\infty}^{x}(x-\xi)^{-\alpha+N-1}f(\xi)\,\mathrm{d}\xi$	$\gamma=\alpha$	$\theta=\alpha$	$x\in\mathbb{R}$
	$D^{\alpha}_{\infty}f(x)=\frac{(-1)^N}{\Gamma(N-\alpha)}\frac{\mathrm{d}^N}{\mathrm{d}x^N}\int_{x}^{+\infty}(\xi-x)^{-\alpha+N-1}f(\xi)\,\mathrm{d}\xi$	$\gamma=\alpha$	$\theta=-\alpha$	$x\in\mathbb{R}$
LC	$D^{\alpha}_{-\infty}f(x)=\frac{1}{\Gamma(N-\alpha)}\int_{-\infty}^{x}(x-\xi)^{-\alpha+N-1}\frac{\mathrm{d}^N f(\xi)}{\mathrm{d}\xi^N}\,\mathrm{d}\xi$	$\gamma=\alpha$	$\theta=\alpha$	$x\in\mathbb{R}$
	$D^{\alpha}_{+\infty}f(x)=\frac{(-1)^N}{\Gamma(N-\alpha)}\int_{x}^{\infty}(x-\xi)^{-\alpha+N-1}\frac{\mathrm{d}^N f(\xi)}{\mathrm{d}\xi^N}\,\mathrm{d}\xi$	$\gamma=\alpha$	$\theta=-\alpha$	$x\in\mathbb{R}$
GL	$D^{\alpha}_{-\infty}f(x)=\lim_{h\to0^+}h^{-\alpha}\sum_{k=0}^{\infty}(-1)^k\binom{\alpha}{k}f(x-kh)$	$\gamma=\alpha$	$\theta=\alpha$	$x\in\mathbb{R}$
	$D^{\alpha}_{\infty}f(x)=\lim_{h\to0^+}h^{-\alpha}\sum_{k=0}^{\infty}(-1)^k\binom{\alpha}{k}f(x+kh)$	$\gamma=\alpha$	$\theta=-\alpha$	$x\in\mathbb{R}$

4.4. RL and C Derivatives

Despite the high degree of generality exhibited by the above derivatives, they are not used in most papers. In fact, such papers use derivatives specialized for functions that are defined on intervals $[a,b],\ -\infty<a<b<\infty$ [4,6]. Usually $a\geq0$ so that right functions are assumed. When $a=0$ we use frequently the designation "causal function". The GL derivatives assume the form

$$ {}^{\mathrm{GL}}D^{\alpha}_{a+}[f(x)]=\lim_{h\to0}\frac{1}{h^{\alpha}}\sum_{k=0}^{\lfloor n\rfloor}(-1)^k\frac{\Gamma(\alpha+1)f(x-kh)}{\Gamma(k+1)\Gamma(\alpha-k+1)},\quad nh=x-a \tag{44}$$

and

$$ {}^{\mathrm{GL}}D^{\alpha}_{b-}[f(x)]=\lim_{h\to0}\frac{1}{h^{\alpha}}\sum_{k=0}^{\lfloor n\rfloor}(-1)^k\frac{\Gamma(\alpha+1)f(x+kh)}{\Gamma(k+1)\Gamma(\alpha-k+1)},\quad nh=b-x. \tag{45}$$

This was the procedure of Grünwald [21] and Letnikov [22]. Therefore, they are nothing else than the application of (27) to bounded support functions and so *they are not new derivatives*. For the integral formulations, the contributions of Liouville and Riemann [23] were joined to obtain the Riemann–Liouville (RL) derivative [4,6]; likewise, from the Liouville–Caputo derivative, the (Dzherbashian-)Caputo (C) derivative [5,6] is obtained. We have:

- Riemann–Liouville (RL) derivatives

$$ {}^{RL}D^{\alpha}_{a+}f(t)=\frac{1}{\Gamma(N-\alpha)}\frac{\mathrm{d}^N}{\mathrm{d}t^N}\int_{a}^{t}f(\tau)(t-\tau)^{N-\alpha-1}\,\mathrm{d}\tau,\quad t>a \tag{46}$$

and

$$ {}^{RL}D^{\alpha}_{b-}f(t)=\frac{(-1)^N}{\Gamma(N-\alpha)}\frac{\mathrm{d}^N}{\mathrm{d}t^N}\int_{t}^{b}f(\tau)(t-\tau)^{N-\alpha-1}\,\mathrm{d}\tau,\quad t>a. \tag{47}$$

- (Dzherbashian-)Caputo (C) derivatives

$$ {}^{C}D^{\alpha}_{a+}f(t)=\frac{1}{\Gamma(N-\alpha)}\int_{a}^{t}f^{(N)}(\tau)(t-\tau)^{N-\alpha-1}\,\mathrm{d}\tau,\quad t>a \tag{48}$$

and

$$ {}^{C}D^{\alpha}_{b-}f(t)=\frac{(-1)^N}{\Gamma(N-\alpha)}\int_{t}^{b}f^{(N)}(\tau)(t-\tau)^{N-\alpha-1}\,\mathrm{d}\tau,\quad t<b. \tag{49}$$

These formulations, collected in Table 2, are the ones most used, although they have several inconveniences, mainly the initial conditions problem they introduce [17]. However, they form the basis for many applications.

Table 2. Classical derivatives.

Name	Definition	Parameters		Domain
RL	$D^{\alpha}_{a^+} f(x) = \dfrac{1}{\Gamma(N-\alpha)} \dfrac{\mathrm{d}^N}{\mathrm{d}x^N} \displaystyle\int_a^x (x-\xi)^{-\alpha+N-1} f(\xi)\,\mathrm{d}\xi$	$\gamma = \alpha$	$\theta = \alpha$	$x \in [a,b]$
	$D^{\alpha}_{b^-} f(x) = \dfrac{(-1)^N}{\Gamma(N-\alpha)} \dfrac{\mathrm{d}^N}{\mathrm{d}x^N} \displaystyle\int_x^b (\xi-x)^{-\alpha+N-1} f(\xi)\,\mathrm{d}\xi$	$\gamma = \alpha$	$\theta = -\alpha$	$x \in [a,b]$
C	$D^{\alpha}_{a^+} f(x) = \dfrac{1}{\Gamma(N-\alpha)} \displaystyle\int_a^x (x-\xi)^{-\alpha+N-1} \dfrac{\mathrm{d}^N f(\xi)}{\mathrm{d}\xi^N}\,\mathrm{d}\xi$	$\gamma = \alpha$	$\theta = \alpha$	$x \in [a,b]$
	$D^{\alpha}_{b^-} f(x) = \dfrac{(-1)^N}{\Gamma(N-\alpha)} \displaystyle\int_x^b (x-\xi)^{-\alpha+N-1} \dfrac{\mathrm{d}^N f(\xi)}{\mathrm{d}\xi^N}\,\mathrm{d}\xi$	$\gamma = \alpha$	$\theta = -\alpha$	$x \in [a,b]$
GL	$D^{\alpha}_{a^+} f(x) = \lim\limits_{h\to 0^+} \dfrac{\sum_{k=0}^{\lfloor \frac{x-a}{h} \rfloor} (-1)^k \binom{\alpha}{k} f(x-kh)}{h^{\alpha}}$	$\gamma = \alpha$	$\theta = \alpha$	$x \in [a,b]$
	$D^{\alpha}_{a^-} f(x) = \lim\limits_{h\to 0^+} \dfrac{\sum_{k=0}^{\lfloor \frac{a-x}{h} \rfloor} (-1)^k \binom{\alpha}{k} f(x+kh)}{h^{\alpha}}$	$\gamma = \alpha$	$\theta = -\alpha$	$x \in [a,b]$

4.5. Multistep Derivatives

The procedure introduced in (36) can be generalized. For example,

$$s^{\alpha} = s^{N-k} s^{\alpha-N} s^{k}, \qquad \alpha \in \mathbb{R}^+, N \in \mathbb{Z}_0^+, \tag{50}$$

with $k < N$ and $N > \alpha$. Then,

$$D^{\alpha} f(t) = D^{N-k} \frac{1}{\Gamma(N-\alpha)} \int_a^t f^{(k)}(\tau)(t-\tau)^{N-\alpha-1}\,\mathrm{d}\tau. \tag{51}$$

This procedure was proposed by Davidson and Essex [8]. However, the derivative they proposed is valid only for functions which are null for $t < 0$. Cavanati [8] introduced an algorithm that is a particular case obtained with $k = N - 1$.

Another similar algorithm was presented by Hilfer [24], also for causal functions. It can be stated as

$$s^{\alpha} = s^{\mu(1-\alpha)} s s^{(1-\mu)(1-\alpha)}, \tag{52}$$

where $0 < \alpha, \mu < 1$. It reads

$$D^{\alpha} f(t) = D^{\mu(1-\alpha)} \frac{1}{\Gamma((1-\mu)(1-\alpha))} \frac{d}{dx} \int_a^t f(\tau)(t-\tau)^{(1-\mu)(1-\alpha)}\,\mathrm{d}\tau. \tag{53}$$

This approach can be generalized. For example, if $1 < \alpha < 2$, then we can set

$$s^{\alpha} = s^{\mu(2-\alpha)} s^2 s^{(1-\mu)(2-\alpha)} \tag{54}$$

or

$$s^{\alpha} = s s^{\mu(2-\alpha)} s s^{(1-\mu)(2-\alpha)}. \tag{55}$$

This kind of reasoning shows that, following this method, we can invent billions of "derivatives". However, it is not clear how this is interesting in real life applications. Indeed, we are not introducing a new derivative, but using repeatedly one basic derivative. Some of the such derivatives were proposed because they introduced more favorable initial conditions. However, this is in general incorrect, since the initial conditions do not depend on the derivatives, but on the physical structure of the system at hand [17,25].

4.6. Second Generation Operators

The derivatives we introduced above are "shift invariant". However, there are other derivatives that do not enjoy such properties, and therefore cannot be obtained from the UFD introduced above. It is the case of the Hadamard derivatives [6] that are "scale invariant". Other examples are the "quantum derivatives" [15]. It is important to refer also the Marchaud derivatives [4] that exhibit a kind of regularization. On the other hand, some modifications and variable changes can be made in the above derivatives, leading to interesting operators that may not necessarily be considered to be derivatives when seen in the light of the above criterion [7,8]. They are introduced in Table 3.

Table 3. Operators (not necessarily derivatives) obtained from modified derivatives.

Name	Definition	Domain
Hadamard	$D^{\alpha}_{0^+}f(x)=\dfrac{1}{\Gamma(-\alpha)}\displaystyle\int_0^x \xi\left(\log\frac{\xi}{x}\right)^{-\alpha-1} f(\xi)\,\mathrm{d}\xi$	$\alpha<0,\ x\geq 0$
	$D^{\alpha}_{0^+}f(x)=\dfrac{1}{\Gamma(1-\alpha)}\displaystyle\int_0^x \xi\left(\log\frac{x}{\xi}\right)^{-\alpha-1} (f(x)-f(\xi))\,\mathrm{d}\xi$	$\alpha>0,\ x\geq 0$
Quantum	$D^{\alpha}_{q}f(x)=x^{-\alpha}\lim\limits_{q\to 1}\dfrac{\sum\limits_{j=0}^{+\infty}\begin{bmatrix}\alpha\\ j\end{bmatrix}_q(-1)^j q^{\frac{j(j+1)}{2}}q^{-j\alpha}f\left(q^j x\right)}{(1-q)^{\alpha}}$	
	$D^{\alpha}_{q^{-1}}f(x)=x^{-\alpha}\lim\limits_{q\to 1}\dfrac{\sum\limits_{j=0}^{+\infty}\begin{bmatrix}\alpha\\ j\end{bmatrix}_q(-1)^j q^{\frac{j(j-1)}{2}}q^{-j\alpha}f\left(q^{-j} x\right)}{(1-q^{-1})^{\alpha}}$	
	$\begin{bmatrix}\alpha\\ j\end{bmatrix}_q=\dfrac{\prod\limits_{i=0}^{j-1}\left(1-q^{\alpha+i}\right)}{\dfrac{1-q^j}{1-q}}$	
Marchaud	$D^{\alpha}_{0^+}f(x)=\dfrac{\alpha}{\Gamma(1-\alpha)}\displaystyle\int_0^{+\infty}\xi^{-\alpha-1}(f(x)-f(x-\xi))\,\mathrm{d}\xi$	$\alpha>0$
	$D^{\alpha}_{0^-}f(x)=\dfrac{\alpha}{\Gamma(1-\alpha)}\displaystyle\int_0^{+\infty}\xi^{-\alpha-1}(f(x)-f(x+\xi))\,\mathrm{d}\xi$	$\alpha<0$
Jumarie	$D^{\alpha}_{0^+}f(x)=\dfrac{1}{\Gamma(N-\alpha)}\dfrac{\mathrm{d}^N}{\mathrm{d}x^N}\displaystyle\int_0^x (x-\xi)^{-\alpha+N-1}(f(\xi)-f(0))\,\mathrm{d}\xi$	$\alpha>0,\ x\geq 0$
Erdélyi-Kober	$D^{\alpha}_{a^+,\sigma,\eta}f(x)=\dfrac{\sigma x^{\sigma(\eta-\alpha)}}{\Gamma(-\alpha)}\displaystyle\int_a^x (x^{\sigma}-\xi^{\sigma})^{-\alpha-1}\xi^{\sigma(1+\eta)-1}f(\xi)\,\mathrm{d}\xi$	$\alpha<0,\ x\geq a$
	$D^{\alpha}_{a^-,\sigma,\eta}f(x)=\dfrac{\sigma x^{-\sigma\alpha}}{\Gamma(-\alpha)}\displaystyle\int_x^a (\xi^{\sigma}-x^{\sigma})^{-\alpha-1}\xi^{\sigma(1+\alpha-\eta)-1}f(\xi)\,\mathrm{d}\xi$	$\alpha<0,\ x\leq a$
	$D^{\alpha}_{a^+,\sigma,\eta}f(x)=x^{-\sigma(\eta+\alpha)}\left(\dfrac{1}{\sigma x^{\sigma-1}}\dfrac{\mathrm{d}}{\mathrm{d}x}\right)^N x^{\sigma(\alpha+N+\eta)}D^{-N-\alpha}_{a^+,\sigma,\eta}f(x)$	$\alpha>0,\ x\geq a$
	$D^{\alpha}_{a^-,\sigma,\eta}f(x)=x^{\sigma\eta}\left(\dfrac{1}{\sigma x^{\sigma-1}}\dfrac{\mathrm{d}}{\mathrm{d}x}\right)^N x^{\sigma(N-\eta)}D^{-N-\alpha}_{a^-,\sigma,\eta}f(x)$	$\alpha>0,\ x\leq a$

Table 3. *Cont.*

Name	Definition	Domain
k-Hilfer	$D^{\alpha}_{k,a^+} f(x) = \dfrac{1}{k^{-\frac{\alpha}{k}}\Gamma(-\alpha)} \int_a^x (x-\xi)^{-\frac{\alpha}{k}-1} f(\xi)\,\mathrm{d}\xi$	$\alpha < 0,\ x \geq a$
	$D^{\mu,\nu}_{k,a^+} f(x) = D^{\nu(\mu-1)}_{k,a^+} \dfrac{\mathrm{d}}{\mathrm{d}x} D^{(1-\nu)(\mu-1)}_{k,a^+} f(x)$	$0 \leq \mu \leq 1,\ 0 < \nu < 1,$ $x \geq a$
Hilfer-Katugampola	$D^{\alpha}_{\rho,a^+} f(x) = \dfrac{\rho^{1+\alpha}}{\Gamma(-\alpha)} \int_a^x (x^{\rho} - \xi^{\rho})^{-\alpha-1} f(\xi)\,\mathrm{d}\xi$	$\rho > 0,\ \alpha < 0,\ x \geq a$
	$D^{\alpha}_{\rho,a^-} f(x) = \dfrac{\rho^{1+\alpha}}{\Gamma(-\alpha)} \int_x^a (x^{\rho} - \xi^{\rho})^{-\alpha-1} f(\xi)\,\mathrm{d}\xi$	$\rho > 0,\ \alpha < 0,\ x \leq a$
	$D^{\alpha,\beta}_{\rho,a^{\pm}} f(x) = \pm D^{\beta(\alpha-1)}_{\rho,a^{\pm}} \left(x^{1-\rho} \dfrac{\mathrm{d}}{\mathrm{d}x} D^{(1-\beta)(\alpha-1)}_{\rho,a^{\pm}} f(x) \right)$	$\rho > 0,\ 1 > \alpha > 0,$ $0 \geq \beta \geq 0$
ψ-Hilfer, with	$D^{\alpha,\psi(x)}_{a^+} f(x) = \dfrac{1}{\Gamma(-\alpha)} \int_a^x \psi'(x)(\psi(x) - \psi(\xi))^{-\alpha-1} f(\xi)\,\mathrm{d}\xi$	$\alpha < 0,\ I = [a,x]$
$\psi(x) \in C^N(I,\mathbb{R})$	$D^{\alpha,\psi(x)}_{a^-} f(x) = \dfrac{1}{\Gamma(-\alpha)} \int_x^a \psi'(x)(\psi(\xi) - \psi(x))^{-\alpha-1} f(\xi)\,\mathrm{d}\xi$	$\alpha < 0,\ I = [x,a]$
$\psi'(x) \neq 0,\ \forall x \in I$	$D^{\alpha,\beta,\psi(x)}_{a^+} f(x) = D^{\beta(\alpha-N),\psi(x)}_{a^+} \left(\dfrac{1}{\psi'(x)} \dfrac{\mathrm{d}}{\mathrm{d}x} \right)^N D^{(1-\beta)(\alpha-N),\psi(x)}_{a^+} f(x)$	$\alpha > 0,\ 0 \leq \beta \leq 1,$ $I = [a,x]$
$x_2 > x_1 \Rightarrow$ $\Rightarrow \psi(x_2) > \psi(x_1)$	$D^{\alpha,\beta,\psi(x)}_{a^-} f(x) = D^{\beta(\alpha-N),\psi(x)}_{a^-} \left(\dfrac{1}{\psi'(x)} \dfrac{\mathrm{d}}{\mathrm{d}x} \right)^N D^{(1-\beta)(\alpha-N),\psi(x)}_{a^+} f(x)$	$\alpha > 0,\ 0 \leq \beta \leq 1,$ $I = [x,a]$

5. Pseudo-Fractional-Derivatives

5.1. Some Comments

In many quarters, the treatment of FDs and systems was taken to be a quite difficult task. Therefore, simplified versions thereof were welcomed, even if some important features that characterize most systems were lost in the process — for example, in modeling natural or human-made systems that are essentially low-pass or bandpass systems. However, some proposed systems are high-pass, and then have limited usefulness; in the following, we describe some of them. We must remark that sometimes the word "fractional" is used as a "trade mark" that helps to "sell a product". It is the case of the "Memory-dependent derivative" [26], that is nothing else than a running average.

5.2. "Derivatives" That Are High-Pass Filters

Consider a simple differential equation:

$$a_1 y'(t) + a_0 y(t) = b_1 x'(t). \tag{56}$$

This equation is recognized easily as the model of the classic high-pass filter [27]. Its transfer function is

$$H(s) = \frac{b_1 s}{a_1 s + a_0} = \frac{b_1}{a_1} \frac{s}{s+\sigma}, \quad \Re(s) > -\sigma = -a_0/a_1. \tag{57}$$

One way of relating the input and output is:

$$y(t) = \frac{b_1}{a_1} \int_0^t e^{-\sigma(t-\tau)} x'(\tau)\,\mathrm{d}\tau. \tag{58}$$

This expression was chosen with this form to remember the Caputo derivative (41). Now set:

$$a_0 = \alpha,\ a_1 = 1-\alpha,\ b_1 = M(\alpha),\ \sigma = \frac{\alpha}{1-\alpha} \tag{59}$$

to obtain

$$y(t) = \frac{M(\alpha)}{1-\alpha} \int_0^t e^{-\frac{\alpha}{1-\alpha}(t-\tau)} x'(\tau)\, \mathrm{d}\tau. \tag{60}$$

This is the expression of the "famous" *fractional Caputo-Fabrizio* "derivative". As is clear, it is neither a derivative, nor fractional.

Let us continue and substitute the exponential in (60) by the Mittag–Leffler function. We obtain also a fractional high-pass filter with TF given by

$$H(s) = \frac{M(\alpha)}{1-\alpha} \frac{s^\alpha}{s^\alpha + \sigma}, \ \Re(s) > 0, \tag{61}$$

that is called Atangana-Baleanu "derivative". Attending to its FT, it is a system, but not a derivative. There are many variations of this operator, but they remain high-pass filters, not derivatives [8].

For a list of these and similar operators, see Table 4.

Table 4. "Derivatives" with non-singular kernel; in all cases, the order verifies $0 < \alpha < 1$, and $M(\alpha)$ verifies $M(0) = M(1) = 1$.

Name	Definition	Domain
Caputo-Fabrizio	$D^\alpha_{a^+} f(x) = \frac{M(\alpha)}{1-\alpha} \int_a^x e^{-\frac{\alpha(x-\xi)}{1-\alpha}} \frac{\mathrm{d}f(\xi)}{\mathrm{d}\xi}\, \mathrm{d}\xi$	$x \geq a$
Yang et al.	$D^\alpha_{a^+} f(x) = \frac{M(\alpha)}{1-\alpha} \frac{\mathrm{d}}{\mathrm{d}x} \int_a^x e^{-\frac{\alpha(x-\xi)}{1-\alpha}} f(\xi)\, \mathrm{d}\xi$	$x \geq a$
Atangana-Baleanu-Caputo	$D^\alpha_{a^+} f(x) = \frac{M(\alpha)}{1-\alpha} \int_a^x E_\alpha\left(-\frac{\alpha(x-\xi)^\alpha}{1-\alpha}\right) \frac{\mathrm{d}f(\xi)}{\mathrm{d}\xi}\, \mathrm{d}\xi$	$x \geq a$
Atangana-Baleanu-Riemann–Liouville	$D^\alpha_{a^+} f(x) = \frac{M(\alpha)}{1-\alpha} \frac{\mathrm{d}}{\mathrm{d}x} \int_a^x E_\alpha\left(-\frac{\alpha(x-\xi)^\alpha}{1-\alpha}\right) f(\xi)\, \mathrm{d}\xi$	$x \geq a$
Generalized Caputo	$D^{\alpha,\beta}_{a^+} f(x) = \frac{M(\alpha)}{1-\alpha} \int_a^x E_\beta\left(-\frac{\alpha(x-\xi)^\beta}{1-\alpha}\right) \frac{\mathrm{d}f(\xi)}{\mathrm{d}\xi}\, \mathrm{d}\xi$	$0 \leq \beta \leq 1,\ x \geq a$
Generalized Riemann–Liouville	$D^{\alpha,\beta}_{a^+} f(x) = \frac{M(\alpha)}{1-\alpha} \frac{\mathrm{d}}{\mathrm{d}x} \int_a^x E_\beta\left(-\frac{\alpha(x-\xi)^\beta}{1-\alpha}\right) f(\xi)\, \mathrm{d}\xi$	$0 \leq \beta \leq 1,\ x \geq a$
Caputo-Fabrizio, Gaussian kernel	$D^\alpha_{a^+} f(x) = \frac{1+\alpha^2}{\sqrt{\pi^\alpha(1-\alpha)}} \int_a^x e^{-\frac{\alpha(x-\xi)^2}{1-\alpha}} \frac{\mathrm{d}f(\xi)}{\mathrm{d}\xi}\, \mathrm{d}\xi$	$f(a) = 0,\ x \geq a$
Sun-Hao-Zhang-Baleanu	$D^\alpha_{a^+} f(x) = \frac{M(\alpha)}{(1-\alpha)^{\frac{1}{\alpha}}} \int_a^x e^{-\frac{\alpha(x-\xi)^\alpha}{1-\alpha}} \frac{\mathrm{d}f(\xi)}{\mathrm{d}\xi}\, \mathrm{d}\xi$	$x \geq a$

5.3. "Disguised" Order 1 Derivatives

The "local fractional derivative" introduced by Kolwankar reads [28]

$$D^\alpha f(t) = \lim_{\tau \to t} D^\alpha_t [f(\tau) - f(t)], \tag{62}$$

where D^α_t is the RL derivative. V. Tarasov [29] showed that, if $\alpha < 1$, this derivative is equivalent to the order 1 derivative, therefore not fractional.

A similar result can be found for the "conformable" derivative [30]

$$D^\alpha f(t) = \lim_{\varepsilon \to 0} \frac{f(t + \varepsilon t^{1-\alpha}) - f(t)}{\varepsilon} \tag{63}$$

and

$$D^\alpha f(t) = \lim_{\varepsilon \to 0} \frac{f\left(t e^{\varepsilon t^{-\alpha}}\right) - f(t)}{\varepsilon}. \tag{64}$$

For differentiable functions and for both derivatives, this results in

$$D^{\alpha} f(t) = t^{1-\alpha} \frac{\mathrm{d}f}{\mathrm{d}t}(t). \tag{65}$$

Several modified versions of these derivatives were proposed [8].
The so-called "fractal derivative" was introduced in [31] and reads

$$\frac{\partial f(t)}{\partial t^{\alpha}} = \lim_{s \to t} \frac{f(t) - f(s)}{t^{\alpha} - s^{\alpha}}. \tag{66}$$

It is a strange derivative that gives ∞ for differentiable functions, unless $\alpha = 1$.
For a list of these and similar operators, see Table 5.

Table 5. Local formulations of derivatives ($\alpha > 0$).

Name	Definition	Domain
Kolwankar	$D^{\alpha} f(x) = \lim_{\xi \to x} D_x^{\alpha}(f(\xi) - f(x))$ where D_x^{α} is the RL derivative	$x \in \mathbb{R}^+$
Chen	$D^{\alpha} f(x) = \lim_{\xi \to x} \frac{f(x) - f(\xi)}{x^{\alpha} - \xi^{\alpha}}$	$x \in \mathbb{R}^+$
Conformable	$D^{\alpha} f(x) = \lim_{\varepsilon \to 0} \frac{f(x + \varepsilon x^{1-\alpha}) - f(x)}{\varepsilon}$	$x \in \mathbb{R}^+$
Katugampola	$D^{\alpha} f(x) = \lim_{\varepsilon \to 0} \frac{f(x\, e^{\varepsilon x^{-\alpha}}) - f(x)}{\varepsilon}$	$x \in \mathbb{R}^+$
M	$D^{\alpha,\beta} f(x) = \lim_{\varepsilon \to 0} \frac{f(x\, E_{\beta}(\varepsilon x^{-\alpha})) - f(x)}{\varepsilon}$	$x \in \mathbb{R}^+, \beta > 0$
Deformable	$D^{\alpha} f(x) = \lim_{\varepsilon \to 0} \frac{(1 + \varepsilon - \varepsilon\alpha) f(x + \varepsilon\alpha) - f(x)}{\varepsilon}$	$x \in \mathbb{R}^+$
Beta	$D^{\alpha} f(x) = \lim_{\varepsilon \to 0} \frac{f\left(x + \varepsilon\left(x + \frac{1}{\Gamma(\alpha)}\right)^{1-\alpha}\right) - f(x)}{\varepsilon}$	$0 < \alpha \leq 1,\ x \in \mathbb{R}^+$
AGO	$D^{\alpha} f(x) = \lim_{\varepsilon \to 0} \frac{f\left(x + \epsilon(\psi(x))^{1-\alpha}\right) - f(x)}{\varepsilon}$ $\psi(x) \in C(\mathbb{R}^+),\ \psi'(x) \neq 0$	$0 < \alpha < 1, x \in \mathbb{R}^+$
Generalized	$D^{\alpha} f(x) = \lim_{\varepsilon \to 0} \frac{f\left(x - \psi(x) + \psi(x)\, e^{\frac{\epsilon(\psi(x))^{-\alpha}}{\psi'(x)}}\right) - f(x)}{\varepsilon}$ $\psi(x) \in C(\mathbb{R}^+),\ \psi'(x) \neq 0$	$0 < \alpha < 1,\ x \in \mathbb{R}^+$
General conformable	$D_{\psi}^{\alpha} f(x) = \lim_{\varepsilon \to 0} \frac{f(x + \varepsilon\, \psi(x, \alpha)) - f(x)}{\varepsilon}$	$\alpha > 0,\ x \in \mathbb{R}^+$
Lazopoulos's Lambda	$D_{\Lambda}^{\alpha} f(x) = \frac{\frac{1}{\Gamma(N-\alpha)} \frac{\mathrm{d}^N}{\mathrm{d}x^N} \int_a^x (x - \xi)^{-\alpha+N-1} f(\xi)\, \mathrm{d}\xi}{\frac{1}{\Gamma(N-\alpha)} \frac{\mathrm{d}^N}{\mathrm{d}x^N} \int_a^x (x - \xi)^{-\alpha+N-1} \xi\, \mathrm{d}\xi}$	$x \in \mathbb{R}^+$

Figures 2–4 illustrate the results obtained with different FD formulations, when applied to compute the $\alpha = 0.5$ order derivative of the function $f(x) = \cos(\omega x)$, for $\omega = \{0.01\pi, 0.1\pi, 100\pi\}$, respectively. The derivatives are compared with the one obtained with the GL operator, which serves as baseline. In all cases we adopt $a = 0$, while FD specific parameters are as presented in the legends of the graphs. We verify that some formulations fail to compute the derivatives accurately, while others diverge.

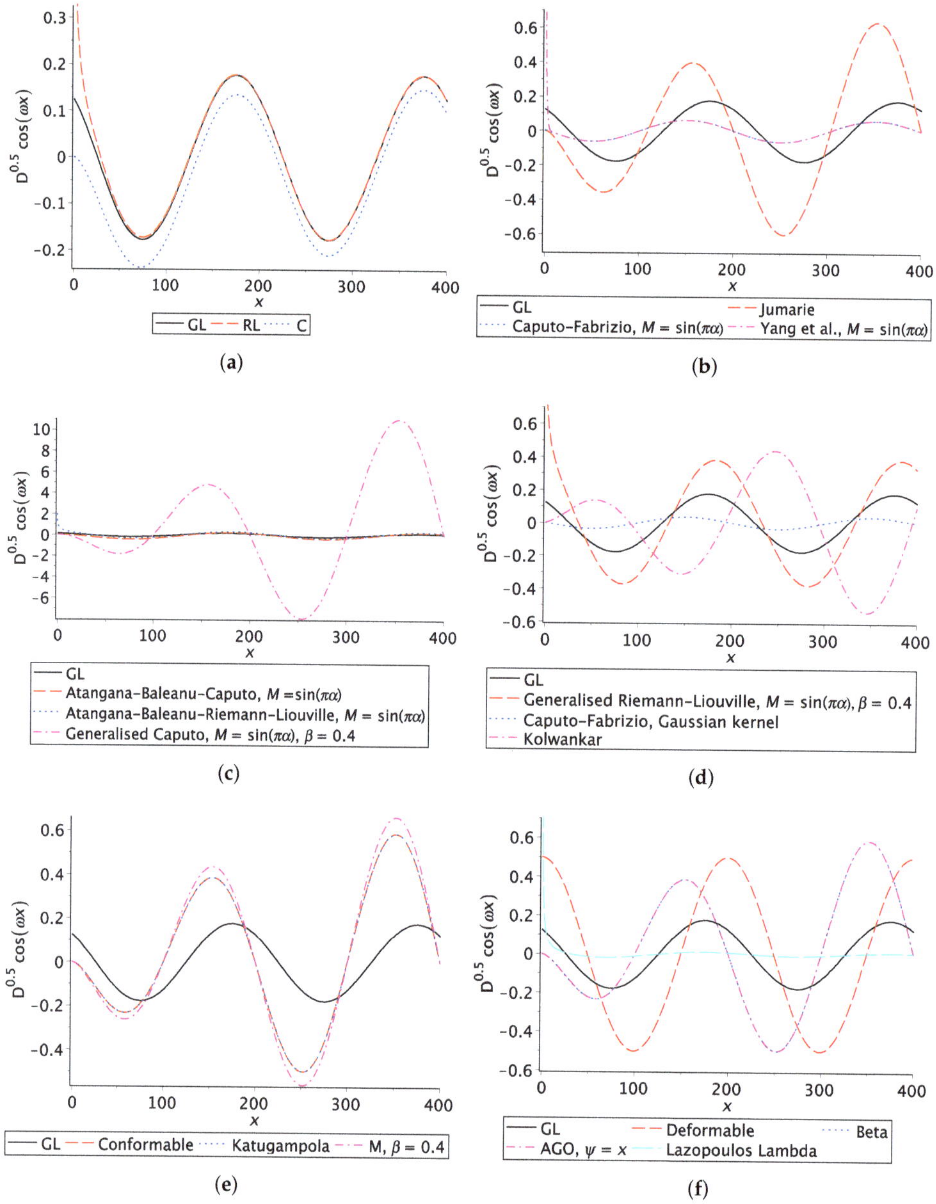

Figure 2. The $\alpha = 0.5$ order derivative of function $f(x) = \cos(0.01\pi x)$ with different FD formulations. See Tables 2–5.

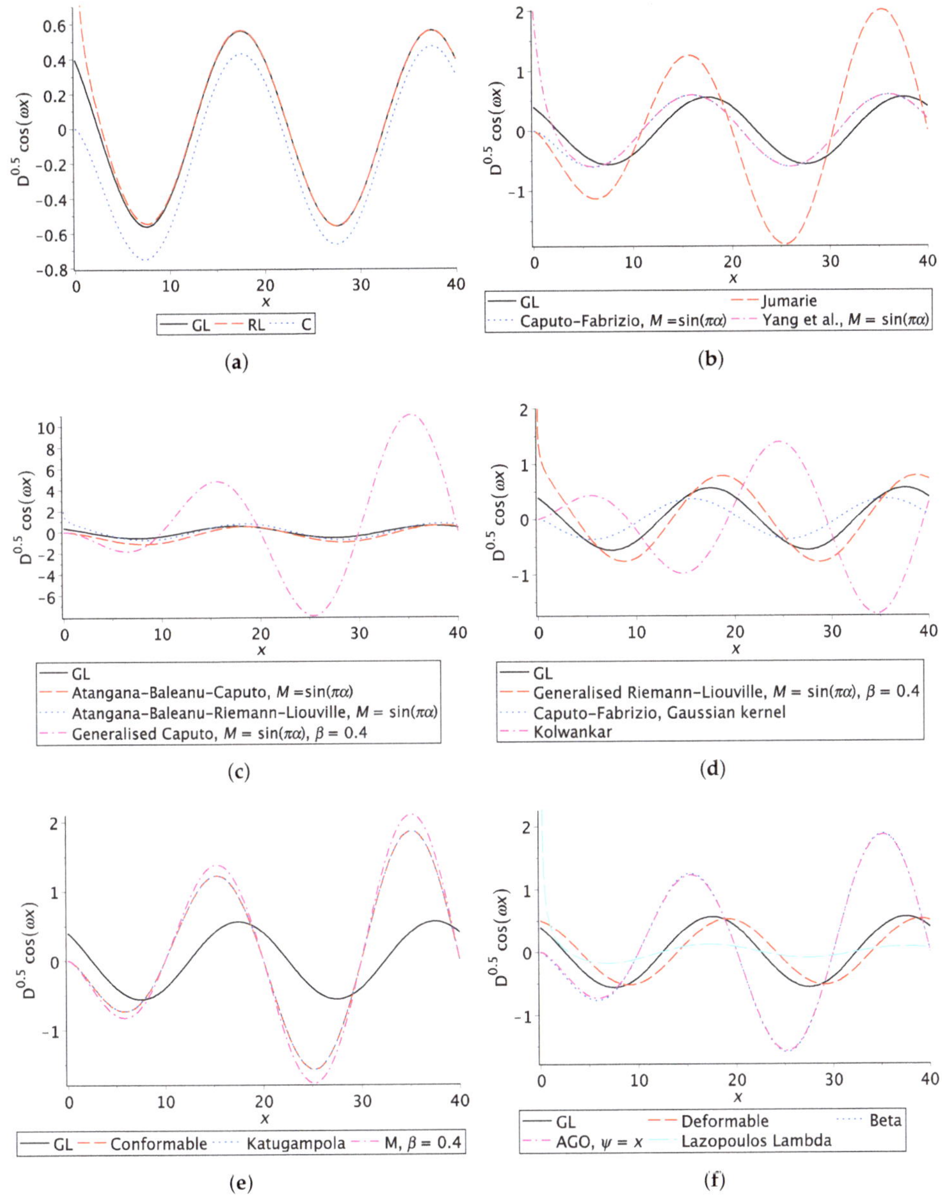

Figure 3. The $\alpha = 0.5$ order derivative of function $f(x) = \cos(0.1\pi x)$ with different FD formulations. See Tables 2–5.

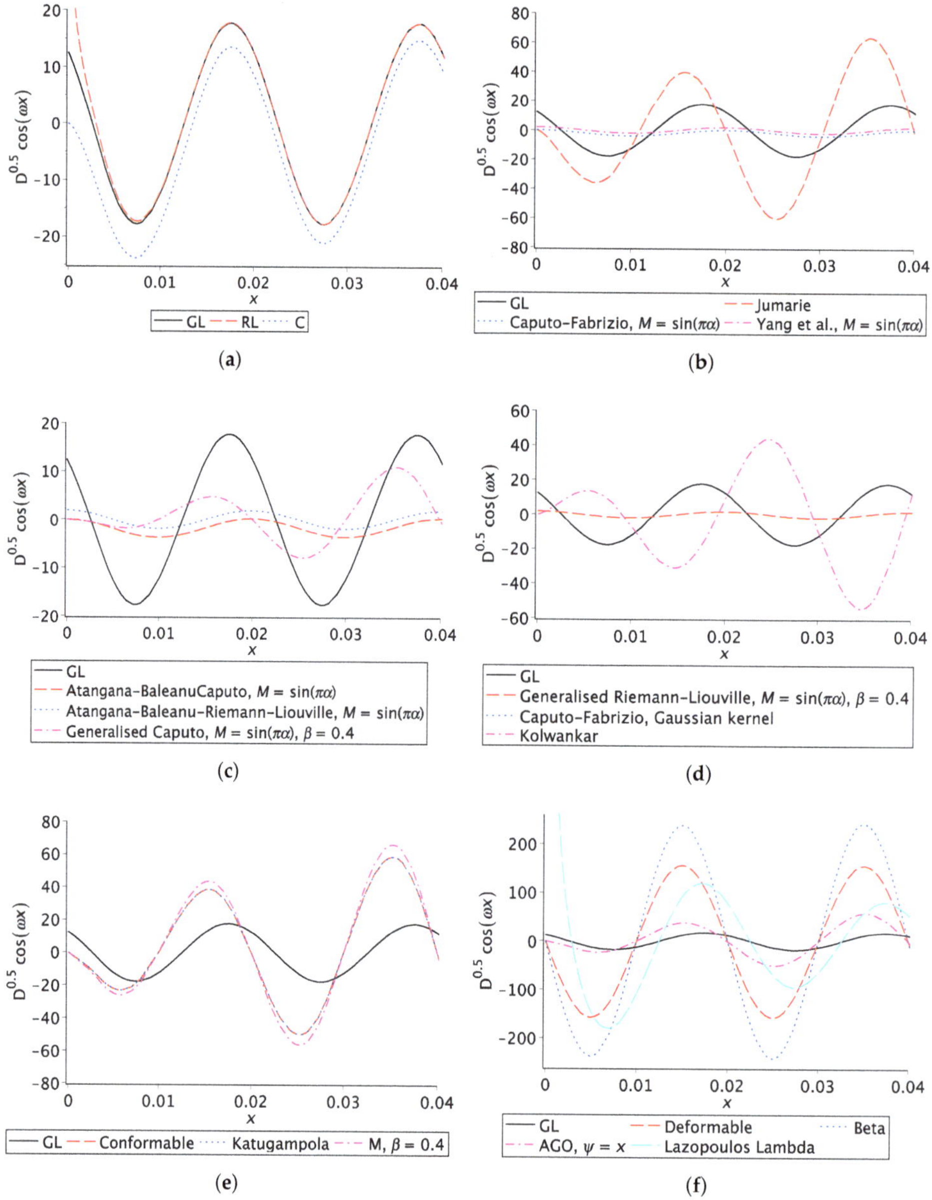

Figure 4. The $\alpha = 0.5$ order derivative of function $f(x) = \cos(100\pi x)$ with different FD formulations. See Tables 2–5.

As is well known, the classic order 1 derivative of a sinusoidal is the same sinusoidal multiplied by the angular frequency and with a change of phase equal to $\pi/2$. Therefore, we expect something similar in the fractional case, at least after passing some time corresponding to the transient regime. This happens with those operators that we classified as derivatives. For the others, we have amplifications/attenuations and, in some cases, modulations, confirming that they are systems (filters), but not derivatives. Some have a non-acceptable behavior: they are unstable.

6. Which Derivatives?

After this journey into the world of FDs, it is important to answer the question: "Do we need such different formulations?"

In previous papers [10–12], some answers to this question were given.

1. In problems involving time, we have almost always to use causal derivatives. Therefore, the GL or one of the integral versions (37), (39), or (41) should be used.
2. In space problems, we can use the above formulæor the corresponding right-side versions, if there is any privileged direction. If this does not happen, we must use a two-sided derivative, preferably (2).

However, most derivatives described in [7,8] are particular cases, and the particularity is introduced by the domain. Therefore, do we need to define a new derivative each time the domain changes? This creates a big difficulty: we cannot keep increasing the number of derivatives, differing only because they are defined on different domains, which constrains the application fields.

Author Contributions: Conceptualization, M.D.O.; methodology, M.D.O.; formal analysis, all authors; writing—original draft preparation, all authors; writing—review and editing, all authors; visualization, A.M.L. All authors have read and agreed to the published version of the manuscript.

Funding: This work was partially funded by National Funds through the FCT-Foundation for Science and Technology within the scope of the CTS Research Unit—Center of Technology and Systems/UNINOVA/FC /NOVA, under the reference UIDB/00066/2020, and also by FCT through IDMEC, under LAETA, project UID/EMS/50022/2020.

Conflicts of Interest: The authors declare no conflict of interest.

References

1. Dugowson, S. Les Différentielles Métaphysiques (Histoire et Philosophie de la Généralisation de l'ordre de Dérivation). Ph.D. Thesis, Université Paris Nord, Paris, France, 1994.
2. Liouville, J. Memóire sur le calcul des différentielles à indices quelconques. *J. L'École Polytech. Paris* **1832**, *13*, 71–162.
3. Liouville, J. Memóire sur quelques questions de Géométrie et de Méchanique, et sur un nouveau genre de calcul pour résoudre ces questions. *J. L'École Polytech. Paris* **1832**, *13*, 1–69.
4. Samko, S.G.; Kilbas, A.A.; Marichev, O.I. *Fractional Integrals and Derivatives;* Gordon and Breach: Yverdon, Zwitserland, 1993.
5. Podlubny, I. *Fractional Differential Equations: An Introduction to Fractional Derivatives, Fractional Differential Equations, to Methods of Their Solution and Some of Their Applications;* Academic Press: San Diego, CA, USA, 1999; pp. 1–3.
6. Kilbas, A.A.; Srivastava, H.M.; Trujillo, J.J. *Theory and Applications of Fractional Differential Equations;* Elsevier: Amsterdam, The Netherlands, 2006.
7. De Oliveira, E.C.; Tenreiro Machado, J.A. A review of definitions for fractional derivatives and integral. *Math. Probl. Eng.* **2014**, *2014*, 238459. [CrossRef]
8. Teodoro, G.S.; Machado, J.T.; De Oliveira, E.C. A review of definitions of fractional derivatives and other operators. *J. Comput. Phys.* **2019**, *388*, 195–208. [CrossRef]
9. Ortigueira, M.D.; Machado, J.T. What is a fractional derivative? *J. Comput. Phys.* **2015**, *293*, 4–13. [CrossRef]
10. Ortigueira, M.D.; Machado, J.A.T. Which Derivative? *Fractal Fract.* **2017**, *1*, 3. [CrossRef]
11. Ortigueira, M.D.; Machado, J.A.T. Fractional Derivatives: The Perspective of System Theory. *Mathematics* **2019**, *7*, 150. [CrossRef]
12. Ortigueira, M.D. Two-sided and regularised Riesz-Feller derivatives. *Math. Methods Appl. Sci.* **2021**, *44*, 8057–8069. [CrossRef]
13. Ortigueira, M.D.; Bengochea, G.; Machado, J.A.T. Substantial, tempered, and shifted fractional derivatives: Three faces of a tetrahedron. *Math. Methods Appl. Sci.* **2021**, *44*, 9191–9209. [CrossRef]
14. Ortigueira, M.D.; Bengochea, G. Bilateral Tempered Fractional Derivatives. *Symmetry* **2021**, *13*, 823. [CrossRef]
15. Ortigueira, M.D. *Fractional Calculus for Scientists and Engineers;* Lecture Notes in Electrical Engineering; Springer: Berlin/Heidelberg, Germany, 2011.
16. Bohner, M.; Peterson, A. *Dynamic Equations on Time Scales: An Introduction with Applications;* Springer Science & Business Media: Berlin/Heidelberg, Germany, 2001.
17. Ortigueira, M.D.; Valério, D. *Fractional Signals and Systems;* De Gruyter: Berlin, Germany, 2020.

18. Ortigueira, M.D.; Machado, J.A.T. New discrete-time fractional derivatives based on the bilinear transformation: Definitions and properties. Recent Advances in the Fractional-Order Circuits and Systems: Theory, Design and Applications. *J. Adv. Res.* **2020**, *25*, 1–10. [CrossRef] [PubMed]
19. Liouville, J. Note sur une formule pour les différentielles à indices quelconques à l'occasion d'un mémoire de M. Tortolini. *J. Math. Pures Appl.* **1855**, *20*, 115–120.
20. Herrmann, R. *Fractional Calculus*, 3rd ed.; World Scientific: Singapore, 2018.
21. Grünwald, A.K. Ueber "begrentz" Derivationen und deren Anwendung. *Z. Math. Phys.* **1867**, *12*, 441–480.
22. Letnikov, A. Note relative à l'explication des principes fondamentaux de la théorie de la différentiation à indice quelconque (A propos d'un mémoire). *Mat. Sb.* **1873**, *6*, 413–445.
23. Riemann, B. Versuch einer allgemeinen Auffassung der Integration und Differentiation. In *Gesammelte Werke*; Cambridge University Press: Cambridge, UK, 1876; Volume 62.
24. Rudolf, H. *Applications of Fractional Calculus in Physics*; World Scientific: Singapore, 2000.
25. Ortigueira, M.D.; Machado, J.T. Revisiting the 1D and 2D Laplace transforms. *Mathematics* **2020**, *8*, 1330. [CrossRef]
26. Wang, J.L.; Li, H.F. Surpassing the fractional derivative: Concept of the memory-dependent derivative. *Comput. Math. Appl.* **2011**, *62*, 1562–1567. [CrossRef]
27. Oppenheim, A.V.; Willsky, A.S.; Hamid, S. *Signals and Systems*, 2nd ed.; Prentice-Hall: Upper Saddle River, NJ, USA, 1997.
28. Kolwankar, K.M.; Gangal, A.D. Local fractional calculus: A calculus for fractal space-time. In *Fractals*; Springer: Berlin/Heidelberg, Germany, 1999; pp. 171–181.
29. Tarasov, V.E. No nonlocality. No fractional derivative. *Commun. Nonlinear Sci. Numer. Simul.* **2018**, *62*, 157–163. [CrossRef]
30. Khalil, R.; Al Horani, M.; Yousef, A.; Sababheh, M. A new definition of fractional derivative. *J. Comput. Appl. Math.* **2014**, *264*, 65–70. [CrossRef]
31. Chen, W. Time–space fabric underlying anomalous diffusion. *Chaos Solitons Fractals* **2006**, *28*, 923–929. [CrossRef]

mathematics

Article

Operational Calculus for the General Fractional Derivatives of Arbitrary Order

Maryam Al-Kandari [1,*], Latif A-M. Hanna [1] and Yuri Luchko [2]

1 Department of Mathematics, Kuwait University, Kuwait City 12037, Kuwait; latif@sci.kuniv.edu.kw
2 Department of Mathematics, Physics, and Chemistry, Berlin University of Applied Sciences and Technology, 10587 Berlin, Germany; luchko@bht-berlin.de
* Correspondence: maryam@sci.kuniv.edu.kw

Abstract: In this paper, we deal with the general fractional integrals and the general fractional derivatives of arbitrary order with the kernels from a class of functions that have an integrable singularity of power function type at the origin. In particular, we introduce the sequential fractional derivatives of this type and derive an explicit formula for their projector operator. The main contribution of this paper is a construction of an operational calculus of Mikusiński type for the general fractional derivatives of arbitrary order. In particular, we present a representation of the m-fold sequential general fractional derivatives of arbitrary order as algebraic operations in the field of convolution quotients and derive some important operational relations.

Keywords: Sonine kernel; general fractional integral; general fractional derivative of arbitrary order; fundamental theorems of fractional calculus; operational calculus; convolution series

MSC: 26A33; 26B30; 33E30; 44A10; 44A35; 44A40; 45D05; 45E10; 45J05

Citation: Al-Kandari, M.; Hanna, L.A.-M.; Luchko, Y. Operational Calculus for the General Fractional Derivatives of Arbitrary Order. *Mathematics* **2022**, *10*, 1590. https://doi.org/10.3390/math10091590

Academic Editors: Alexandra M.S.F. Galhano, António M. Lopes and Carlo Cattani

Received: 26 March 2022
Accepted: 29 April 2022
Published: 7 May 2022

1. Introduction

Starting from the early phases in development of calculus, several attempts of interpretation of derivatives and integrals as some purely algebraic symbols were undertaken. Leibniz, Euler, Lagrange, and other founders of calculus introduced and employed the algebraic rules for manipulation with the integral and differential operators. In most of the cases, these rules led to correct results. However, they were just formal algorithms without a strict mathematical background. The most prominent example of this procedure were the works by Oliver Heaviside who systematically employed an algebraic approach to investigation of a number of practical problems including differential equations of electromagnetism and theory of oscillators. Because of importance of Heaviside's approach for engineers, mathematicians started to think about its rigorous mathematical background. In the works by Bromwich, Carson, Van der Pol, Doetsch, and other mathematicians, the methods of Heaviside were justified in terms of the Laplace transform and its modifications. However, a requirement of existence of the Laplace transform led to some conditions on behavior of the functions in infinity that restricted applicability of the Heaviside operational calculus.

In the 1950s, a cardinal return to the original ideas of operational calculus was proposed in the works by Jan Mikusiński and his co-authors (see [1] and the references therein), where an operational calculus for the first order derivative has been developed and applied for a number of mathematical and real world problems. The main components of this approach are an interpretation of the Laplace convolution as a multiplication in the ring of functions continuous on the real positive semi-axis and an extension of this ring to the field of convolution quotients. Later on, the Mikusiński scheme was employed by several mathematicians for development of operational calculi for some special differential operators with the variable coefficients (see, for example, [2–4]) that turned out to be particular cases of the hyper-Bessel differential operator in the form

$$(By)(t) = t^{-\beta}\prod_{j=1}^{m}\left(\gamma_j + \frac{1}{\beta}t\frac{d}{dt}\right)y(t),\ \beta > 0,\ \gamma_j \in \mathbb{R},\ j = 1,\ldots,m. \tag{1}$$

An operational calculus of Mikusiński type for the hyper-Bessel differential operator (1) was constructed by Dimovski in [5].

A new stage in further development of operational calculi of Mikusiński type was initiated in the works by Luchko and his co-authors, where operational calculi for different fractional derivatives were constructed and applied for derivation of the closed form solution formulas for the fractional integral and differential equations. The first operational calculus for a fractional derivative has been proposed in [6,7], where an operational calculus of Mikusiński type for the multiple Erdélyi-Kober fractional derivative was developed. The two prominent particular cases of this fractional derivative are the hyper-Bessel differential operator (1) and the Riemann–Liouville fractional derivative. An extended version of the operational calculus for the Riemann–Liouville fractional derivative was suggested in [8,9]. An operational calculus for another basic fractional derivative, the Caputo derivative, was developed in [10]. It is worth mentioning that in [10], this operational calculus was applied for derivation of the closed form solutions formulas to the multi-term fractional differential equations involving the Caputo fractional derivatives with the commensurate and non-commensurate orders. An operational calculus of Mikusiński type for the Hilfer fractional derivative was worked out in [11]. In [12], the case of the Caputo-type fractional Erdélyi-Kober derivative was treated. In [13], a Mikusiński type operational calculus for the general fractional derivative (GFD) of the "generalized order" form the interval $(0,\ 1)$ in the Caputo sense was developed. This GFD is a composition of a Laplace convolution integral with a Sonine kernel with an integrable singularity of power function type at the point zero and the first order derivative. In [13], this calculus was applied for derivation of the closed form solution formulas for the initial-value problems for the multi-term fractional differential equations with the sequential fractional derivatives of this type. The case of the GFD of arbitrary order in the Riemann–Liouville sense was treated in the very recent paper [14], where the corresponding operational calculus was applied to solve the multi-term fractional differential equations with these derivatives and the suitably formulated initial conditions. In [15,16], a survey of the operational calculi for several different fractional derivatives was provided.

The rest of this paper is organized as follows: In Section 2, we present an overview of some important properties of the general fractional integrals (GFI) and the GFD of arbitrary order. Then we introduce the sequential GFDs, prove the 1st and the 2nd fundamental theorem of fractional calculus (FC) for these derivatives, and derive an explicit form for their projector operators. Section 3 is devoted to construction of an operational calculus of the Mikusiński type for the GFD of arbitrary order. In particular, the sequential GFDs of arbitrary order are represented as multiplication with certain elements of the constructed field of convolution quotients. The developed operational calculus can be applied for derivation of the closed form solution formulas for the ordinary and partial fractional differential equations with the sequential GFDs of arbitrary order.

2. General Fractional Derivatives of Arbitrary Order

The first publication devoted to the GFI and GFD was the paper [17] by Sonine published in 1884, even if no fractional integrals and fractional derivatives were explicitly mentioned there. In [17], Sonine generalized the solution method employed by Abel in [18,19] for the integro-differential equation (in slightly different notations)

$$f(t) = \frac{1}{\Gamma(1-\alpha)}\int_0^t \frac{g'(\tau)\,d\tau}{(t-\tau)^{\alpha}},\ 0 < \alpha < 1 \tag{2}$$

to the case of more general integral equations. In his derivations, Abel essentially used the relation

$$(h_\alpha * h_{1-\alpha})(t) = \{1\},\ t > 0,\ 0 < \alpha < 1, \tag{3}$$

where h_α denotes a power function

$$h_\alpha(t) := \frac{t^{\alpha-1}}{\Gamma(\alpha)},\ t > 0,\ \alpha > 0, \tag{4}$$

the operation $*$ stands for the Laplace convolution

$$(f_1 \, * \, f_2)(t) = \int_0^t f_1(t-\tau) f_2(\tau)\, d\tau, \tag{5}$$

and $\{1\}$ is the function that is identically equal to 1 for $t > 0$. Abel's solution to the integro-differential Equation (2) under the condition $g(0) = 0$ is nowadays well-known as the Riemann–Liouville fractional integral:

$$g(t) = \frac{1}{\Gamma(\alpha)} \int_0^t (t-\tau)^{\alpha-1} f(\tau)\, d\tau,\ t > 0. \tag{6}$$

The brilliant idea of Sonine was to replace the power functions h_α and $h_{1-\alpha}$ in the relation (3) with the arbitrary functions $\kappa,\ k$ that satisfy the same relation:

$$(\kappa \, * \, k)(t) = \{1\},\ t > 0. \tag{7}$$

Nowadays such functions are called the Sonine kernels and the condition (7) is refereed to as the Sonine condition.

By employing the Abel method, Sonine solved the convolution type integral equation

$$f(t) = \int_0^t \kappa(t-\tau) g(\tau)\, d\tau \ = \ (\kappa \, * \, g)(t) \tag{8}$$

in explicit form:

$$g(t) = \frac{d}{dt} \int_0^t k(t-\tau) f(\tau)\, d\tau \ = \ \frac{d}{dt}(k \, * \, f)(t), \tag{9}$$

provided the kernels $\kappa,\ k$ satisfy the Sonine condition (7).

It is worth mentioning that derivations of both Abel and Sonine were not rigorous from the modern viewpoint because they did not introduce the suitable spaces of functions and did not provide conditions for validity of their formal manipulations with integrals and derivatives. Only recently, the operators (8) and (9) became a subject of active research in FC and nowadays their mathematical theory is under intensive construction, see [13,20–28].

In this paper, we deal with the GFI and GFD of arbitrary order introduced in [24] for the first time. It is well-known that both the Riemann–Liouville fractional integral

$$(I^\alpha_{0+} f)(t) = \frac{1}{\Gamma(\alpha)} \int_0^t (t-\tau)^{\alpha-1} f(\tau)\, d\tau,\ \alpha > 0,\ t > 0 \tag{10}$$

and the Riemann–Liouville and Caputo fractional derivatives

$$(D^\alpha_{0+} f)(t) = \frac{d^n}{dt^n}(I^{n-\alpha}_{0+} f)(t),\ n-1 \le \alpha < n,\ n \in \mathbb{N}, \tag{11}$$

$$({}_*D^\alpha_{0+} f)(t) = \left(D^\alpha_{0+} \left(f(\cdot) - \sum_{j=0}^{n-1} f^{(j)}(0) h_{j+1}(\cdot) \right) \right)(t),\ t > 0,\ n-1 \le \alpha < n,\ n \in \mathbb{N} \tag{12}$$

are well-defined for any order $\alpha > 0$.

However, the relation (3) is valid only under the restriction $0 < \alpha < 1$. Similarly, the Sonine operators defined by the right-hand sides of the Formulas (8) and (9) (GFI and GFD in the modern FC notations) have a "generalized order" between zero and one because of the Sonine condition (7). To define the GFI and GFD of arbitrary positive order, in [24], the Sonine condition (7) was extended and specified for the kernels from some suitable spaces of functions:

Definition 1 ([24]). *Let the functions κ and k satisfy the condition*

$$(\kappa * k)(t) = \{1\}^{<n>}(t),\ n \in \mathbb{N},\ t > 0, \tag{13}$$

where

$$\{1\}^{<n>}(t) := (\underbrace{\{1\} * \ldots * \{1\}}_{n\ times})(t) = h_n(t) = \frac{t^{n-1}}{(n-1)!}$$

and the inclusions $\kappa \in C_{-1}(0, +\infty)$ and $k \in C_{-1,0}(0, +\infty)$ hold true, where

$$C_{-1}(0, +\infty) := \{f : f(t) = t^p f_1(t),\ t > 0,\ p > -1,\ f_1 \in C([0, +\infty))\}, \tag{14}$$

$$C_{-1,0}(0, +\infty) = \{f : f(t) = t^p f_1(t),\ t > 0,\ -1 < p < 0,\ f_1 \in C([0, +\infty))\}. \tag{15}$$

The set of pairs (κ, k) of such kernels is denoted by $\mathcal{L}_n$.

For a detailed treatment of the case $n = 1$, i.e., for a theory of the GFI and GFD with the kernels from $\mathcal{L}_1$ we refer to [23]. In this paper, our focus will be on the case $n > 1$, even if the case $n = 1$ is also included in all formulations and derivations.

It is worth mentioning that one cannot interchange the kernels κ and k in Definition 1 with $n > 1$ because of the non-symmetrical inclusions $\kappa \in C_{-1}(0, +\infty)$ and $k \in C_{-1,0}(0, +\infty)$ (in the case $n = 1$, Definition 1 is symmetrical and one can interchange the kernels κ and k). However, the kernel $\kappa(t) = h_\alpha(t),\ \alpha > 0$ of the Riemann–Liouville integral (10) and the kernel $k(t) = h_{n-\alpha}(t)$ of the Riemann–Liouville and Caputo fractional derivatives (11) and (12) of order $\alpha,\ n-1 < \alpha < n,\ n \in \mathbb{N}$ can be also not interchanged in the case $n > 1$. Evidently, the kernels $\kappa(t) = h_\alpha(t),\ \alpha > 0$ and $k(t) = h_{n-\alpha}(t),\ n-1 < \alpha < n,\ n \in \mathbb{N}$ belong to the kernel set $\mathcal{L}_n$. However, they are the Sonine kernels only in the case $n = 1$, i.e., only in the case of the fractional derivatives orders less than one.

For other examples of the kernels from $\mathcal{L}_n,\ n > 1$ and for the procedures how to construct them starting from the known Sonine kernels from the set $\mathcal{L}_1$ we refer to [24,28].

Now we define the GFI and GFD of arbitrary order, present their known properties and derive some new ones.

Definition 2 ([24]). *Let $(\kappa, k) \in \mathcal{L}_n$. The GFI with the kernel κ and the GFD of arbitrary order with the kernel k are defined as follows:*

$$(\mathbb{I}_{(\kappa)} f)(t) := \int_0^t \kappa(t-\tau) f(\tau)\, d\tau,\ t > 0, \tag{16}$$

$$({}_*\mathbb{D}_{(k)} f)(t) := \left(\mathbb{D}_{(k)}\left(f(\cdot) - \sum_{j=0}^{n-1} f^{(j)}(0) h_{j+1}(\cdot)\right)\right)(t),\ t > 0, \tag{17}$$

where the GFD $\mathbb{D}_{(k)}$ defined in the Riemann–Liouville sense is given by the relation

$$(\mathbb{D}_{(k)} f)(t) := \frac{d^n}{dt^n} \int_0^t k(t-\tau) f(\tau)\, d\tau,\ t > 0. \tag{18}$$

As already mentioned, the kernels of the Riemann–Liouville fractional integral (10) and the Caputo fractional derivative (12) belong to the kernel set $\mathcal{L}_n$ and thus these FC operators are particular cases of the GFI (16) and the GFD (17), respectively.

Another interesting particular case of the GFI (16) and the GFD (17) was presented in [24]. Let the condition $n-2 < \nu < n-1,\ n \in \mathbb{N}$ holds true. Then the operator

$$(\mathbb{I}_{(\kappa)} f)(t) = \int_0^t (t-\tau)^{\nu/2} J_\nu(2\sqrt{t-\tau}) f(\tau) d\tau,\ t > 0, \tag{19}$$

is a particular case of the GFI (16) and the corresponding GFD of arbitrary order takes the following form:

$$({}_*\mathbb{D}_{(k)} f)(t) := \left(\mathbb{D}_{(k)} \left(f(\cdot) - \sum_{j=0}^{n-1} f^{(j)}(0) h_{j+1}(\cdot)\right)\right)(t),\ t > 0, \tag{20}$$

where

$$(\mathbb{D}_{(k)} f)(t) = \frac{d^n}{dt^n} \int_0^t (t-\tau)^{n/2-\nu/2-1} I_{n-\nu-2}(2\sqrt{t-\tau}) f(\tau)\, d\tau,\ t > 0 \tag{21}$$

and the functions J_ν and I_ν are the Bessel and the modified Bessel functions, respectively.

In this paper, we investigate the GFI and the GFD of arbitrary order on the space $C_{-1}(0,+\infty)$ defined by the Formula (14) and its sub-spaces. In particular, the sub-spaces

$$C_{-1}^m(0,+\infty) := \{f :\ f^{(m)} \in C_{-1}(0,+\infty)\}, m \in \mathbb{N}_0 = \mathbb{N} \cup \{0\}$$

will be often used. These sub-spaces were introduced and investigated in [10] in connection with construction of an operational calculus of Mikusiński type for the Caputo fractional derivative.

As mentioned in [24], if the inclusion $k \in C_{-1}^{n-1}(0,+\infty)$ holds valid, the GFD (17) can be represented as follows:

$$({}_*\mathbb{D}_{(k)} f)(t) = (\mathbb{D}_{(k)} f)(t) - \sum_{j=0}^{n-1} f^{(j)}(0) \frac{d^{n-j-1}}{dt^{n-j-1}} k(t),\ t > 0, \tag{22}$$

where $\mathbb{D}_{(k)}$ is given by the relation (18). Moreover, for $f \in C_{-1}^n(0,+\infty)$, the GFD (17) takes the form

$$({}_*\mathbb{D}_{(k)} f)(t) = \int_0^t k(t-\tau) f^{(n)}(\tau)\, d\tau,\ t > 0 \tag{23}$$

that is often used in the case of the power law kernel $k(\tau) = h_{n-\alpha}(\tau),\ n-1 < \alpha < n,\ n \in \mathbb{N}$ of the Caputo fractional derivative.

The basic properties of the GFI (16) of arbitrary order on the space $C_{-1}(0,+\infty)$ immediately follow from the well-known properties of the Laplace convolution [23]:

$$\mathbb{I}_{(\kappa)} : C_{-1}(0,+\infty) \to C_{-1}(0,+\infty) \text{ (mapping property)}, \tag{24}$$

$$\mathbb{I}_{(\kappa_1)} \mathbb{I}_{(\kappa_2)} = \mathbb{I}_{(\kappa_2)} \mathbb{I}_{(\kappa_1)} \text{ (commutativity law)}, \tag{25}$$

$$\mathbb{I}_{(\kappa_1)} \mathbb{I}_{(\kappa_2)} = \mathbb{I}_{(\kappa_1 * \kappa_2)} \text{ (index law)}. \tag{26}$$

We also mention the following important theorem:

Theorem 1 ([10])**.** *The triple $\mathcal{R}_{-1} = (C_{-1}(0,+\infty), +, *)$ with the usual addition $+$ and multiplication $*$ in form of the Laplace convolution is a commutative ring without unity with respect to multiplication and without divisors of zero.*

Furthermore, in [24], two fundamental theorems of FC for the GFI and the GFD of arbitrary order have been proved. The formulations of these theorems are provided below, for the proofs see [24].

Theorem 2 ([24])**.** *Let* $(\kappa,\, k) \in \mathcal{L}_n$.

The GFD (17) *of arbitrary order is a left inverse operator to the GFI* (16) *on the space* $C_{-1,(k)}(0,+\infty)$:

$$({}_*\mathbb{D}_{(k)}\, \mathbb{I}_{(\kappa)}\, f)(t) = f(t),\ f \in C_{-1,(k)}(0,+\infty),\ t>0, \tag{27}$$

where $C_{-1,(k)}(0,+\infty) := \{f:\ f(t) = (\mathbb{I}_{(k)}\, \phi)(t),\ \phi \in C_{-1}(0,+\infty)\}$.

Theorem 3 ([24])**.** *Let* $(\kappa,\, k) \in \mathcal{L}_n$.

Then the relation

$$(\mathbb{I}_{(\kappa)}\, {}_*\mathbb{D}_{(k)}\, f)(t) = f(t) - \sum_{j=0}^{n-1} f^{(j)}(0)\, h_{j+1}(t) \tag{28}$$

holds true on the space $C_{-1}^{n}(0,+\infty)$.

It is worth mentioning that the result of Theorem 3 (2nd fundamental theorem of FC for the GFD of arbitrary order) can be reformulated in terms of the so-called projector operator P of the GFD (17):

$$(P\,f)(t) := f(t) - (\mathbb{I}_{(\kappa)}\, {}_*\mathbb{D}_{(k)}\, f)(t) = \sum_{j=0}^{n-1} f^{(j)}(0)\, h_{j+1}(t). \tag{29}$$

The form of the projector operator P determines the natural initial conditions for the initial-value problems for the fractional differential equations with the GFD (17). According to the representation (29), they are provided in terms of the integer order derivatives $f^{(j)}(0),\ j = 0,1,\ldots,n-1$ of the unknown function as it is the case for the ordinary differential equations and for the fractional differential equations with the Caputo derivatives.

In the rest of this section, we define the m-fold sequential GFIs and GFDs of arbitrary order with the kernels $(\kappa,\, k) \in \mathcal{L}_n,\ n \in \mathbb{N}$ and investigate their properties. In the case $n = 1$, the m-fold GFIs and the m-fold sequential GFDs were introduced and studied in [25,29]. In [14], the m-fold sequential GFDs in the Riemann–Liouville sense with the kernels $(\kappa,\, k) \in \mathcal{L}_n$ have been considered based on the GFD of arbitrary order defined in the Riemann–Liouville sense by the Formula (18). In this paper, we deal with the case of the GFD of arbitrary order in form (17).

First we define the convolution powers $f^{<m>},\ m \in \mathbb{N}_0$ of a function f as follows:

$$f^{<m>}(t) := \begin{cases} \{1\}, & m = 0, \\ f(t), & m = 1, \\ (\underbrace{f * \ldots * f}_{m\ times})(t), & m = 2,3,\ldots. \end{cases} \tag{30}$$

Definition 3. *Let* $(\kappa,\, k) \in \mathcal{L}_n,\ n \in \mathbb{N}$.

The m-fold GFI is defined as a composition of m GFIs with the kernel κ:

$$(\mathbb{I}_{(\kappa)}^{<m>}\, f)(t) := (\underbrace{\mathbb{I}_{(\kappa)} \cdots \mathbb{I}_{(\kappa)}}_{m\ times}\, f)(t) = (\kappa^{<m>} * f)(t),\ t > 0. \tag{31}$$

The corresponding m-fold sequential GFD of arbitrary order is defined as follows:

$$({}_*\mathbb{D}_{(k)}^{<m>} f)(t) := (\underbrace{{}_*\mathbb{D}_{(k)} \ldots {}_*\mathbb{D}_{(k)}}_{m \ times} f)(t),\ t > 0. \tag{32}$$

In analogy to the case of the Riemann–Liouville fractional integral and the Caputo fractional derivative, the operators $\mathbb{I}_{(\kappa)}^{<0>}$ and ${}_*\mathbb{D}_{(k)}^{<0>}$ are interpreted as the identity operator Id.

Due to Theorem 1, the kernel $\kappa^{<m>},\ m \in \mathbb{N}$ from the Formula (31) is from the space $C_{-1}(0,+\infty)$ and thus the m-fold GFI can be represented as a GFI with the kernel $\kappa^{<m>}$:

$$(\mathbb{I}_{(\kappa)}^{<m>} f)(t) = (\kappa^{<m>} * f)(t) = (\mathbb{I}_{(\kappa)^{<m>}} f)(t),\ t > 0. \tag{33}$$

The m-fold sequential GFD (32) of arbitrary order is a direct generalization of the sequential fractional derivative in the Caputo sense to the case of the integro-differential operators with the kernels from $\mathcal{L}_n$.

The 1st fundamental theorem of FC (Theorem 2) for the GFI (16) and the GFD (17) of arbitrary order immediately leads to the following important result:

Theorem 4 (1st fundamental theorem of FC for the m-fold sequential GFD of arbitrary order)**.** *Let* $(\kappa,\ k) \in \mathcal{L}_n,\ n \in \mathbb{N}$.

The m-fold sequential GFD (32) *of arbitrary order is a left inverse operator to the m-fold GFI* (31) *on the space* $C_{-1,(k)}(0,+\infty)$:

$$(\mathbb{D}_{(k)}^{<m>}\ \mathbb{I}_{(\kappa)}^{<m>} f)(t) = f(t),\ f \in C_{-1,(k)}(0,+\infty),\ t > 0. \tag{34}$$

To generalize Theorem 3 to the case of the m-fold sequential GFDs, we first introduce the suitable spaces of functions in the form

$$C_{-1,(k)}^{n,m}(0,+\infty) := \{f \in C_{-1}^{n}(0,+\infty) :\ {}_*D_{(k)}^{<i>} f \in C_{-1}^{n}(0,+\infty),\ i - 1,\ldots,m-1\}. \tag{35}$$

For $m = 1$, we set $C_{-1,(k)}^{n,1}(0,+\infty) := C_{-1}^{n}(0,+\infty)$.

Theorem 5 (2nd fundamental theorem of FC for the m-fold sequential GFD of arbitrary order)**.** *Let* $(\kappa,\ k) \in \mathcal{L}_n$.

Then the relation

$$(\mathbb{I}_{(\kappa)}^{<m>}\ {}_*\mathbb{D}_{(k)}^{<m>} f)(t) = f(t) - \sum_{j=0}^{n-1} \frac{d^j f}{dt^j}(0)\, h_{j+1}(t) - \sum_{j=0}^{n-1}\sum_{i=1}^{m-1} \left(\frac{d^j}{dt^j}\, {}_*\mathbb{D}_{(k)}^{<i>} f\right)(0)\, (\kappa^{<i>} * h_{j+1})(t) \tag{36}$$

holds true on the space $C_{-1,(k)}^{n,m}(0,+\infty)$.

Proof. In the case $m = 1$, the statement of Theorem 5 is Theorem 3 that was proved in [24]. Now we proceed with the case $m = 2$. Then we have the inclusion $f \subset C_{-1,(k)}^{n,2}(0,+\infty)$ and the representation

$$(\mathbb{I}_{(\kappa)}^{<2>}\ {}_*\mathbb{D}_{(k)}^{<2>} f)(t) = (\mathbb{I}_{(\kappa)}\mathbb{I}_{(\kappa)}\ {}_*\mathbb{D}_{(k)}\ {}_*\mathbb{D}_{(k)} f)(t) =$$

$$(\mathbb{I}_{(\kappa)}(\mathbb{I}_{(\kappa)}\ {}_*\mathbb{D}_{(k)}(\ {}_*\mathbb{D}_{(k)}f)))(t).$$

For a function $f \in C_{-1,(k)}^{n,2}(0,+\infty)$, the inclusion ${}_*D_{(k)}f \in C_{-1}^{n}(0,+\infty)$ holds true and thus we can apply Theorem 3 to the inner composition $\mathbb{I}_{(\kappa)}\ {}_*\mathbb{D}_{(k)}$ acting on the function ${}_*\mathbb{D}_{(k)} f$ and get the formula

$$(\mathbb{I}_{(\kappa)}^{<2>} \,_*\mathbb{D}_{(k)}^{<2>} f)(t) = \left(\mathbb{I}_{(\kappa)} \left[(\,_*\mathbb{D}_{(k)} f)(t) - \sum_{j=0}^{n-1} \left(\frac{d^j}{dt^j} \,_*\mathbb{D}_{(k)} f \right)(0)\, h_{j+1})(t) \right]\right)(t) =$$

$$(\mathbb{I}_{(\kappa)} \,_*\mathbb{D}_{(k)} f)(t) - \sum_{j=0}^{n-1} \left(\frac{d^j}{dt^j} \,_*\mathbb{D}_{(k)} f \right)(0)\, (\kappa * h_{j+1})(t).$$

The final result immediately follows by applying Theorem 3 to the composition $\mathbb{I}_{(\kappa)} \,_*\mathbb{D}_{(k)} f$ at the right-hand side of the last formula:

$$(\mathbb{I}_{(\kappa)}^{<2>} \mathbb{D}_{(k)}^{<2>} f)(t) = f(t) - \sum_{j=0}^{n-1} \frac{d^j f}{dt^j}(0)\, h_{j+1}(t) - \sum_{j=0}^{n-1} \left(\frac{d^j}{dt^j} \,_*\mathbb{D}_{(k)} f \right)(0)\, (\kappa * h_{j+1})(t).$$

For $m = 3, 4, \ldots$, we employ the recurrent formula

$$(\mathbb{I}_{(\kappa)}^{<m>} \,_*\mathbb{D}_{(k)}^{<m>} f)(t) = (\mathbb{I}_{(\kappa)}^{<m-1>} (\mathbb{I}_{(\kappa)} \,_*\mathbb{D}_{(k)} (\,_*\mathbb{D}_{(k)}^{<m-1>} f)))(t) =$$

$$\left(\mathbb{I}_{(\kappa)}^{<m-1>} \left[(\,_*\mathbb{D}_{(k)}^{<m-1>} f)(t) - \sum_{j=0}^{n-1} \left(\frac{d^j}{dt^j} \,_*\mathbb{D}_{(k)}^{<m-1>} f \right)(0)\, h_{j+1}(t) \right]\right)(t) =$$

$$\left(\mathbb{I}_{(\kappa)}^{<m-1>} \,_*\mathbb{D}_{(k)}^{<m-1>} f\right)(t) - \sum_{j=0}^{n-1} \left(\frac{d^j}{dt^j} \,_*\mathbb{D}_{(k)}^{<m-1>} f \right)(0)\, (\kappa^{<m-1>} * h_{j+1})(t)$$

and the principle of the mathematical induction to complete the proof of the representation (36). □

Remark 1. *The Formula* (36) *can be rewritten in terms of the projector operator P_m of the m-fold sequential GFD* (32) *as follows:*

$$(P_m f)(t) := f(t) - (\mathbb{I}_{(\kappa)}^{<m>} \,_*\mathbb{D}_{(k)}^{<m>} f)(t) =$$

$$\sum_{j=0}^{n-1} \frac{d^j f}{dt^j}(0)\, h_{j+1}(t) + \sum_{j=0}^{n-1} \sum_{i=1}^{m-1} \left(\frac{d^j}{dt^j} \,_*\mathbb{D}_{(k)}^{<i>} f \right)(0)\, (\kappa^{<i>} * h_{j+1})(t),\ t > 0. \tag{37}$$

According to Theorem 5, the Formula (37) *is valid on the space $C_{-1,(k)}^{n,m}(0, +\infty)$.*

It is worth mentioning that the Formula (37) *specifies the natural form of the initial conditions while dealing with the fractional differential equations that contain the sequential GFDs of the orders up to m. These initial conditions should be formulated in terms of the values of the integer order derivatives applied to the sequential GFDs of the orders up to $(m-1)$ evaluated at the point zero:*

$$\frac{d^j f}{dt^j}(0) = a_{j0},\ j = 0, \ldots, n-1,\ \left(\frac{d^j}{dt^j} \,_*\mathbb{D}_{(k)}^{<i>} f \right)(0) = a_{ji},\ j = 0, \ldots, n-1,\ i = 1, \ldots, m-1.$$

Remark 2. *Evidently, the identity operator Id on the space $C_{-1,(k)}^{n,m}(0, +\infty)$ can be represented as follows:*

$$\begin{aligned} Id =& (Id - \mathbb{I}_{(\kappa)} \,_*\mathbb{D}_{(k)}) + (\mathbb{I}_{(\kappa)} \,_*\mathbb{D}_{(k)} - \mathbb{I}_{(\kappa)}^{<2>} \,_*\mathbb{D}_{(k)}^{<2>}) + \ldots \\ &+ (\mathbb{I}_{(\kappa)}^{<m-1>} \,_*\mathbb{D}_{(k)}^{<m-1>} - \mathbb{I}_{(\kappa)}^{<m>} \,_*\mathbb{D}_{(k)}^{<m>}) + \mathbb{I}_{(\kappa)}^{<m>} \,_*\mathbb{D}_{(k)}^{<m>} \\ =& (Id - \mathbb{I}_{(\kappa)} \,_*\mathbb{D}_{(k)}) + (\mathbb{I}_{(\kappa)} (\,_*\mathbb{D}_{(k)} - (\mathbb{I}_{(\kappa)} \,_*\mathbb{D}_{(k)}) \,_*\mathbb{D}_{(k)})) + \ldots \\ &+ (\mathbb{I}_{(\kappa)}^{<m-1>} (\,_*\mathbb{D}_{(k)}^{<m-1>} - (\mathbb{I}_{(\kappa)} \,_*\mathbb{D}_{(k)}) \,_*\mathbb{D}_{(k)}^{<m-1>})) + \mathbb{I}_{(\kappa)}^{<m>} \,_*\mathbb{D}_{(k)}^{<m>} \\ =& P + (\mathbb{I}_{(\kappa)} P(\,_*\mathbb{D}_{(k)})) + \cdots + (\mathbb{I}_{(\kappa)}^{<m-1>} P(\,_*\mathbb{D}_{(k)}^{<m-1>})) + \mathbb{I}_{(\kappa)}^{<m>} \,_*\mathbb{D}_{(k)}^{<m>}, \end{aligned}$$

where P stands for the projector operator of the GFD (17) given by the Formula (29).

As a consequence, the projector operator P_m of the m-fold sequential GFD (32) can be also represented in a more compact form:

$$(P_m f)(t) = \sum_{i=0}^{m-1} (\mathbb{I}_{(\kappa)}^{<i>} (P ({}_*\mathbb{D}_{(k)}^{<i>}) f))(t),\ t > 0. \tag{38}$$

It is worth mentioning that for any function $f \in C_{-1,(k)}^{n,m}(0, +\infty)$, its image $P_m f$ by the projector operator is from the kernel of the m-fold sequential GFD:

$$\begin{aligned}({}_*\mathbb{D}_{(k)}^{<m>} (P_m f))(t) =& ({}_*\mathbb{D}_{(k)}^{<m>} f)(t) - ({}_*\mathbb{D}_{(k)}^{<m>} \mathbb{I}_{(\kappa)}^{<m>} {}_*\mathbb{D}_{(k)}^{<m>} f)(t) \\ =& ({}_*\mathbb{D}_{(k)}^{<m>} f)(t) - ({}_*\mathbb{D}_{(k)}^{<m>} f)(t) = 0.\end{aligned}$$

We also mention the inclusions

$$\text{Ker}\,{}_*\mathbb{D}_{(k)} \subset \text{Ker}\,{}_*\mathbb{D}_{(k)}^{<2>} \subset \cdots \subset \text{Ker}\,{}_*\mathbb{D}_{(k)}^{<m>}$$

that immediately follow from the definition of the m-fold sequential GFD.

3. Operational Calculus for the GFD of Arbitrary Order

As already mentioned in Introduction, an operational calculus of the Mikusiński type for the GFI and GFD with the kernels $(\kappa,\ k) \in \mathcal{L}_1$ has been constructed in [13]. This case corresponds to the GFI and GFD of the "generalized order" between zero and one. In this paper, we extend the constructions presented in [13] to the case of the kernels $(\kappa,\ k) \in \mathcal{L}_n,\ n \in \mathbb{N}$, i.e., for the GFI and GFD of arbitrary order.

The first important component of any operational calculus of Mikusiński type is a suitable ring of functions. For the operational calculus for the GFI and GFD with the kernels $(\kappa,\ k) \in \mathcal{L}_n$, this ring is described in Theorem 1 that says that the triple $\mathcal{R}_{-1} = (C_{-1}(0, +\infty), +, *)$ with the usual addition $+$ and multiplication $*$ in form of the Laplace convolution is a commutative ring without divisors of zero.

Furthermore, Definition 1 ensures that the kernels κ and k from $\mathcal{L}_n$ are elements of this ring:

$$\kappa \in \mathcal{R}_{-1},\ k \in \mathcal{R}_{-1} \text{ if } (\kappa,\ k) \in \mathcal{L}_n.$$

In particular, this means that the GFI with the kernel κ is reduced to a conventional multiplication on the ring $\mathcal{R}_{-1}$:

$$(\mathbb{I}_{(\kappa)} f)(t) = (\kappa * f)(t),\ t > 0,\ f \in \mathcal{R}_{-1}. \tag{39}$$

As to the GFD of arbitrary order, it cannot be reduced to the algebraic operations on the ring $\mathcal{R}_{-1}$. The reason is that the GFD is a left-inverse operator to the GFI and the ring $\mathcal{R}_{-1}$ does not possess a unity element with respect to multiplication. Thus, no inverse element to $\kappa \in \mathcal{R}_{-1}$ does exist in $\mathcal{R}_{-1}$.

To demonstrate the last statement, let us assume that the GFD (17) of arbitrary order can be represented as a convolution with a certain element $\kappa^{-1} \in \mathcal{R}_{-1}$:

$$({}_*\mathbb{D}_{(k)} f)(t) = (\kappa^{-1} * f)(t),\ t > 0,\ f \in \mathcal{R}_{-1}.$$

According to Theorem 2, the GFD (17) of arbitrary order is a left inverse operator to the GFI (16). Combining the last equation with the representation (39), we arrive at the relation

$$({}_*\mathbb{D}_{(k)}\, \mathbb{I}_{(\kappa)} f)(t) = (\kappa^{-1} * (\kappa * f))(t) = ((\kappa^{-1} * \kappa) * f)(t) = f(t),\ t > 0,\ f \in \mathcal{R}_{-1}$$

that contradicts to the fact that the ring $\mathcal{R}_{-1}$ does not possess a unity element with respect to multiplication.

To resolve the problem mentioned above, the ring $\mathcal{R}_{-1}$ is extended to a field of convolution quotients. Let us remind that $\mathcal{R}_{-1}$ does not have any divisors of zero (Theorem 1) and thus this extension follows a standard procedure, see, e.g., [10,13].

First, an equivalence relation on the set

$$C_{-1}^2(0,+\infty) := C_{-1}(0,+\infty) \times (C_{-1}(0,+\infty) \setminus \{0\})$$

is introduced:

$$(f_1, g_1) \sim (f_2, g_2) \Leftrightarrow (f_1 * g_2)(t) = (f_2 * g_1)(t),\ (f_1, g_1),\ (f_2, g_2) \in C_{-1}^2(0,+\infty).$$

Then we consider the equivalences classes $C_{-1}^2(0,+\infty)/\sim$ and denote them as quotients:

$$\frac{f}{g} := \{(f_1, g_1) \in C_{-1}^2(0,+\infty) :\ (f_1, g_1) \sim (f, g)\}.$$

On the space $C_{-1}^2(0,+\infty)/\sim$, usual operations of addition and multiplication are introduced:

$$\frac{f_1}{g_1} + \frac{f_2}{g_2} := \frac{f_1 * g_2 + f_2 * g_1}{g_1 * g_2},$$

$$\frac{f_1}{g_1} \cdot \frac{f_2}{g_2} := \frac{f_1 * f_2}{g_1 * g_2}.$$

These operations are correctly defined (they do not depend on the representatives of the equivalence classes). Theorem 1 and the definitions provided above immediately lead to the following important result:

Theorem 6 ([10]). *The triple $\mathcal{F}_{-1} = (C_{-1}^2(0,+\infty)/\sim,\ +,\ \cdot)$ is a field that is usually referred to as the field of convolution quotients.*

As usual, the ring $\mathcal{R}_{-1}$ can be embedded into the field $\mathcal{F}_{-1}$:

$$f \mapsto \frac{f * \kappa}{\kappa}, \tag{40}$$

where κ is the kernel of the GFI (16).

The set $C_{-1}^2(0,+\infty)/\sim$ of the equivalence classes is also a vector space with the addition operation mentioned above and multiplication with a scalar $\lambda \in \mathbb{R}$ or $\lambda \in \mathbb{C}$ defined as follows [10]:

$$\lambda \frac{f}{g} := \frac{\lambda f}{g},\ \frac{f}{g} \in C_{-1}^2(0,+\infty)/\sim.$$

On the other hand, the constant function $\{\lambda\}$ (the function that takes the value λ for any $t \geq 0$) is from the space $C_{-1}(0,+\infty)$ and thus it is an element of the ring $\mathcal{R}_{-1}$. According to our definitions, multiplication with $\{\lambda\}$ in the field $\mathcal{F}_{-1}$ is given by the following expression:

$$\{\lambda\} \cdot \frac{f}{g} = \frac{\{\lambda\} * f}{g},\ \frac{f}{g} \in \mathcal{F}_{-1}.$$

As we see, one has to differentiate between multiplication of an element from $\mathcal{F}_{-1}$ with a scalar λ and with the constant function $\{\lambda\}$.

Even if the ring $\mathcal{R}_{-1}$ is embedded into the field $\mathcal{F}_{-1}$, some elements of the field of convolution quotients cannot be reduced to the conventional functions from the ring. One of them is the unity element $I = \frac{\kappa}{\kappa}$ of the field $\mathcal{F}_{-1}$ with respect to multiplication (see [13]). Such elements can be interpreted as a kind of generalized functions (hyper-functions in the terminology of [30]). The inverse element to the kernel κ of the GFI (16) is another important hyper-function.

Definition 4 ([13]). *The inverse element to the kernel $\kappa \in \mathcal{R}_{-1}$ in the field $\mathcal{F}_{-1}$ in form*

$$S_\kappa := \frac{\kappa}{\kappa * \kappa} \tag{41}$$

is called an algebraic inverse element to the GFI (16).

By definition, the relation

$$\kappa \cdot S_\kappa = \frac{\kappa * \kappa}{\kappa} \cdot \frac{\kappa}{\kappa * \kappa} = \frac{\kappa^{<3>}}{\kappa^{<3>}} = I \tag{42}$$

holds true, where I is the unity of the field $\mathcal{F}_{-1}$ with respect to multiplication.

Then, following [13], we introduce an important notion of an algebraic GFD of arbitrary order.

Definition 5. *Let $(\kappa, k) \in \mathcal{L}_n$, $n \in \mathbb{N}$ and $f \in C^{n-1}[0, +\infty)$.*
The algebraic GFD of arbitrary order is defined as follows:

$${}_*\mathbb{D}_{(k)}\, f = S_\kappa \cdot f - S_\kappa \cdot (P\, f), \tag{43}$$

where the function f and the projector operator $P\,f$ given by the Formula (29) *are interpreted as elements of the convolution quotients field $\mathcal{F}_{-1}$.*

As we see, the algebraic GFD of arbitrary order is defined for any function $f \in C^{n-1}[0, +\infty)$. However, the GFD of arbitrary order given by the Formula (17) evidently does not always exist on the space $C^{n-1}[0, +\infty)$. Thus, the algebraic GFD (43) can be interpreted as a kind of a generalized derivative that assigns a certain element of the field $\mathcal{F}_{-1}$ to any function $f \in C^{n-1}[0, +\infty)$. However, the algebraic GFD (43) coincides with the GFD (17) on the space $C^n_{-1}(0, +\infty)$ (the inclusion $C^n_{-1}(0, +\infty) \subset C^{n-1}[0, +\infty)$ holds valid, see [10]).

Theorem 7. *Let $(\kappa, k) \in \mathcal{L}_n$ and $f \in C^n_{-1}(0, +\infty)$.*
Then the algebraic GFD (43) *coincides with the GFD* (17) *of arbitrary order:*

$$({}_*\mathbb{D}_{(k)}\, f)(t) \;=\; {}_*\mathbb{D}_{(k)}\, f = S_\kappa \cdot f - S_\kappa \cdot (P\, f), \tag{44}$$

where the projector operator $P\,f$ is defined by the Formula (29) *and the functions f, ${}_*\mathbb{D}_{(k)}\, f$, and $P\,f$ are interpreted as elements of the field $\mathcal{F}_{-1}$.*

Proof. On the space $C^n_{-1}(0, +\infty)$, the projector operator P is given by the Formula (29). Thus, we have the following representation:

$$(\mathbb{I}_{(\kappa)}\, {}_*\mathbb{D}_{(k)}\, f)(t) = (\kappa * ({}_*\mathbb{D}_{(k)}\, f))(t) = f(t) - (P\, f)(t),\; t > 0. \tag{45}$$

For any $f \in C^n_{-1}(0, +\infty)$, both the function at the right- and the function at the left-hand side of the Formula (45) evidently belong to the space $C_{-1}(0, +\infty)$. Because the ring $\mathcal{R}_{-1}$ is embedded into the convolution quotients field $\mathcal{F}_{-1}$, the Formula (45) can be interpreted as an equality of two elements from $\mathcal{F}_{-1}$. By multiplying this equality with the element S_κ and using the relation (42) we immediately arrive at the Formula (44). □

Remark 3. *In the case of the kernels $\kappa(t) = h_\alpha(t)$, $\alpha > 0$ and $k(t) = h_{n-\alpha}(t)$, $n - 1 < \alpha < n$, $n \in \mathbb{N}$, the GFI* (16) *is the well-known Riemann–Liouville fractional integral and the GFD* (17) *is the Caputo fractional derivative of order α. In* [10]*, a formula of type* (44) *for the Caputo fractional derivative has been derived for the first time.*

The constructions presented above can be extended to the case of the m-fold sequential GFD (32) of arbitrary order.

Definition 6. *Let $(\kappa, k) \in \mathcal{L}_n$, $n \in \mathbb{N}$ and ${}_*\mathbb{D}_{(k)}^{<i>} f \in C^{n-1}[0, +\infty)$, $i = 0, \ldots, m-1$.*
The m-fold sequential algebraic GFD of arbitrary order is defined by the expression

$$
{}_*\mathbb{D}_{(k)}^{<m>} f = S_\kappa^m \cdot f - S_\kappa^m \cdot (P_m f), \tag{46}
$$

where the function f and the projector operator $P_m f$ given by the Formula (37) are interpreted as the elements of the convolution quotients field $\mathcal{F}_{-1}$.

According to the Formula (37), the m-fold sequential algebraic GFD of arbitrary order is well defined for any function f that satisfies the conditions ${}_*\mathbb{D}_{(k)}^{<i>} f \in C^{n-1}[0, +\infty)$, $i = 0, \ldots, m-1$. Evidently, these conditions do not ensure existence of the m-fold sequential GFD (32) of arbitrary order. Therefore, the m-fold sequential algebraic GFD of arbitrary order can be interpreted as a kind of a generalized derivative. Still, for the functions from the space $C_{-1,(k)}^{n,m}(0, +\infty)$ given by the Formula (35), Theorem 5 and the same arguments that were used in the case $m = 1$ (Theorem 7 and its proof) lead to the following important result:

Theorem 8. *Let $(\kappa, k) \in \mathcal{L}_n$, $n \in \mathbb{N}$ and $f \in C_{-1,(k)}^{n,m}(0, +\infty)$.*
Then the m-fold sequential algebraic GFD (46) coincides with the m-fold sequential GFD (32):

$$
({}_*\mathbb{D}_{(k)}^{<m>} f)(t) = {}_*\mathbb{D}_{(k)}^{<m>} f = S_\kappa^m \cdot f - S_\kappa^m \cdot (P_m f), \tag{47}
$$

where the projector operator P_m is defined by the Formula (37) and the functions f, $\mathbb{D}_{(k)}^{<m>} f$, and $P_m f$ are interpreted as the elements of the field $\mathcal{F}_{-1}$.

Remark 4. *The representation (37) of the projector operator P_m along with the formulas $(I_{(\kappa)}^{<i>} f)(t) = (\kappa^{<i>} * f)(t)$ and $S_\kappa \cdot \kappa = I$ lead to another form of the Formula (47):*

$$
({}_*\mathbb{D}_{(k)}^{<m>} f)(t) = {}_*\mathbb{D}_{(k)}^{<m>} f = S_\kappa^m \cdot f - \sum_{i=0}^{m-1} S_\kappa^{m-i} \cdot (P\,({}_*\mathbb{D}_{(k)}^{<i>} f)), \tag{48}
$$

where the projector operator P is given by Equation (29).

According to the representations (44) and (47) (or (48)), the GFD (17) of arbitrary order and the m-fold sequential GFD (32) of arbitrary order can be represented as algebraic operations (multiplications) on the field $\mathcal{F}_{-1}$ of convolution quotients. These representations can be used to reduce the initial-value problems for the fractional differential equations with the GFDs of arbitrary order and the m-fold sequential GFDs of arbitrary order to some algebraic equations on the field $\mathcal{F}_{-1}$ of convolution quotients. The solutions to these equations are some elements of $\mathcal{F}_{-1}$ that in general are generalized functions or hyperfunctions. However, usually one looks for solutions in form of conventional functions, say, from the space $C_{-1}(0, +\infty)$. That is why the so-called operational relations (representations of some elements of the field $\mathcal{F}_{-1}$ as conventional functions from the ring $\mathcal{R}_{-1}$) are another important component of any operational calculus of Mikusiński type. In the rest of this section, we provide operational relations for the elements of the field $\mathcal{F}_{-1}$ in form of the rational functions $R(S_\kappa) = Q(S_\kappa)/P(S_\kappa)$ with $\deg(Q) < \deg(P)$. The result formulated in the next theorem is a basis for derivation of these operational relations.

Theorem 9 ([14]). *Let a function $\kappa \in C_{-1}(0, +\infty)$ be represented in the form*

$$
\kappa(t) = h_p(t)\kappa_1(t),\ t > 0,\ p > 0,\ \kappa_1 \in C[0, +\infty) \tag{49}
$$

and the convergence radius of the power series

$$\Sigma(z) = \sum_{j=0}^{+\infty} a_j\, z^j,\ a_j \in \mathbb{C},\ z \in \mathbb{C} \tag{50}$$

be non-zero. Then the convolution series

$$\Sigma_\kappa(t) = \sum_{j=0}^{+\infty} a_j\, \kappa^{<j+1>}(t) \tag{51}$$

is convergent for all $t > 0$ and defines a function from the space $C_{-1}(0,+\infty)$. Moreover, the series

$$t^{1-\alpha}\, \Sigma_\kappa(t) = \sum_{j=0}^{+\infty} a_j\, t^{1-\alpha}\, \kappa^{<j+1>}(t),\quad \alpha = \min\{p,\, 1\} \tag{52}$$

is uniformly convergent for $t \in [0,\, T]$ for any $T > 0$.

It is worth mentioning that the convolution series (51) is a far reaching generalization of the power series (50) that is also a convolution series generated by the kernel $\kappa = \{1\}$ (Mikusiński's operational calculus for the first order derivative).

As an example, let us consider the geometric series

$$\Sigma(t) = \sum_{j=1}^{+\infty} \lambda^{j-1} t^j,\ \lambda \in \mathbb{C},\ \lambda \neq 0,\ t \in \mathbb{C}. \tag{53}$$

Its convergence radius is $r = 1/|\lambda| > 0$. Theorem 9 ensures that the convolution series

$$l_{\kappa,\lambda}(t) = \sum_{j=1}^{+\infty} \lambda^{j-1} \kappa^{<j>}(t),\ \lambda \in \mathbb{C},\ t > 0 \tag{54}$$

is a function that belongs to the space $C_{-1}(0,+\infty)$.

In the framework of the Mikusiński operational calculus for the first order derivative, the kernel function $\kappa = \{1\}$ was employed. It is easy to verify that for this kernel the relation $\kappa^{<j>}(t) = \{1\}^{<j>}(t) = h_j(t)$ holds valid. Thus, the convolution series (54) is an exponential function:

$$l_{\kappa,\lambda}(t) = \sum_{j=1}^{+\infty} \lambda^{j-1} h_j(t) = \sum_{j=0}^{+\infty} \frac{(\lambda\, t)^j}{j!} = e^{\lambda\, t}. \tag{55}$$

Another important example is the operational calculus of Mikusiński type for the Caputo fractional derivative ([10]). In this case, the kernel κ is the power function h_α and the formula $\kappa^{<j>}(t) = h_\alpha^{<j>}(t) = h_{j\alpha}(t)$ holds valid. Thus, the convolution series (54) takes the form

$$l_{\kappa,\lambda}(t) = \sum_{j=1}^{+\infty} \lambda^{j-1} h_{j\alpha}(t) = t^{\alpha-1} \sum_{j=0}^{+\infty} \frac{\lambda^j\, t^{j\alpha}}{\Gamma(j\alpha+\alpha)} = t^{\alpha-1} E_{\alpha,\alpha}(\lambda\, t^\alpha), \tag{56}$$

where the two-parameters Mittag–Leffler function $E_{\alpha,\beta}$ is defined by the following absolutely convergent series:

$$E_{\alpha,\beta}(z) = \sum_{j=0}^{+\infty} \frac{z^j}{\Gamma(\alpha\, j + \beta)},\ \Re(\alpha) > 0,\ z, \beta \in \mathbb{C}. \tag{57}$$

Some other particular cases of the convolution series (54) were presented in [13].

A very important operational relation in terms of the convolution series $l_{\kappa,\lambda}$ is presented in the next theorem.

Theorem 10. *For any $\lambda \in \mathbb{C}$, the operational relation*

$$\frac{I}{S_\kappa - \lambda} = l_{\kappa,\lambda}(t),\ t > 0 \tag{58}$$

holds true, where by $\frac{I}{e}$ we denote the element of the field $\mathcal{F}_{-1}$ inverse to the element $e \in \mathcal{F}_{-1}$.

Proof. In the case $\lambda = 0$, the convolution series (54) is just the kernel function $\kappa(t)$ and the operational relation (58) in the form $\frac{I}{S_\kappa} = \kappa$ holds true by definition of S_κ (see the Formula (42)).

For $\lambda \neq 0$, the GFI $\mathbb{I}_{(\kappa)}$ can be applied to the convolution series $l_{\kappa,\lambda}$ term by term due to Theorem 9 and we get the following chain of equations:

$$(I - \lambda\kappa) \cdot l_{\kappa,\lambda} = l_{\kappa,\lambda} - \lambda(\mathbb{I}_{(\kappa)} \sum_{j=1}^{+\infty} \lambda^{j-1}\kappa^{j})(t) = l_{\kappa,\lambda} - \lambda \sum_{j=1}^{+\infty} \lambda^{j-1}(\mathbb{I}_{(\kappa)}\, \kappa^{j})(t) =$$

$$l_{\kappa,\lambda} - \sum_{j=1}^{+\infty} \lambda^{j}\kappa^{j+1}(t) = \sum_{j=1}^{+\infty} \lambda^{j-1}\kappa^{j}(t) - \sum_{j=2}^{+\infty} \lambda^{j-1}\kappa^{j}(t) = \kappa.$$

Otherwise, the element $I - \lambda\kappa$ of the field $\mathcal{F}_{-1}$ can be represented as follows:

$$I - \lambda\kappa = \kappa \cdot S_\kappa - \lambda\kappa = \kappa \cdot (S_\kappa - \lambda).$$

The last two representations lead to the relation

$$\kappa \cdot (S_\kappa - \lambda) \cdot l_{\kappa,\lambda} = (I - \lambda\kappa) \cdot l_{\kappa,\lambda} = \kappa.$$

Thus, the element $(S_\kappa - \lambda) \cdot l_{\kappa,\lambda}$ is the unity I of the field $\mathcal{F}_{-1}$ and the proof of the operational relation (58) is completed. □

In the case of the kernels $(\kappa, k) \in \mathcal{L}_1$, Theorem (10) has been formulated and proved in [13].

As an example, we consider the kernel $\kappa = \{1\}$ (Mikusiński's operational calculus for the first order derivative). According to the Formula (55), the operational relation (58) can be represented in the well-known form:

$$\frac{I}{S_\kappa - \lambda} = l_{\kappa,\lambda}(t) = e^{\lambda t}. \tag{59}$$

In the case of the kernel $\kappa(t) = h_\alpha(t),\ t > 0$ (operational calculus for the Caputo fractional derivative [10]), the Formula (56) leads to the following operational relation:

$$\frac{I}{S_\kappa - \lambda} = l_{\kappa,\lambda}(t) = t^{\alpha-1}E_{\alpha,\alpha}(\lambda\, t^{\alpha}), \tag{60}$$

where the two-parameters Mittag–Leffler function $E_{\alpha,\beta}$ is defined by (57). This operational relation has been deduced in [6] for the fist time.

Other particular cases of the operational relation (58) have been presented in [13].

As already mentioned, the operational relation (58) can be used to deduce other useful operational relations. The embedding of the ring $\mathcal{R}_{-1}$ into the field $\mathcal{F}_{-1}$ means in particular that a convolution of any ring elements complies with multiplication of the corresponding elements of the field of convolution quotients. Thus, we get the following operational relation:

$$\frac{I}{(S_\kappa - \lambda)^m} = l_{\kappa,\lambda}^{<m>}(t),\ t > 0,\ m \in \mathbb{N}. \tag{61}$$

For a representation of the convolution powers $l_{\kappa,\lambda}^{<m>}$ in terms of the convolution series see [13].

As an example, we consider the kernel $\kappa = \{1\}$ (Mikusiński's operational calculus for the first order derivative). The operational relation (61) takes the well-known form [1]:

$$\frac{I}{(S_\kappa - \lambda)^m} = h_m(t)\, e^{\lambda t}. \tag{62}$$

In the case of the kernel $\kappa(t) = h_\alpha(t),\ t > 0$ (operational calculus for the Caputo fractional derivative), we get the following operational relation [6,10]:

$$\frac{I}{(S_\kappa - \lambda)^m} = t^{m\alpha-1} E^m_{\alpha,m\alpha}(\lambda t^\alpha),\ t > 0,\ m \in \mathbb{N}, \tag{63}$$

where the Mittag–Leffler type function $E^m_{\alpha,\beta}$ is defined via the following convergent series:

$$E^m_{\alpha,\beta}(z) := \sum_{j=0}^{\infty} \frac{(m)_j z^j}{j!\Gamma(\alpha j + \beta)},\ \alpha, \beta > 0,\ z \in \mathbb{C},\ (m)_j = \prod_{i=0}^{j-1}(m+i).$$

Combining the operational relations (58) and (61), we deduce another important operational relation.

Let $R(S_\kappa) = Q(S_\kappa)/P(S_\kappa)$, where Q and P are polynomials and $\deg(Q) < \deg(P)$. In this case, the rational function $R(S_\kappa)$ can be represented as a sum of the partial fractions:

$$R(S_\kappa) = \sum_{j=1}^{J}\sum_{i=1}^{m_j} \frac{a_{ij}}{(S_\kappa - \lambda_j)^i},\ \sum_{j=1}^{J} m_j = \deg(P). \tag{64}$$

Then the operational relation

$$R(S_\kappa) = \sum_{j=1}^{J}\sum_{i=1}^{m_j} a_{ij}\, l_{\kappa,\lambda_j}^{<i>}(t),\ t > 0,\ \sum_{j=1}^{J} m_j = \deg(P), \tag{65}$$

holds true, where the constants λ_j and m_j, $j = 1, \ldots, J$ are uniquely determined by representation of the rational function $R(S_\kappa)$ as a sum of the partial fractions in form (64).

The operational relation (65) is a direct consequence from the Formula (64) and the operational relation (61).

4. Discussion

In this paper, we first discussed some important properties of the general fractional integrals (GFI) and the GFD of arbitrary order introduced recently in the works of the third named co-author. The new objects defined for the first time in this paper are the m-fold GFIs and the sequential GFDs of arbitrary order. For the m-fold GFIs and the sequential GFDs of arbitrary order we proved the 1st and the 2nd fundamental theorems of FC and derived an explicit form for their projector operator. The main contribution of this paper is an operational calculus of Mikusiński type for the GFDs of arbitrary order. In the field of the convolution quotients, the GFDs of arbitrary order and the sequential GFDs of arbitrary order are represented as multiplication with certain elements of the field. We also derived several important operational relations that provide useful representations of some field elements as conventional functions expressed in terms of the so-called convolution series.

In conclusion, we mention that the operational calculus for the GFDs of arbitrary order that we constructed in this paper can be applied for derivation of the closed form formulas for the solutions to the ordinary and partial fractional differential equations containing the m-fold sequential GFDs. These matters will be discussed elsewhere.

Author Contributions: Conceptualization, M.A.-K., L.A.-M.H. and Y.L.; Data curation, M.A.-K. and L.A.-M.H.; Formal analysis, M.A.-K.; Investigation, M.A.-K. and L.A.-M.H.; Methodology, M.A.-K. and Y.L.; Resources, L.A.-M.H. and Y.L.; Supervision, Y.L.; Validation, M.A.-K.; Writing—original draft, M.A.-K. and Y.L.; Writing—review & editing, M.A.-K., L.A.-M.H. and Y.L. All authors have read and agreed to the published version of the manuscript.

Funding: This research was funded by the Kuwait University in the framework of the research project No. SM01/20 "Operational method in Fractional Calculus".

Institutional Review Board Statement: Not applicable.

Informed Consent Statement: Not applicable.

Data Availability Statement: Not applicable.

Acknowledgments: The authors acknowledge the support of the Kuwait University for their joint research project No. SM01/20 "Operational method in Fractional Calculus".

Conflicts of Interest: The authors declare no conflict of interest.

References

1. Mikusiński, J. *Operational Calculus*; Pergamon Press: New York, NJ, USA, 1959.
2. Ditkin, V.A. The theory of operator calculus. *Dokl. Akad. Nauk SSSR* **1957**, *116*, 15–17. (In Russian)
3. Ditkin, V.A.; Prudnikov, A.P. The theory of operational calculus, which is generated by Bessel equation. *J. Vichisl. Mat. Mat. Fiz.* **1963**, *3*, 223–238. (In Russian)
4. Meller, N.A. On an operational calculus for the operator $B_\alpha = t^{-\alpha}Dt^{\alpha+1}D$. *J. Vychisl. Mat. Mat. Fiz.* **1960**, *6*, 161–180. (In Russian)
5. Dimovski, I.H. Operational calculus for a class of differential operators. *Compt. Rend. Acad. Bulg. Sci.* **1996**, *19*, 1111–1114.
6. Luchko, Y. Theory of the Integral Transformations with the Fox H-Function as a Kernel and Some of Its Applications Including Operational Calculus. Ph.D. Thesis, Belorussian State University, Minsk, Russia, 1993.
7. Luchko, Y.; Yakubovich, S. An operational method for solving some classes of integro-differential equations. *Differ. Equ.* **1994**, *30*, 247–256.
8. Luchko, Y.; Srivastava, H.M. The exact solution of certain differential equations of fractional order by using operational calculus. *Comput. Math. Appl.* **1995**, *29*, 73–85. [CrossRef]
9. Hadid, S.B.; Luchko, Y. An operational method for solving fractional differential equations of an arbitrary real order. *Panam. Math. J.* **1996**, *6*, 57–73.
10. Luchko, Y.; Gorenflo, R. An operational method for solving fractional differential equations. *Acta Math. Vietnam.* **1999**, *24*, 207–234.
11. Hilfer, R.; Luchko, Y.; Tomovski, Z. Operational method for the solution of fractional differential equations with generalized Riemann–Liouville fractional derivatives. *Fract. Calc. Appl. Anal.* **2009**, *12*, 299–318.
12. Hanna, L.A.-M.; Luchko, Y. Operational calculus for the Caputo-type fractional Erdélyi-Kober derivative and its applications. *Integral Transform. Spec. Funct.* **2014**, *25*, 359–373. [CrossRef]
13. Luchko, Y. Operational Calculus for the general fractional derivatives with the Sonine kernels. *Fract. Calc. Appl. Anal.* **2021**, *24*, 338–375. [CrossRef]
14. Luchko, Y. Fractional Differential Equations with the General Fractional Derivatives of Arbitrary Order in the Riemann–Liouville Sense. *Mathematics* **2022**, *10*, 849. [CrossRef]
15. Luchko, Y. Operational method in fractional calculus. *Fract. Calc. Appl. Anal.* **1999**, *2*, 463–489.
16. Luchko, Y. Operational method for fractional ordinary differential equations. In *Handbook of Fractional Calculus with Applications. Volume 1: Basic Theory*; Kochubei, A., Luchko, Y., Eds.; De Gruyter: Berlin, Germany, 2019; pp. 91–118.
17. Sonine, N. Sur la généralisation d'une formule d'Abel. *Acta Math.* **1884**, *4*, 171–176. [CrossRef]
18. Abel, N.H. Oplösning af et par opgaver ved hjelp af bestemte integraler. *Mag. Naturvidenskaberne* **1823**, *2*, 2.
19. Abel, N.H. Auflösung einer mechanischen Aufgabe. *J. Die Reine Angew. Math.* **1826**, *1*, 153–157.
20. Kochubei, A.N. General fractional calculus, evolution equations, and renewal processes. *Integr. Equ. Oper. Theory* **2011**, *71*, 583–600. [CrossRef]
21. Kochubei, A.N. General fractional calculus. In *Handbook of Fractional Calculus with Applications. Volume 1: Basic Theory*; Kochubei, A., Luchko, Y., Eds.; De Gruyter: Berlin, Germany, 2019; pp. 111–126.
22. Kochubei, A.N. Equations with general fractional time derivatives. Cauchy problem. In *Handbook of Fractional Calculus with Applications. Volume 2: Fractional Differential Equations*; Kochubei, A., Luchko, Y., Eds.; De Gruyter: Berlin, Germany, 2019; pp. 223–234.
23. Luchko, Y. General Fractional Integrals and Derivatives with the Sonine Kernels. *Mathematics* **2021**, *9*, 594. [CrossRef]

24. Luchko, Y. General Fractional Integrals and Derivatives of Arbitrary Order. *Symmetry* **2021**, *13*, 755. [CrossRef]
25. Luchko, Y. Special Functions of Fractional Calculus in the Form of Convolution Series and Their Applications. *Mathematics* **2021**, *9*, 2132. [CrossRef]
26. Luchko, Y.; Yamamoto, M. General time-fractional diffusion equation: Some uniqueness and existence results for the initial-boundary-value problems. *Fract. Calc. Appl. Anal.* **2016**, *19*, 675–695. [CrossRef]
27. Luchko, Y.; Yamamoto, M. The general fractional derivative and related fractional differential equations. *Mathematics* **2020**, *8*, 2115. [CrossRef]
28. Tarasov, V.E. General Fractional Calculus: Multi-Kernel Approach. *Mathematics* **2021**, *9*, 1501. [CrossRef]
29. Luchko, Y. Convolution series and the generalized convolution Taylor formula. *Fract. Calc. Appl. Anal.* **2022**, *25*, 207–228. [CrossRef]
30. Yosida, K. *Operational Calculus: A Theory of Hyperfunctions*; Springer: Berlin/Heidelberg, Germany, 1984.

mathematics

MDPI

Article

Monotonicity Results for Nabla Riemann–Liouville Fractional Differences

Pshtiwan Othman Mohammed [1,*], Hari Mohan Srivastava [2,3,4,5], Dumitru Baleanu [6,7,8,*], Rashid Jan [9] and Khadijah M. Abualnaja [10]

1. Department of Mathematics, College of Education, University of Sulaimani, Sulaimani 46001, Iraq
2. Department of Mathematics and Statistics, University of Victoria, Victoria, BC V8W 3R4, Canada; harimsri@math.uvic.ca
3. Department of Medical Research, China Medical University Hospital, China Medical University, Taichung 40402, Taiwan
4. Department of Mathematics and Informatics, Azerbaijan University, 71 Jeyhun Hajibeyli Street, AZ1007 Baku, Azerbaijan
5. Section of Mathematics, International Telematic University Uninettuno, I-00186 Rome, Italy
6. Department of Mathematics, Cankaya University, Ankara 06530, Turkey
7. Institute of Space Sciences, R76900 Magurele-Bucharest, Romania
8. Department of Natural Sciences, School of Arts and Sciences, Lebanese American University, Beirut 11022801, Lebanon
9. Department of Mathematics, University of Swabi, Swabi 23430, KPK, Pakistan; rashidjan@uoswabi.edu.pk
10. Department of Mathematics and Statistics, College of Science, Taif University, P.O. Box 11099, Taif 21944, Saudi Arabia; Kh.abualnaja@tu.edu.sa

* Correspondence: pshtiwan.muhammad@univsul.edu.iq (P.O.M.); dumitru@cankaya.edu.tr (D.B.)

Abstract: Positivity analysis is used with some basic conditions to analyse monotonicity across all discrete fractional disciplines. This article addresses the monotonicity of the discrete nabla fractional differences of the Riemann–Liouville type by considering the positivity of $\left({}^{RL}_{b_0}\nabla^{\theta}g\right)(z)$ combined with a condition on $g(b_0+2)$, $g(b_0+3)$ and $g(b_0+4)$, successively. The article ends with a relationship between the discrete nabla fractional and integer differences of the Riemann–Liouville type, which serves to show the monotonicity of the discrete fractional difference $\left({}^{RL}_{b_0}\nabla^{\theta}g\right)(z)$.

Keywords: discrete fractional calculus; discrete nabla Riemann–Liouville fractional differences; monotonicity analysis

MSC: 26A48; 26A51; 33B10; 39A12; 39B62

check for updates

Citation: Mohammed, P.O.; Srivastava, H.M.; Baleanu, D.; Jan, R.; Abualnaja, K.M. Monotonicity Results for Nabla Riemann–Liouville Fractional Differences. *Mathematics* **2022**, *10*, 2433. https://doi.org/10.3390/math10142433

Academic Editors: Alexandra M.S.F. Galhano, António M. Lopes, Carlo Cattani and Hsien-Chung Wu

Received: 8 June 2022
Accepted: 9 July 2022
Published: 12 July 2022

1. Introduction

In recent years, discrete fractional operators (sums and differences) have turned out to be modern tools in the modelling of many phenomena of mathematical analysis [1–3], electric circuits [4,5], medical sciences [6], material sciences and mechanics (see, for details, [7–9]). From among the several meanings and avenues of such studies, we choose to mention the discrete delta/nabla fractional operators of the Riemann–Liouville, Liouville–Caputo or other types (see, for details, [10,11]), from singular and non-singular kernel operators to the definitions based upon the time-scale theory. Some of these definitions are equivalent, even though they seem to be completely different, and they have been established by different authors (see, for example, [12–15]).

The positivity and monotonicity analyses have proven to be useful tools in discrete fractional calculus theory:

For the set $\{b_0, b_0+1, b_0+2, \ldots\}$ denoted by $\mathbb{J}_{b_0}$ with $b_0 \in \mathbb{R}$, let g be defined on $\mathbb{J}_{b_0}$. Then, the function g will be monotonically increasing if $(\nabla g)(z)$ is positive; that is:

$$(\nabla g)(z) := g(z) - g(z-1) \geqq 0,$$

for each $z \in \mathbb{J}_{b_0+1}$.

In the context of discrete fractional calculus, the development of new positivity and monotonicity analyses is a source of interesting mathematical problems (see, for example, [16–21]). In recent years, several papers have been published devoted exclusively to the study of the problem of the monotonicity of discrete nabla/delta fractional operators with a certain kernel (and often under additional assumptions about the function). The interested reader may be referred, for example, to the developments reported in [22–28].

Our results in this paper concern the analysis of monotonicity for the discrete nabla fractional differences of Riemann–Liouville-type under the conditions that $\left({}^{RL}_{b_0}\nabla^{\theta} g\right)(z) \geqq 0$ and the ones coming from one of the following:

- $g(b_0+2) \geqq \frac{\theta}{\ell-1} g(b_0+1)$ for $\ell \in \mathbb{J}_3$ in Theorem 1.
- $g(b_0+3) \geqq \frac{\theta}{\ell-2} g(b_0+2) + \frac{\theta(\ell-\theta-2)}{(\ell-1)(\ell-2)} g(b_0+1)$ for $\ell \in \mathbb{J}_4$ in Theorem 2.
- $g(b_0+4) \geqq \frac{\theta}{\ell-2} g(b_0+3) + \frac{\theta(\ell-\theta-3)}{(\ell-2)(\ell-3)} g(b_0+2) + \frac{\theta(\ell-\theta-2)(\ell-\theta-3)}{(\ell-1)(\ell-2)(\ell-3)} g(b_0+1)$ for $\ell \in \mathbb{J}_5$ in Theorem 3.

Furthermore, we will show that the discrete nabla fractional Riemann–Liouville difference $\left({}^{RL}_{b_0}\nabla^{\theta} g\right)(z) \geqq 0$ for each $z \in \mathbb{J}_{b_0+1}$ by considering the relationship between the discrete nabla fractional and integer differences. It is worth mentioning that our results are motivated by the results in [29], wherein somewhat analogous results were investigated for the discrete delta Riemann–Liouville fractional differences.

The paper is divided into another three sections as follows. Section 2 considers preliminaries on discrete fractional operators of the Riemann–Liouville type and a main lemma, which we need in the next section. Our main results are presented in Section 3, which is separated into two subsections: In Section 3.1, we consider our three main theorems, which establish the monotonicity analysis of the discrete fractional operators. The relationship between the discrete nabla fractional and integer differences will be examined in Section 3.2, which will show how it will be used to establish the positivity of the discrete nabla fractional Riemann–Liouville operators. At the end of each theorem, a corollary is made. Section 4 provides a specific example, which confirms the applicability of our results. The article ends with concluding remarks and brief considerations of several discrete fractional modelling extensions, which can be applicable in the future to obtain monotonicity analysis for other types of discrete fractional operators in Section 5.

2. Preliminaries and a Lemma

Here, we provide some background material regarding the discrete nabla fractional operators toward the proof of our main achievements. As such, the main lemma is given.

Definition 1 (see [13,14,30]). *Let us denote the set $\{b_0, b_0+1, b_0+2, \ldots\}$ by $\mathbb{J}_{b_0}$ and with the starting point $a \in \mathbb{R}$. Assume that g is defined on $\mathbb{J}_{b_0}$. Then, the ∇ Riemann–Liouville fractional sum of order θ (> 0) is expressed as follows:*

$$\left({}_{b_0}\nabla^{-\theta} g\right)(z) = \sum_{r=b_0+1}^{z} \frac{(z-r+1)^{\overline{[\theta-1]}}}{\Gamma(\theta)} g(r) \quad \text{for } z \text{ in } \mathbb{J}_{b_0+1}, \tag{1}$$

where $z^{(\theta)}$ is defined by

$$z^{\overline{[\theta]}} = \frac{\Gamma(z+\theta)}{\Gamma(z)} \quad \text{for } z \text{ and } \theta \text{ in } \mathbb{R}, \tag{2}$$

and it yields zero at a pole. It is also worth recalling that

$$\nabla z^{[\theta]} = \theta z^{[\theta-1]}. \tag{3}$$

Definition 2 (see [30]). *Let g be defined on $\mathbb{J}_{b_0}$. Then the ∇ Riemann–Liouville fractional difference of order θ ($\ell-1<\theta<\ell$) is defined by*

$$\begin{aligned}\left({}^{RL}_{b_0}\nabla^{\theta} g\right)(z) &= \left(\nabla^{\ell}\,{}_{b_0}\nabla^{-(\ell-\theta)} g\right)(z)\\ &= \nabla^{\ell}\left[\sum_{r=b_0+1}^{z} \frac{(z-r+1)^{[\ell-\theta-1]}}{\Gamma(\ell-\theta)} g(r)\right] \quad \text{for } z \text{ in } \mathbb{J}_{b_0+1},\ \ell \text{ in } \mathbb{J}_1.\end{aligned}$$

Recently, Liu et al. [31] established an equivalent definition to Definition 2, as follows.

Definition 3 (see [31]). *Let $\ell-1<\theta<\ell$. Then the ∇ Riemann–Liouville fractional difference of order θ can be expressed as follows:*

$$\left({}^{RL}_{b_0}\nabla^{\theta} g\right)(z) = \sum_{r=b_0+1}^{z} \frac{(z-r+1)^{[-\theta-1]}}{\Gamma(-\theta)} g(r) \quad \text{for } z \text{ in } \mathbb{J}_{b_0+\ell},\ \ell \text{ in } \mathbb{J}_1.$$

In order to begin our work later, we state and prove the following main lemma.

Lemma 1. *For g defined on $\mathbb{J}_{b_0}$, the ∇ Riemann–Liouville fractional difference of order θ ($1<\theta<2$) can be expressed as follows:*

$$\begin{aligned}\left({}^{RL}_{b_0}\nabla^{\theta} g\right)(z) &= (\nabla g)(z) + \frac{(z-b_0)^{[-\theta]}}{\Gamma(1-\theta)} g(b_0+1)\\ &\quad + \sum_{r=b_0+2}^{z-1} \frac{(z-r+1)^{[-\theta]}}{\Gamma(1-\theta)} (\nabla g)(r) \quad \text{for } z \text{ in } \mathbb{J}_{b_0+3}.\end{aligned} \tag{4}$$

In addition, it is essential to observe that

$$\frac{(z-r+1)^{[-\theta]}}{\Gamma(1-\theta)} < 0, \tag{5}$$

for each $r=b_0+2, b_0+3, \dots, z-1$ and $z \in \mathbb{J}_{b_0+3}$.

Proof. According to Definition 3, we note for $1<\theta<2$ and $z \in \mathbb{J}_{b_0+2}$ that

$$\begin{aligned}\left({}^{RL}_{b_0}\nabla^{\theta} g\right)(z) &= \sum_{r=b_0+1}^{z} \frac{(z-r+1)^{[-\theta-1]}}{\Gamma(-\theta)} g(r)\\ &\overset{by}{\underset{(3)}{=}} \sum_{r=b_0+1}^{z} \frac{\nabla (z-r+1)^{[-\theta]}}{\Gamma(1-\theta)} g(r)\\ &= \sum_{r=b_0+1}^{z} \frac{(z-r+1)^{[-\theta]}}{\Gamma(1-\theta)} g(r) - \sum_{r=b_0+1}^{z-1} \frac{(z-r)^{[-\theta]}}{\Gamma(1-\theta)} g(r)\\ &= \frac{1}{\Gamma(1-\theta)}\left[(z-b_0)^{[-\theta]} g(b_0+1) + \sum_{r=b_0+2}^{z} (z-r+1)^{[-\theta]} g(r)\right.\end{aligned}$$

$$- \sum_{r=b_0+2}^{z} (z-r+1)^{[-\theta]} g(r-1) \Big]$$
$$= (\nabla g)(z) + \frac{(z-b_0)^{[-\theta]}}{\Gamma(1-\theta)} g(b_0+1) + \sum_{r=b_0+2}^{z-1} \frac{(z-r+1)^{[-\theta]}}{\Gamma(1-\theta)} (\nabla g)(r),$$

which completes the proof of (4). For $r = b_0+2, b_0+3, \ldots, z-1$ with $z \in \mathbb{J}_{b_0+3}$, we have

$$\frac{(z-r+1)^{[-\theta]}}{\Gamma(1-\theta)} = \frac{\Gamma(z-r-\theta+1)}{\Gamma(1-\theta)\Gamma(z-r+1)} = \frac{(z-r-\theta)(z-r-\theta-1)\cdots(2-\theta)(1-\theta)}{(z-r)!},$$

which is clearly positive for $\theta \in (1,2)$. Therefore, the second part of the lemma is proved. Hence, the proof of the lemma is complete. □

3. Main Results

This section is divided into two main subsections.

3.1. Monotonicity Results

This section is devoted to the study of the monotonicity analysis of the discrete fractional Riemann–Liouville differences.

Theorem 1. *Suppose that $g : \mathbb{J}_{b_0} \longrightarrow \mathbb{R}$ satisfies each of the following conditions:*

$$\text{(i)} \quad \left({}^{RL}_{b_0}\nabla^{\theta} g\right)(z) \geqq 0 \quad \text{for each } z \in \mathbb{J}_{b_0+3},$$
$$\text{(ii)} \quad g(b_0+2) \geqq \frac{\theta}{\ell-1} g(b_0+1) \quad \text{for } \ell \in \mathbb{J}_3,$$

for $\theta \in (1,2)$. Then $(\nabla g)(z) \geqq 0$ for $z \in \mathbb{J}_{b_0+3}$.

Proof. According to the assumption that $\left({}^{RL}_{b_0}\nabla^{\theta} g\right)(z) \geqq 0$ and the identity (4), one can see, for $z \in \mathbb{J}_{b_0+3}$, that

$$(\nabla g)(z) \geqq -\frac{(z-b_0)^{[-\theta]}}{\Gamma(1-\theta)} g(b_0+1) - \sum_{r=b_0+2}^{z-1} \frac{(z-r+1)^{[-\theta]}}{\Gamma(1-\theta)} (\nabla g)(r).$$

If we set $z := b_0 + \ell$ for $\ell \in \mathbb{J}_3$, it follows that

$$(\nabla g)(b_0+\ell) \geqq -\frac{\ell^{[-\theta]}}{\Gamma(1-\theta)} g(b_0+1) - \sum_{r=b_0+2}^{b_0+\ell-1} \frac{(b_0+\ell-r+1)^{[-\theta]}}{\Gamma(1-\theta)} (\nabla g)(r). \tag{6}$$

We proceed with the proof using the principle of mathematical induction on ℓ for the inequality (6). Indeed, for $\ell = 3$, we have

$$
\begin{aligned}
(\nabla g)(b_0+3) &\geqq -\frac{3^{[-\theta]}}{\Gamma(1-\theta)}g(b_0+1)-\frac{2^{[-\theta]}}{\Gamma(1-\theta)}(\nabla g)(b_0+2)\\
&=\frac{-1}{\Gamma(1-\theta)}\left[\frac{\Gamma(3-\theta)}{\Gamma(3)}g(b_0+1)+\frac{\Gamma(2-\theta)}{\Gamma(2)}(\nabla g)(b_0+2)\right]\\
&=\frac{-\Gamma(2-\theta)}{\Gamma(1-\theta)}\left[\frac{2-\theta}{2}g(b_0+1)+g(b_0+2)-g(b_0+1)\right]\\
&=\underbrace{(\theta-1)}_{>0}\underbrace{\left[\frac{-\theta}{2}g(b_0+1)+g(b_0+2)\right]}_{\geqq 0 \text{ per condition (ii)}}\geqq 0.
\end{aligned}
$$

Suppose that $(\nabla g)(b_0+\jmath)\geqq 0$ for $\jmath=3,4,\ldots,\ell-1$ and $\ell\in\mathbb{J}_4$. Then, we shall show that $(\nabla g)(b_0+\ell)\geqq 0$. By making use of (6), we obtain

$$
\begin{aligned}
(\nabla g)(b_0+\ell) &\geqq -\frac{\ell^{[-\theta]}}{\Gamma(1-\theta)}g(b_0+1)-\sum_{r=b_0+2}^{b_0+\ell-1}\frac{(b_0+\ell-r+1)^{[-\theta]}}{\Gamma(1-\theta)}(\nabla g)(r)\\
&=-\frac{\ell^{[-\theta]}}{\Gamma(1-\theta)}g(b_0+1)-\frac{(\ell-1)^{[-\theta]}}{\Gamma(1-\theta)}(\nabla g)(b_0+2)\\
&\quad-\sum_{r=b_0+3}^{b_0+\ell-1}\underbrace{\frac{(b_0+\ell-r+1)^{[-\theta]}}{\Gamma(1-\theta)}}_{<0 \text{ per (5)}}\underbrace{(\nabla g)(r)}_{\geqq 0 \text{ per our claim}}\\
&\geqq -\frac{\ell^{[-\theta]}}{\Gamma(1-\theta)}g(b_0+1)-\frac{(\ell-1)^{[-\theta]}}{\Gamma(1-\theta)}(\nabla g)(b_0+2)\\
&=\frac{-1}{\Gamma(1-\theta)}\left[\frac{\Gamma(\ell-\theta)}{\Gamma(\ell)}g(b_0+1)+\frac{\Gamma(\ell-\theta-1)}{\Gamma(\ell-1)}(\nabla g)(b_0+2)\right]\\
&=\frac{-\Gamma(\ell-\theta-1)}{\Gamma(1-\theta)\Gamma(\ell-1)}\underbrace{\left[-\frac{\theta}{\ell-1}g(b_0+1)+g(b_0+2)\right]}_{\geqq 0 \text{ per condition (ii)}}\geqq 0,
\end{aligned}
$$

where we have used that

$$
\frac{-\Gamma(\ell-\theta-1)}{\Gamma(1-\theta)\Gamma(\ell-1)}=-\frac{(\ell-\theta-2)(\ell-\theta-3)\cdots(2-\theta)(1-\theta)}{(\ell-2)!}>0, \tag{7}
$$

for $\theta\in(1,2)$ and $\ell\geqq 3$. Thus, the proof is complete. □

Corollary 1. *If the function* $g:\mathbb{J}_{b_0}\longrightarrow\mathbb{R}$ *satisfies*

$$
\begin{aligned}
&\text{(i)}\quad \left({}^{RL}_{b_0}\nabla^{\theta}g\right)(z)\leqq 0 \quad \text{for each } z\in\mathbb{J}_{b_0+3},\\
&\text{(ii)}\quad g(b_0+2)\leqq\frac{\theta}{\ell-1}g(b_0+1) \quad \text{for } \ell\in\mathbb{J}_3,
\end{aligned}
$$

for $\theta\in(1,2)$*, then,* $(\nabla g)(z)\leqq 0$ *for* $z\in\mathbb{J}_{b_0+2}$.

Proof. Define $h:=-g$. Thus, the proof follows immediately from Theorem 1 applying for the function g. □

Theorem 2. *Assume that $g : \mathbb{J}_{b_0} \longrightarrow \mathbb{R}$ satisfies each of the following conditions:*

$$\text{(i)} \quad \left({}^{RL}_{b_0}\nabla^{\theta} g\right)(z) \geqq 0 \quad \text{for each } z \in \mathbb{J}_{b_0+3},$$

$$\text{(ii)} \quad g(b_0+3) \geqq \frac{\theta}{\ell-2} g(b_0+2) + \frac{\theta(\ell-\theta-2)}{(\ell-1)(\ell-2)} g(b_0+1) \quad \text{for } \ell \in \mathbb{J}_4,$$

for $\theta \in (1,2)$. Then $(\nabla g)(z) \geqq 0$ for $z \in \mathbb{J}_{b_0+4}$.

Proof. The proof will make use of the principle of mathematical induction on ℓ for the inequality (6). In fact, for $\ell = 4$, it follows that

$$\begin{aligned}
&(\nabla g)(b_0+4)\\
&\geqq -\frac{4^{[-\theta]}}{\Gamma(1-\theta)} g(b_0+1) - \frac{3^{[-\theta]}}{\Gamma(1-\theta)} (\nabla g)(b_0+2) - \frac{2^{[-\theta]}}{\Gamma(1-\theta)} (\nabla g)(b_0+3)\\
&= \frac{-1}{\Gamma(1-\theta)} \left[\frac{\Gamma(4-\theta)}{\Gamma(4)} g(b_0+1) + \frac{\Gamma(3-\theta)}{\Gamma(3)} (\nabla g)(b_0+2) + \frac{\Gamma(2-\theta)}{\Gamma(2)} (\nabla g)(b_0+3) \right]\\
&= \frac{-\Gamma(2-\theta)}{\Gamma(1-\theta)} \left[\frac{(2-\theta)(3-\theta)}{6} g(b_0+1) + \frac{2-\theta}{2} \{ g(b_0+2) - g(b_0+1) \} \right.\\
&\left. + g(b_0+3) - g(b_0+2) \right]\\
&= \underbrace{(\theta-1)}_{>0} \underbrace{\left[-\frac{\theta(2-\theta)}{6} g(b_0+1) - \frac{\theta}{2} g(b_0+2) + g(b_0+3) \right]}_{\geqq 0 \text{ per condition (ii)}} \geqq 0.
\end{aligned}$$

If we let $(\nabla g)(b_0+\jmath) \geqq 0$ for $\jmath = 4, 5, \ldots, \ell-1$ and $\ell \in \mathbb{J}_5$, then, we try to show that $(\nabla g)(b_0+\ell) \geqq 0$, with the help of (6), we can deduce

$$\begin{aligned}
&(\nabla g)(b_0+\ell)\\
&\geqq -\frac{\ell^{[-\theta]}}{\Gamma(1-\theta)} g(b_0+1) - \frac{(\ell-1)^{[-\theta]}}{\Gamma(1-\theta)} (\nabla g)(b_0+2) - \frac{(\ell-2)^{[-\theta]}}{\Gamma(1-\theta)} (\nabla g)(b_0+3)\\
&\underbrace{- \sum_{r=b_0+4}^{b_0+\ell-1} \frac{(b_0+\ell-r+1)^{[-\theta]}}{\Gamma(1-\theta)}}_{\geqq 0 \text{ per (5) and our claim}}\\
&\geqq -\frac{\ell^{[-\theta]}}{\Gamma(1-\theta)} g(b_0+1) - \frac{(\ell-1)^{[-\theta]}}{\Gamma(1-\theta)} (\nabla g)(b_0+2) - \frac{(\ell-2)^{[-\theta]}}{\Gamma(1-\theta)} (\nabla g)(b_0+3)\\
&= \frac{-1}{\Gamma(1-\theta)} \left[\frac{\Gamma(\ell-\theta)}{\Gamma(\ell)} g(b_0+1) + \frac{\Gamma(\ell-\theta-1)}{\Gamma(\ell-1)} (\nabla g)(b_0+2) \right.\\
&\left. + \frac{\Gamma(\ell-\theta-2)}{\Gamma(\ell-2)} (\nabla g)(b_0+3) \right]\\
&= \underbrace{\frac{-\Gamma(\ell-\theta-2)}{\Gamma(1-\theta)\Gamma(\ell-2)}}_{>0 \text{ per (7)}} \underbrace{\left[-\frac{\theta(\ell-\theta-2)}{(\ell-1)(\ell-2)} g(b_0+1) - \frac{\theta}{\ell-2} g(b_0+2) + g(b_0+3) \right]}_{\geqq 0 \text{ per condition (ii)}} \geqq 0,
\end{aligned}$$

which completes the proof. □

Corollary 2. *If the function* $g : \mathbb{J}_{b_0} \longrightarrow \mathbb{R}$ *satisfies*

$$\text{(i)} \quad \left({}^{RL}_{b_0}\nabla^{\theta} g\right)(z) \leqq 0 \quad \text{for each } z \in \mathbb{J}_{b_0+3},$$

$$\text{(ii)} \quad g(b_0+3) \leqq \frac{\theta}{\ell-2} g(b_0+2) + \frac{\theta(\ell-\theta-2)}{(\ell-1)(\ell-2)} g(b_0+1) \quad \text{for } \ell \in \mathbb{J}_4,$$

for $\theta \in (1,2)$*, then,* $(\nabla g)(z) \leqq 0$ *for* $z \in \mathbb{J}_{b_0+5}$.

Proof. The proof follows immediately from Theorem 2 applied to the function $h := -g$. □

Theorem 3. *Assume that* $g : \mathbb{J}_{b_0} \longrightarrow \mathbb{R}$ *satisfies each of the following conditions:*

$$\text{(i)} \quad \left({}^{RL}_{b_0}\nabla^{\theta} g\right)(z) \geqq 0 \quad \text{for each } z \in \mathbb{J}_{b_0+3},$$

$$\begin{aligned} \text{(ii)} \quad g(b_0+4) \geqq{}& \frac{\theta}{\ell-2} g(b_0+3) + \frac{\theta(\ell-\theta-3)}{(\ell-2)(\ell-3)} g(b_0+2) \\ &+ \frac{\theta(\ell-\theta-2)(\ell-\theta-3)}{(\ell-1)(\ell-2)(\ell-3)} g(b_0+1) \quad \text{for } \ell \in \mathbb{J}_5, \end{aligned}$$

for $\theta \in (1,2)$*. Then* $(\nabla g)(z) \geqq 0$ *for* $z \in \mathbb{J}_{b_0+5}$.

Proof. Again, we will prove this theorem using the principle of mathematical induction induction on ℓ. Thus, (6) at $\ell = 5$ leads to

$$\begin{aligned} &(\nabla g)(b_0+5) \\ &\geqq -\frac{5^{[-\theta]}}{\Gamma(1-\theta)} g(b_0+1) - \frac{4^{[-\theta]}}{\Gamma(1-\theta)} (\nabla g)(b_0+2) - \frac{3^{[-\theta]}}{\Gamma(1-\theta)} (\nabla g)(b_0+3) \\ &- \frac{2^{[-\theta]}}{\Gamma(1-\theta)} (\nabla g)(b_0+4) \\ &= \frac{-\Gamma(2-\theta)}{\Gamma(1-\theta)} \left[\frac{(2-\theta)(3-\theta)(4-\theta)}{24} g(b_0+1) + \frac{(2-\theta)(3-\theta)}{6} (\nabla g)(b_0+2) \right. \\ &\left. + \frac{2-\theta}{2} (\nabla g)(b_0+3) + (\nabla g)(b_0+4) \right] \\ &= \underbrace{(\theta-1)}_{>0} \underbrace{\left[-\frac{\theta(2-\theta)(3-\theta)}{24} g(b_0+1) - \frac{\theta(2-\theta)}{6} g(b_0+2) - \frac{\theta}{2} g(b_0+3) + g(b_0+4) \right]}_{\geqq 0 \text{ per condition (ii)}} \geqq 0. \end{aligned}$$

Now, we assume that $(\nabla g)(b_0+j) \geqq 0$ for $j = 5, 6, \ldots, \ell-1$ and $\ell \in \mathbb{J}_6$. Then, we have to show that $(\nabla g)(b_0+\ell) \geqq 0$. In view of (6), we can deduce

$$\begin{aligned} &(\nabla g)(b_0+\ell) \\ &\geqq -\frac{\ell^{[-\theta]}}{\Gamma(1-\theta)} g(b_0+1) - \frac{(\ell-1)^{[-\theta]}}{\Gamma(1-\theta)} (\nabla g)(b_0+2) - \frac{(\ell-2)^{[-\theta]}}{\Gamma(1-\theta)} (\nabla g)(b_0+3) \\ &- \frac{(\ell-3)^{[-\theta]}}{\Gamma(1-\theta)} (\nabla g)(b_0+4) - \underbrace{\sum_{r=b_0+5}^{b_0+\ell-1} \frac{(b_0+\ell-r+1)^{[-\theta]}}{\Gamma(1-\theta)}}_{\geqq 0 \text{ per (5) and our claim}} \end{aligned}$$

$$\geqq -\frac{\ell^{[-\theta]}}{\Gamma(1-\theta)}g(b_0+1) - \frac{(\ell-1)^{[-\theta]}}{\Gamma(1-\theta)}(\nabla g)(b_0+2)$$
$$-\frac{(\ell-2)^{[-\theta]}}{\Gamma(1-\theta)}(\nabla g)(b_0+3) - \frac{(\ell-3)^{[-\theta]}}{\Gamma(1-\theta)}(\nabla g)(b_0+4)$$
$$= \frac{-1}{\Gamma(1-\theta)}\left[\frac{\Gamma(\ell-\theta)}{\Gamma(\ell)}g(b_0+1) + \frac{\Gamma(\ell-\theta-1)}{\Gamma(\ell-1)}(\nabla g)(b_0+2)\right.$$
$$\left.+\frac{\Gamma(\ell-\theta-2)}{\Gamma(\ell-2)}(\nabla g)(b_0+3) + \frac{\Gamma(\ell-\theta-3)}{\Gamma(\ell-3)}(\nabla g)(b_0+4)\right]$$
$$= \underbrace{\frac{-\Gamma(\ell-\theta-3)}{\Gamma(1-\theta)\Gamma(\ell-3)}}_{>0 \text{ per } (7)} \underbrace{\left[\begin{array}{c} -\frac{\theta(\ell-\theta-2)(\ell-\theta-3)}{(\ell-1)(\ell-2)(\ell-3)}g(b_0+1) \\ -\frac{\theta(\ell-\theta-3)}{(\ell-2)(\ell-3)}g(b_0+2) - \frac{\theta}{\ell-2}g(b_0+3) + g(b_0+4) \end{array}\right]}_{\geqq 0 \text{ per condition (ii)}} \geqq 0,$$

which completes the proof. □

Corollary 3. *If the function $g : \mathbb{J}_{b_0} \longrightarrow \mathbb{R}$ satisfies*

(i) $\left({}^{RL}_{b_0}\nabla^{\theta} g\right)(z) \leqq 0$ for each $z \in \mathbb{J}_{b_0+3}$,

(ii) $$g(b_0+4) \geqq \frac{\theta}{\ell-2}g(b_0+3) + \frac{\theta(\ell-\theta-3)}{(\ell-2)(\ell-3)}g(b_0+2) + \frac{\theta(\ell-\theta-2)(\ell-\theta-3)}{(\ell-1)(\ell-2)(\ell-3)}g(b_0+1) \quad \text{for } \ell \in \mathbb{J}_5,$$

for $\theta \in (1,2)$, then, $(\nabla g)(z) \leqq 0$ for $z \in \mathbb{J}_{b_0+4}$.

Proof. This proof follows directly from Theorem 3 applied to the function $h := -g$. □

3.2. Discrete Nabla Fractional and Integer Differences

Based on Lemma 1, we can now establish a relationship between the discrete nabla fractional and integer differences of the Riemann–Liouville type, and we immediately present the final monotonicity result of this study.

Theorem 4. *Let g be defined on $\mathbb{J}_{b_0}$ and $N-1 < \theta < N$ with $N \in \mathbb{J}_1$. Then*

$$\left({}^{RL}_{b_0}\nabla^{\theta} g\right)(b_0+N+\ell) = \sum_{\imath=0}^{N-1} \frac{(N+\ell)^{[-\theta+\imath]}}{\Gamma(1-\theta-\imath)}(\nabla^{\imath} g)(b_0+1)$$
$$+\frac{1}{\Gamma(N-\theta)}\sum_{\imath=1}^{\ell-1}(\ell-\imath+1)^{[-\theta+N-1]}\left(\nabla^N g\right)(b_0+\imath+N) + \left(\nabla^N g\right)(b_0+\ell+N),$$

for all $\ell \in \mathbb{J}_0$. In addition, we have

$$\frac{(\ell-\imath+1)^{[-\theta+N-1]}}{\Gamma(N-\theta)} = \frac{(N-\theta+\ell-\imath-1)(N-\theta+\ell-\imath-2)\cdots(N+1-\theta)(N-\theta)}{(\ell-\imath)!} > 0, \tag{8}$$

for $\imath = 1,2,\ldots,\ell-1$ with $\ell \in \mathbb{J}_1$.

Proof. For $N = 1$, from Lemma 1, we find for $z \in \mathbb{J}_{b_0+2}$ that

$$\left({}^{RL}_{b_0}\nabla^{\theta} g\right)(z) = \frac{(z-b_0)^{[-\theta]}}{\Gamma(1-\theta)} g(b_0+1) + \sum_{r=b_0+2}^{z} \frac{(z-r+1)^{[-\theta]}}{\Gamma(1-\theta)} (\nabla g)(r). \tag{9}$$

Now, by the same technique used in Lemma 1, for $N = 2$, we have:

$$\begin{aligned}
&\left({}^{RL}_{b_0}\nabla^{\theta} g\right)(z) \\
&= \nabla^2 \left[\sum_{r=b_0+1}^{z} \frac{(z-r+1)^{[1-\theta]}}{\Gamma(2-\theta)} g(r)\right] = \nabla\left[\nabla\left(\sum_{r=b_0+1}^{z} \frac{(z-r+1)^{[1-\theta]}}{\Gamma(2-\theta)} g(r)\right)\right] \\
&= \nabla\left[\frac{(z-b_0)^{[1-\theta]}}{\Gamma(2-\theta)} g(b_0+1) + \sum_{r=b_0+2}^{z} \frac{(z-r+1)^{[1-\theta]}}{\Gamma(2-\theta)} (\nabla g)(r)\right] \\
&= \nabla\left[\frac{(z-b_0)^{[1-\theta]}}{\Gamma(2-\theta)} g(b_0+1)\right] + \nabla\left[\sum_{r=b_0+2}^{z} \frac{(z-r+1)^{[1-\theta]}}{\Gamma(2-\theta)} (\nabla g)(r)\right] \\
&= \frac{(z-b_0)^{[-\theta]}}{\Gamma(1-\theta)} g(b_0+1) + \frac{(z-b_0)^{[1-\theta]}}{\Gamma(2-\theta)} g(b_0+1) \\
&+ \sum_{r=b_0+3}^{z} \frac{(z-r+1)^{[1-\theta]}}{\Gamma(2-\theta)} (\nabla^2 g)(r),
\end{aligned} \tag{10}$$

for $z \in \mathbb{J}_{b_0+3}$, where we have used the following fact

$$\nabla\,(z-b_0)^{[1-\theta]} = (z-b_0)^{[1-\theta]} - (z-1-b_0)^{[1-\theta]} = (1-\theta)(z-b_0)^{[-\theta]}.$$

We can continue by the same process to obtain

$$\begin{aligned}
\left({}^{RL}_{b_0}\nabla^{\theta} g\right)(z) = \sum_{\imath=0}^{N-1} \frac{(z-b_0)^{[-\theta+\imath]}}{\Gamma(1-\theta+\imath)} (\nabla^{\imath} g)(b_0+1) \\
+ \sum_{r=b_0+N+1}^{z} \frac{(z-r+1)^{[N-\theta-1]}}{\Gamma(N-\theta)} (\nabla^{N} g)(r),
\end{aligned} \tag{11}$$

for $z \in \mathbb{J}_{b_0+N+1}$. Now, we define $z := b_0 + N + \ell$ for $\ell \in \mathbb{J}_1$ to obtain

$$\begin{aligned}
&\left({}^{RL}_{b_0}\nabla^{\theta} g\right)(b_0+N+\ell) \\
&= \sum_{\imath=0}^{N-1} \frac{(N+\ell)^{[-\theta+\imath]}}{\Gamma(1-\theta+\imath)} (\nabla^{\imath} g)(b_0+1) + \sum_{r=b_0+N+1}^{b_0+N+\ell} \frac{(b_0+N+\ell-r+1)^{[N-\theta-1]}}{\Gamma(N-\theta)} (\nabla^{N} g)(r) \\
&= \sum_{\imath=0}^{N-1} \frac{(N+\ell)^{[-\theta+\imath]}}{\Gamma(1-\theta+\imath)} (\nabla^{\imath} g)(b_0+1) + \sum_{\imath=1}^{\ell} \frac{(\ell-\imath+1)^{[N-\theta-1]}}{\Gamma(N-\theta)} (\nabla^{N} g)(b_0+\imath+N) \\
&= \sum_{\imath=0}^{N-1} \frac{(N+\ell)^{[-\theta+\imath]}}{\Gamma(1-\theta+\imath)} (\nabla^{\imath} g)(b_0+1) \\
&+ \sum_{\imath=1}^{\ell-1} \frac{(\ell-\imath+1)^{[N-\theta-1]}}{\Gamma(N-\theta)} (\nabla^{N} g)(b_0+\imath+N) + (\nabla^{N} g)(b_0+\ell+N),
\end{aligned}$$

which completes the proof of the first part. For the second part of the theorem, we see that

$$\begin{aligned}\frac{(\ell-\imath+1)^{[-\theta+N-1]}}{\Gamma(N-\theta)} &= \frac{\Gamma(\ell-\imath-\theta+N)}{\Gamma(\ell-\imath+1)\Gamma(N-\theta)} \\ &= \frac{(N-\theta+\ell-\imath-1)(N-\theta+\ell-\imath-2)\cdots(N+1-\theta)(N-\theta)}{(\ell-\imath)!} > 0,\end{aligned}$$

for $N-1<\theta<N$ and $\imath=1,2,\ldots,\ell-1$. Hence, the proof is complete. □

Our final result is on the positivity of $\left({}^{RL}_{b_0}\nabla^{\theta}g\right)(z)$, as follows.

Theorem 5. *Let $g:\mathbb{J}_{b_0}\longrightarrow\mathbb{R}$ be a function, $N-1<\theta<N$ with $N\in\mathbb{J}_1$, $(\nabla^N g)(z)\geqq 0$ for $z\in\mathbb{J}_{b_0+1}$, $(-1)^{N-\imath}(\nabla^N g)(b_0+1)\leqq 0$ for $\imath=0,1,\ldots,N-1$. Then $\left({}^{RL}_{b_0}\nabla^{\theta}g\right)(z)\geqq 0$ for each $z\in\mathbb{J}_{b_0+1}$.*

Proof. Let ℓ be a fixed but arbitrary element in $\mathbb{J}_1$. We can then see that

$$\begin{aligned}\frac{(N+\ell)^{[-\theta+\imath]}}{\Gamma(1-\theta+\imath)} &= \frac{\Gamma(N+\ell-\theta+\imath)}{\Gamma(1-\theta+\imath)\Gamma(N+\ell)} \\ &= \frac{(N+\ell-\theta+\imath-1)(N+\ell-\theta+\imath-2)\cdots(2-\theta+\imath)(1-\theta+\imath)}{(N+\ell-1)!}.\end{aligned}$$

Now, if $N-\imath-1$ is even, then we obtain

$$\frac{(N+\ell)^{[-\theta+\imath]}}{\Gamma(1-\theta+\imath)}>0 \quad\text{and}\quad \left(\nabla^N g\right)(b_0+1)\geqq 0,$$

but if $N-\imath-1$ is odd, then we obtain

$$\frac{(N+\ell)^{[-\theta+\imath]}}{\Gamma(1-\theta+\imath)}<0 \quad\text{and}\quad \left(\nabla^N g\right)(b_0+1)\leqq 0.$$

These provide that

$$\sum_{\imath=0}^{N-1}\frac{(N+\ell)^{[-\theta+\imath]}}{\Gamma(1-\theta+\imath)}(\nabla^{\imath}g)(b_0+1)\geqq 0. \tag{12}$$

Hence, according to Theorems 4, (8) and (12) and the assumption that $(\nabla^N g)(z)\geqq 0$, we can deduce

$$\begin{aligned}\left({}^{RL}_{b_0}\nabla^{\theta}g\right)(b_0+N+\ell) &= \sum_{\imath=0}^{N-1}\frac{(N+\ell)^{[-\theta+\imath]}}{\Gamma(1-\theta+\imath)}(\nabla^{\imath}g)(b_0+1) \\ &\quad+\sum_{\imath=1}^{\ell-1}\frac{(\ell-\imath+1)^{[N-\theta-1]}}{\Gamma(N-\theta)}(\nabla^N g)(b_0+\imath+N)+(\nabla^N g)(b_0+\ell+N) \\ &\geqq \sum_{\imath=1}^{\ell-1}\frac{(\ell-\imath+1)^{[N-\theta-1]}}{\Gamma(N-\theta)}(\nabla^N g)(b_0+\imath+N)\geqq 0.\end{aligned}$$

By putting $z:=b_0+N+\ell$ for arbitrary $\ell\in\mathbb{J}_1$, we obtain $\left({}^{RL}_{b_0}\nabla^{\theta}g\right)(z)\geqq 0$ for each $z\in\mathbb{J}_{b_0+1}$ as desired. □

Corollary 4. *Let $g : \mathbb{J}_{b_0} \longrightarrow \mathbb{R}$ be a function, $N-1 < \theta < N$ with $N \in \mathbb{J}_1$, $(\nabla^N g)(z) \leqq 0$ for $z \in \mathbb{J}_{b_0+1}$, $(-1)^{N-\imath}(\nabla^N g)(b_0+1) \geqq 0$ for $\imath = 0,1,\ldots,N-1$. Then $\left({}^{RL}_{b_0}\nabla^\theta g\right)(z) \leqq 0$ for each $z \in \mathbb{J}_{b_0+1}$.*

4. Application: A Specific Example

In this section, we provide a specific example to illustrate our results.

Consider the function

$$g(z) = \left(\frac{8}{3}\right)^z \quad \text{for } z \in \mathbb{J}_{b_0+2}.$$

At first, we will try to show that $\left({}^{RL}_{b_0}\nabla^\theta g\right)(z) \geq 0$ for $z \in \{b_0+3, b_0+4\}$, $\theta = \frac{3}{2}$ and $b_0 = 0$. From Definition 3 at $z = b_0+3$, we have

$$\begin{aligned}
&\left({}^{RL}_{b_0}\nabla^\theta g\right)(b_0+3) \\
&= \frac{1}{\Gamma(-\theta)} \sum_{r=b_0+1}^{b_0+3} (b_0+3-r+1)^{[-\theta-1]} g(r) \\
&= \frac{1}{\Gamma(-\theta)} \left\{ (3)^{[-\theta-1]} g(b_0+1) + (2)^{[-\theta-1]} g(b_0+2) + (1)^{[-\theta-1]} g(b_0+3) \right\} \\
&= \frac{-\theta(1-\theta)}{2} g(b_0+1) - \theta\, g(b_0+2) + g(b_0+3) \\
&= \frac{251}{27} \geq 0,
\end{aligned}$$

which leads to

$$g(b_0+3) \geq \frac{\theta(1-\theta)}{2} g(b_0+1) + \theta\, g(b_0+2). \tag{13}$$

In addition, Definition 3 at $z = b_0+4$ gives

$$\begin{aligned}
&\left({}^{RL}_{b_0}\nabla^\theta g\right)(b_0+4) \\
&= \frac{1}{\Gamma(-\theta)} \sum_{r=b_0+1}^{b_0+4} (b_0+4-r+1)^{[-\theta-1]} g(r) \\
&= \frac{1}{\Gamma(-\theta)} \Big\{ (4)^{[-\theta-1]} g(b_0+1) + (3)^{[-\theta-1]} g(b_0+2) \\
&\qquad + (2)^{[-\theta-1]} g(b_0+3) + (1)^{[-\theta-1]} g(b_0+4) \Big\} \\
&= \frac{-\theta(1-\theta)(2-\theta)}{6} g(b_0+1) - \frac{\theta(1-\theta)}{2} g(b_0+2) - \theta\, g(b_0+3) + g(b_0+4) \\
&= \frac{3179}{162} \geq 0,
\end{aligned}$$

which implies that

$$g(b_0+4) \geq \frac{\theta(1-\theta)(2-\theta)}{6} g(b_0+1) + \frac{\theta(1-\theta)}{2} g(b_0+2) + \theta\, g(b_0+3). \tag{14}$$

On the other hand, we consider the condition:

$$g(b_0+2) \geq \frac{\theta}{\ell-1} g(b_0+1),$$

at $\ell = 3,4$. At $\ell = 3$, it follows that

$$\left(\frac{8}{3}\right)^2 = g(b_0+2) \geq \frac{\theta}{2} g(b_0+1) = \frac{3}{4}\left(\frac{8}{3}\right),$$

which means that

$$g(b_0+2) \geq \frac{\theta}{2} g(b_0+1). \tag{15}$$

In addition, at $\ell = 4$, it follows that

$$\begin{aligned}\left(\frac{8}{3}\right)^2 &= g(b_0+2)\\ &\geq \frac{\Gamma(\theta+2)}{\Gamma(\theta)\Gamma(3)} g(b_0+\theta h) = \frac{\theta}{3} g(b_0+1) = \frac{1}{2}\left(\frac{8}{3}\right),\end{aligned}$$

which is equivalent to

$$g(b_0+2) \geq \frac{\theta}{3} g(b_0+1). \tag{16}$$

Thus, we can conclude from the inequalities (13)–(14) that

$$\begin{aligned}(\nabla g)(b_0+3) &= g(b_0+3) - g(b_0+2)\\ &\geq \frac{\theta(1-\theta)}{2} g(b_0+1) + \theta\, g(b_0+2) - g(b_0+2)\\ &= (\theta-1)\left\{\frac{-\theta}{2} g(b_0+1) + \theta\, g(b_0+2)\right\} \geq 0.\end{aligned}$$

Also, from the inequalities (15)–(16) we can conclude that

$$\begin{aligned}(\nabla g)(b_0+4) &= g(b_0+4) - g(b_0+3)\\ &\geq \frac{\theta(1-\theta)(2-\theta)}{6} g(b_0+1) + \frac{\theta(1-\theta)}{2} g(b_0+2) + \theta\, g(b_0+3) - g(b_0+3)\\ &= \frac{(\theta-1)(2-\theta)}{2}\left\{\frac{-\theta}{3} g(b_0+1) + \theta\, g(b_0+2)\right\} \geq 0.\end{aligned}$$

These inequalities imply that g is non-decreasing in the time set $\{b_0+3, b_0+4\}$.

5. Conclusions and Future Directions

In this paper, we studied the monotonicity analysis for the discrete nabla fractional differences of the Riemann–Liouville type. The first three main results were dedicated to the positivity of $(\nabla g)(z)$ by assuming that $\left({}^{RL}_{b_0}\nabla^{\theta} g\right)(z) \geqq 0$ combined with the condition that $g(b_0+2) \geqq \frac{\theta}{\ell-1} g(b_0+1)$ for $\ell \in \mathbb{J}_3$ in Theorem 1, $g(b_0+3) \geqq \frac{\theta}{\ell-2} g(h_0+2) + \frac{\theta(\ell-\theta-2)}{(\ell-1)(\ell-2)} g(b_0+1)$ for $\ell \in \mathbb{J}_4$ in Theorem 2, and $g(b_0+4) \geqq \frac{\theta}{\ell-2} g(b_0+3) + \frac{\theta(\ell-\theta-3)}{(\ell-2)(\ell-3)} g(b_0+2) + \frac{\theta(\ell-\theta-2)(\ell-\theta-3)}{(\ell-1)(\ell-2)(\ell-3)} g(b_0+1)$ for $\ell \in \mathbb{J}_5$ in Theorem 3.

On the other hand, the relationship between the discrete nabla fractional and integer differences of the Riemann–Liouville type has been made. From which the positivity of the discrete nabla fractional differences of the Riemann–Liouville type has been established. In addition, some particular results have been obtained in the corollaries, which showed the negativity (decreasing) of the function.

There is vast room for monotonicity analysis to be explored in this fertile field of discrete fractional operators, for example, discrete Caputo–Fabrizio and Atangana–Baleanu fractional operators (see [30,32,33] for information about these discrete operators).

Author Contributions: Conceptualisation, P.O.M., H.M.S., D.B. and R.J.; Data curation, P.O.M. and K.M.A.; Formal analysis, D.B. and K.M.A.; Funding acquisition, D.B. and K.M.A.; Investigation, P.O.M., H.M.S., D.B., R.J. and K.M.A.; Methodology, R.J. and K.M.A.; Project administration, H.M.S. and D.B.; Resources, R.J.; Software, P.O.M.; Supervision, H.M.S. and D.B.; Validation, R.J.; Visualisation, R.J.; Writing–original draft, P.O.M.; Writing–review and editing, H.M.S. and K.M.A. All authors have read and agreed to the published version of the manuscript.

Funding: This research received no external funding.

Institutional Review Board Statement: Not applicable.

Informed Consent Statement: Not applicable.

Data Availability Statement: Not applicable.

Conflicts of Interest: The authors declare no conflict of interest.

References

1. Atici, F.M.; Atici, M.; Belcher, M.; Marshall, D. A new approach for modeling with discrete fractional equations. *Fund. Inform.* **2017**, *151*, 313–324.
2. Atici, F.; Sengul, S. Modeling with discrete fractional equations. *J. Math. Anal. Appl.* **2010**, *369*, 1–9. [CrossRef]
3. Chen, C.R.; Bohner, M.; Jia, B.G. Ulam-hyers stability of Caputo fractional difference equations. *Math. Meth. Appl. Sci.* **2019**, *42*, 7461–7470. [CrossRef]
4. Lizama, C. The Poisson distribution, abstract fractional difference equations, and stability. *Proc. Amer. Math. Soc.* **2017**, *145*, 3809–3827. [CrossRef]
5. Silem, A.; Wu, H.; Zhang, D.-J. Discrete rogue waves and blow-up from solitons of a nonisospectral semi-discrete nonlinear Schrödinger equation. *Appl. Math. Lett.* **2021**, *116*, 107049. [CrossRef]
6. Atici, F.M.; Atici, M.; Nguyen, N.; Zhoroev, T.; Koch, G. A study on discrete and discrete fractional pharmacokinetics-pharmacodynamics models for tumor growth and anti-cancer effects. *Comput. Math. Biophys.* **2019**, *7*, 10–24. [CrossRef]
7. Goodrich, C.S. On discrete sequential fractional boundary value problems. *J. Math. Anal. Appl.* **2012**, *385*, 111–124. [CrossRef]
8. Ferreira, R.A.C.; Torres, D.F.M. Fractional h-difference equations arising from the calculus of variations. *Appl. Anal. Discrete Math.* **2011**, *5*, 110–121. [CrossRef]
9. Wu, G.; Baleanu, D. Discrete chaos in fractional delayed logistic maps. *Nonlinear Dyn.* **2015**, *80*, 1697–1703. [CrossRef]
10. Srivastava, H.M. An introductory overview of fractional-calculus operators based upon the Fox-Wright and related higher transcendental functions. *J. Adv. Engrg. Comput.* **2021**, *5*, 135–166. [CrossRef]
11. Srivastava, H.M. Some parametric and argument variations of the operators of fractional calculus and related special functions and integral transformations. *J. Nonlinear Convex Anal.* **2021**, *22*, 1501–1520.
12. Goodrich, C.S.; Peterson, A.C. *Discrete Fractional Calculus*; Springer: Berlin/Heidelberg, Germany, 2015.
13. Abdeljawad, T. On delta and nabla caputo fractional differences and dual identities. *Discrete Dyn. Nat. Soc.* **2013**, *2013*, 12. [CrossRef]
14. Abdeljawad, T.; Jarad, F.; Atangana, A.; Mohammed, P.O. On a new type of fractional difference operators on h-step isolated time scales. *J. Fract. Calc. Nonlinear Syst.* **2021**, *1*, 46–74.
15. Abdeljawad, T.; Atici, F. On the definitions of nabla fractional operators. *Abstr. Appl. Anal.* **2012**, *2012*, 1–13. [CrossRef]
16. Dahal, R.; Goodrich, C.S. Mixed order monotonicity results for sequential fractional nabla differences. *J. Differ. Equ. Appl.* **2019**, *25*, 837–854. [CrossRef]
17. Du, F.; Jia, B.; Erbe, L.; Peterson, A. Monotonicity and convexity for nabla fractional (q,h)-differences. *J. Differ. Equ. Appl.* **2016**, *22*, 1224–1243. [CrossRef]
18. Mohammed, P.O.; Almutairi, O.; Agarwal, R.P.; Hamed, Y.S. On convexity, monotonicity and positivity analysis for discrete fractional operators defined using exponential kernels. *Fractal Fract.* **2022**, *6*, 55. [CrossRef]
19. Dahal, R.; Goodrich, C.S.; Lyons, B. Monotonicity results for sequential fractional differences of mixed orders with negative lower bound. *J. Differ. Equ. Appl.* **2021**, *27*, 1574–1593. [CrossRef]
20. Goodrich, C.S. A note on convexity, concavity, and growth conditions in discrete fractional calculus with delta difference. *Math. Inequal. Appl.* **2016**, *19*, 769–779. [CrossRef]
21. Goodrich, C.S. A sharp convexity result for sequential fractional delta differences. *J. Differ. Equ. Appl.* **2017**, *23*, 1986–2003. [CrossRef]
22. Dahal, R.; Goodrich, C.S. A monotonicity result for discrete fractional difference operators. *Arch. Math.* **2014**, *102*, 293–299. [CrossRef]
23. Atici, F.; Uyanik, M. Analysis of discrete fractional operators. *Appl. Anal. Discrete Math.* **2015**, *9*, 139–149. [CrossRef]
24. Mohammed, P.O.; Abdeljawad, T.; Hamasalh, F.K. On discrete delta Caputo-Fabrizio fractional operators and monotonicity analysis. *Fractal Fract.* **2021**, *5*, 116. [CrossRef]

25. Mohammed, P.O.; Abdeljawad, T.; Hamasalh, F.K. On Riemann–Liouville and Caputo fractional forward difference monotonicity analysis. *Mathematics* **2021**, *9*, 1303. [CrossRef]
26. Abdeljawad, T.; Baleanu, D. Monotonicity analysis of a nabla discrete fractional operator with discrete Mittag-Leffler kernel. *Chaos Solit. Fract.* **2017**, *116*, 1–5. [CrossRef]
27. Goodrich, C.S.; Lizama, C. Positivity, monotonicity, and convexity for convolution operators. *Discrete Contin. Dyn. Syst.* **2020**, *40*, 4961–4983. [CrossRef]
28. Goodrich, C.S.; Lyons, B. Positivity and monotonicity results for triple sequential fractional differences via convolution. *Analysis* **2020**, *40*, 89–103. [CrossRef]
29. Erbe, L.; Goodrich, C.S.; Jia, B.; Peterson, A. Monotonicity results for delta fractional differences revisited. *Math. Slovaca* **2017**, *67*, 895–906. [CrossRef]
30. Abdeljawad, T. Different type kernel h-fractional differences and their fractional h-sums. *Chaos Solit. Fract.* **2018**, *116*, 146–156. [CrossRef]
31. Liu, X.; Du, F.; Anderson, D.; Jia, B. Monotonicity results for nabla fractional h-difference operators. *Math. Meth. Appl. Sci.* **2021**, *44*, 1207–1218. [CrossRef]
32. Abdeljawad, T.; Al-Mdallal, Q.M.; Hajji, M.A. Arbitrary order fractional difference operators with discrete exponential kernels and applications. *Discrete Dyn. Nat. Soc.* **2017**, *2017*, 4149320. [CrossRef]
33. Abdeljawad, T.; Madjidi, F. Lyapunov-type inequalities for fractional difference operators with discrete Mittag-Leffler kernel of order $2 < \alpha < 5/2$. *Eur. Phys. J. Spec. Top.* **2017**, *226*, 3355–3368.

mathematics

MDPI

Article

Nonlinear Differential Equations with Distributed Delay: Some New Oscillatory Solutions

Barakah Almarri [1], Ali Hasan Ali [2,3], António M. Lopes [4,*] and Omar Bazighifan [5,6]

[1] Department of Mathematical Sciences, College of Sciences, Princess Nourah bint Abdulrahman University, P.O. Box 84428, Riyadh 11671, Saudi Arabia; bjalmarri@pnu.edu.sa
[2] Department of Mathematics, College of Education for Pure Sciences, University of Basrah, Basrah 61001, Iraq; ali.hasan@science.unideb.hu
[3] Doctoral School of Mathematical and Computational Sciences, University of Debrecen, H-4002 Debrecen, Hungary
[4] LAETA/INEGI, Faculty of Engineering, University of Porto, 4099-002 Porto, Portugal
[5] Department of Mathematics, Faculty of Science, Hadhramout University, Mukalla 50512, Yemen; o.bazighifan@gmail.com
[6] Section of Mathematics, International Telematic University Uninettuno, CorsoVittorio Emanuele II, 39, 00186 Roma, Italy
* Correspondence: aml@fe.up.pt

Abstract: The oscillation of a class of fourth-order nonlinear damped delay differential equations with distributed deviating arguments is the subject of this research. We propose a new explanation of the fourth-order equation oscillation in terms of the oscillation of a similar well-studied second-order linear differential equation without damping. The extended Riccati transformation, integral averaging approach, and comparison principles are used to provide some additional oscillatory criteria. An example demonstrates the efficacy of the acquired criteria.

Keywords: oscillation; fourth-order; damping term; Riccati transformation; comparison theorem; distributed deviating arguments

MSC: 34C10; 34K11

check for updates

Citation: Almarri, B.; Ali, A.H.; Lopes, A.M.; Bazighifani, O. Nonlinear Differential Equations with Distributed Delay: Some New Oscillatory Solutions. *Mathematics* **2022**, *10*, 995. https://doi.org/10.3390/math10060995

Academic Editor: Ioannis G. Stratis

Received: 9 March 2022
Accepted: 17 March 2022
Published: 19 March 2022

1. Introduction

In our current study, we take into consideration the following fourth-order nonlinear damped delay differential equations with distributed deviating arguments:

$$\left(x_2(t)\left(x_1(t)\left(u''(t)\right)^{\alpha}\right)'\right)' + p(t)\left(u''(\delta(t))\right)^{\alpha} + \int_c^d q(t,\varrho)f(t,u(g(t,\varrho)))d\varrho = 0, \quad (1)$$

where $\alpha \geq 1$ is a ratio of odd non-negative natural numbers and $c < d$. We consider the below assertions all through this article:

$$\begin{cases} x_1, x_2, p, \delta \in C(I,[0,\infty)) \text{ and } x_1, x_2 > 0, \text{ where } I = [t_0, +\infty); \\ q, g \in C[I \times [c,d],[0,\infty)), \delta(t) \leq t, \lim_{t\to+\infty}\delta(t) = \infty,\ g(t,\varrho) \text{ is a non-decreasing} \\ \text{function for } \varrho \in [c,d] \text{ satisfying } g(t,\varrho) \leq t \text{ and } \lim_{t\to+\infty} g(t,\varrho) = \infty; \\ f \in C(\mathbb{R},\mathbb{R}),\ \text{there is a constant } k_1 > 0 \text{ such that } f(t,\ u(t))/u^{\beta} \geq k_1. \end{cases}$$

We define the operators,

$$L^{[0]}u = u,\ L^{[1]}u = u',\ L^{[2]}u = x_1\left(\left(L^{[0]}u\right)''\right)^{\alpha},\ L^{[3]}u = x_2\left(L^{[2]}u\right)' \quad \text{as well as} \quad L^{[4]}u = \left(L^{[3]}u\right)'.$$

The meaning of having a solution to Equation (1) is the function $u(t)$ in $C^2[T_u,\infty)$, for which $L^{[2]}u$, $L^{[4]}u$ is in $C^1[T_u,\infty)$, and Equation (1) holds on $[T_u,\infty)$, such that $T_u \geq t_0$. We only take into consideration the solutions $u(t)$ when $\sup\{|u(t)| : t \geq T\} > 0$ for every $T \geq T_u$. On one hand, such a solution to Equation (1) is termed oscillatory when this solution is not eventually negative and, at the same time, not eventually positive on the interval $[T_u,\infty)$. On the other hand, the same solution is termed non-oscillatory if it is eventually negative or eventually positive. Finally, when every solution is oscillating, the equation is said to be oscillatory.

We define

$$\begin{aligned}
A_1(t_1,t) &= \int_{t_1}^{t} x_1^{-1/\alpha}(s)ds,\\
A_2(t_1,t) &= \int_{t_1}^{t} x_2^{-1}(s)ds,\\
A_3(t_1,t) &= \int_{t_1}^{t} \left((x_1(s))^{-1} A_2(t_1,s) \right)^{1/\alpha} ds,\\
A_4(t_1,t) &= \int_{t_1}^{t}\int_{t_1}^{u} \left((x_1(s))^{-1} A_2(t_1,s) \right)^{1/\alpha} ds\, du,
\end{aligned}$$

for $t_0 \leq t_1 \leq t < \infty$ and assume that

$$A_1(t_1,t) \to \infty, \quad A_2(t_1,t) \to \infty \quad \text{as} \quad t \to \infty. \tag{2}$$

Fourth-order differential equations are often used in mathematical models of a wide range of physical, chemical, and biological processes [1–4]. Problems with elasticity, structural deformation, and soil settling are examples of applications of this type of equation. In addition, in mechanical and engineering fields, questions about the presence of oscillatory and non-oscillatory solutions are mostly arising, and the solutions require the presence of the same mentioned equation [5]. Many researchers have intensively studied the topic of oscillation of fourth or higher order differential equations in depth, and many strategies for establishing oscillatory criteria for fourth or higher order differential equations have been developed. Several works, see [6–18], contain extremely interesting results linked to oscillatory features of solutions of neutral differential equations and damped delay differential equations with or without distributed deviating arguments.

In fact, for the following equation, Bazighifan et al. [19] have developed some oscillation criteria

$$\left(r(t)\left(N_x'''(t)\right)^{\beta}\right)' + \int_a^b q(t,\varrho)x^{\beta}(g(t,\varrho))d\varrho = 0.$$

Moreover, Dzurina et al. [20] introduced some oscillation findings of the below fourth-order equation

$$\left(r_3(t)\left(r_2(t)\left(r_1(t)y'(t)\right)\right)'\right)' + p(t)y'(t) + q(t)y(\tau(t)) = 0.$$

More specifically, there are no requirements for the oscillation of Equation (1) in the previous studies.

By the motivations above, our contribution would be giving certain adequate conditions that ensure that every solution to Equation (1) oscillates, utilizing proper Riccati-type transformation, integral averaging condition, and comparison technique, when the following second-order equation

$$(x_2(t)z'(t))' + \frac{p(t)}{x_1(\delta(t))}z(t) = 0, \tag{3}$$

is oscillatory or non-oscillatory.

2. Basic Lemmas

We state in the current section several Lemmas along with their proofs, which are mostly needed in the rest of this study.

Lemma 1 ([8]). *Assume that Equation* (3) *is non-oscillatory. If Equation* (1) *has a non-oscillatory solution* $u(t)$ *on* I, $t_1 \geq t_0$, *then there is a* $t_2 \in I$ *in a way that* $u(t)L^{[2]}u(t) > 0$ *or* $u(t)L^{[2]}u(t) < 0$ *for* $t \geq t_2$.

Lemma 2. *If the Equation* (1) *has a non-oscillatory solution* $u(t)$ *that satisfies* $u(t)L^{[2]}u(t) > 0$ *in Lemma 1 for* $t \geq t_1 \geq t_0$, *then*

$$L^{[2]}u(t) > A_2(t_1,t)\, L^{[3]}u(t), \quad t \geq t_1, \tag{4}$$

$$L^{[1]}u(t) > A_3(t_1,t)\, \left(L^{[3]}u(t)\right)^{1/\alpha}, \quad t \geq t_1, \tag{5}$$

and

$$u(t) > A_4(t_1,t)\, \left(L^{[3]}u(t)\right)^{1/\alpha}, \quad t \geq t_1. \tag{6}$$

Proof. We suppose that there is a $t_1 \geq t_0$ in a way that $u(t) > 0$ and $u(g(t,\varrho)) > 0$ for $t \geq t_1$. From Equation (1), we have

$$L^{[4]}u(t) = -\Big(\frac{p(t)}{x_1(\delta(t))}\Big) L^{[2]}u(\delta(t)) - k_1 \int_c^d q(t,\varrho) u^\beta(g(t,\varrho))d\varrho \leq 0,$$

and $L^{[3]}u(t)$ is non increasing on I, we obtain

$$L^{[2]}u(t) \geq \int_{t_1}^t (L^{[2]}u(s))'\, ds = \int_{t_1}^t (x_2(s))^{-1} L^{[3]}u(s)\, ds \geq A_2(t_1,t)\, L^{[3]}u(t),$$

which implies that

$$u''(t) \geq \Big(L^{[3]}u(t)\Big)^{1/\alpha} \Big((x_1(t))^{-1} A_2(t_1,t)\Big)^{1/\alpha}.$$

Now, twice integrating above from t_1 to t and using $L^{[3]}u(t) \leq 0$, we find

$$u'(t) \geq \Big(L^{[3]}u(t)\Big)^{1/\alpha} \int_{t_1}^t \Big((x_1(s))^{-1} A_2(t_1,s)\Big)^{1/\alpha} ds$$

and

$$u(t) \geq \Big(L^{[3]}u(t)\Big)^{1/\alpha} \int_{t_1}^t \int_{t_1}^u \Big((x_1(s))^{-1} A_2(t_1,s)\Big)^{1/\alpha} ds\, du \quad \text{for } t \leq t_1.$$

□

Lemma 3 ([10]). *Let* $\xi \in C^1(I,\mathbb{R}^+)$, $\xi(t) \leq t$, $\xi'(t) \geq 0$ *and* $G(t) \in C(I,\mathbb{R}^+)$ *for* $t \geq t_0$. *Assume that* $y(t)$ *is a bounded solution of a second-order delay differential equation:*

$$(x_2(t)\, y'(t))' - \Theta(t)\, y(\xi(t)) = 0. \tag{7}$$

If

$$\limsup_{t\to\infty} \int_{\xi(t)}^t \Theta(s)\, A_2(\xi(t), \xi(s))\, ds > 1 \tag{8}$$

or

$$\limsup_{t\to\infty}\int_{\xi(t)}^{t}\Big((x_2(t))^{-1}\int_{u}^{t}\Theta(s)\,ds\Big)du>1, \tag{9}$$

where $x_2(t)$ is as in Equation (1); thus, the solutions of Equation (7) are oscillatory.

3. Oscillation—Comparison Principle Method

In this section, we shall establish some oscillation criteria for Equation (1). For convenience, we denote

$$Q(t)=\Big(\frac{p(t)}{x_1(\delta(t))}\Big)A_2(t_1,\delta(t)),\quad \psi(t)=\exp\Big(\int_{t_1}^{t}Q(s)ds\Big),$$
$$\widetilde{q}(t,\varrho)=\int_{c}^{d}q(t,\varrho)\,d\varrho,\quad \Theta^*(t)=k_1\,\widetilde{q}(t,\varrho)\Big(A_4(t_1,g(t,d))\Big)^{\beta}.$$

Theorem 1. *Assume that $\alpha\geq\beta$ and the conditions in Equation (2) hold, and Equation (3) is non-oscillatory. Suppose there exists a $\xi\in C^1(I,\mathbb{R})$ such that*

$$g(t,\varrho)\leq\xi(t)\leq\delta(t)\leq t,\quad \xi'(t)\geq 0\quad \textit{for } t\geq t_1,$$

and Equations (8) or (9) holds with

$$\Theta(t)=\ell_*\,k_1\widetilde{q}(t,\varrho)g^{\beta}(t,d)\big(A_1(\xi(t),g(t,d))\big)^{\beta}-\frac{p(t)}{x_1(\delta(t))}\geq 0,\quad t\geq t_1,$$

for constant $\ell_*>0$. Moreover, suppose that every solution of the first-order delay equation

$$z'(t)+\psi^{1-\frac{\beta}{\alpha}}(g(t,d))\,\Theta^*(t)\,z^{\frac{\beta}{\alpha}}(g(t,d))=0. \tag{10}$$

Then, every solution of Equation (1) is oscillatory.

Proof. Let Equation (1) have a non-oscillatory solution $u(t)$. Assume there exists a $t\geq t_1$ such that $u(t)>0$ and $u(g(t,\varrho))>0$ for some $t\geq t_0$. From Lemma 1, $u(t)$ has the conditions either $L^{[2]}u(t)>0$ or $L^{[2]}u(t)<0$ for $t\geq t_1$.

Assume that $u(t)$ has the condition $L^{[2]}u(t)>0$ for $t\geq t_1$, then one can easily see that $L^{[3]}u(t)>0$ for $t\geq t_1$. We can choose $t_2\geq t_1$ such that $g(t,\varrho)\geq t_1$ for $t\geq t_2$, $g(t,\varrho)\to\infty$ as $t\to\infty$, and we have Equation (6),

$$u(g(t,d))>A_4(t_1,g(t,d))\,\big(L^{[3]}u(g(t,d))\big)^{1/\alpha},\quad t\geq t_2. \tag{11}$$

By substituting Equations (4) and (11) into Equation (1) and when $L^{[3]}u(t)$ is decreasing,

$$\begin{aligned}&\Big(L^{[3]}u(t)\Big)'+\Big(\frac{p(t)}{x_1(\delta(t))}\Big)L^{[3]}u(t)A_2(t_1,\delta(t))\\&\qquad+k_1\,\widetilde{q}(t,\varrho)\Big(A_4(t_1,g(t,d))\Big)^{\beta}\Big(L^{[3]}u(g(t,d))\Big)^{\beta/\alpha}\leq 0.\end{aligned} \tag{12}$$

Taking $\phi=L^{[3]}u$, we have

$$\phi'(t)+Q(t)\phi(t)+\Theta^*(t)\phi^{\frac{\beta}{\alpha}}(g(t,d))\leq 0 \tag{13}$$

or

$$\Big(\psi(t)\,\phi(t)\Big)'+\psi(t)\Theta^*(t)\phi^{\frac{\beta}{\alpha}}(g(t,d))\leq 0,\quad \text{for } t\geq t_2. \tag{14}$$

Next, setting $z = \psi\phi > 0$ and $\psi(g(t,d)) \le \phi(t)$, we have

$$z'(t) + \psi^{1-\frac{\beta}{\alpha}}(g(t,d))\Theta^*(t)z^{\frac{\beta}{\alpha}}(g(t,d)) \le 0. \tag{15}$$

This means Equation (15) is positive for this inequality. Furthermore, by ([21], Corollary 2.3.5), it can be seen that Equation (1) has a positive solution, a contradiction.

Next, assume $u(t)$ has the condition $L^{[2]}u(t) < 0$, for $t \ge t_1$, then one can easily see that $L^{[1]}u(t) \ge 0$, $L^{[3]}u(t) > 0$ for $t \ge t_3(\ge t_2)$. Using the monotonicity of $u'(t)$ and mean value property of differentiation, there exists a $\theta \in (0,1)$ such that

$$u(t) \ge \theta\, t\, u'(t), \qquad \text{for } t \ge t_3. \tag{16}$$

Set $w(t) = L^{[1]}u(t)$, then $w'(t) = u''(t) < 0$. Using Equation (16) in Equation (1), we obtain

$$\left(x_2(t)\left(x_1(t)\left[w'(t)\right]^{\alpha}\right)'\right)' + p(t)(w'(\delta(t)))^{\alpha} + k_1\,(t\theta)^{\beta}\,\widetilde{q}(t,\varrho)w^{\beta}(g(t,d)) \le 0,$$

and so $\left(x_1(t)\left[w'(t)\right]^{\alpha}\right) < 0$, we have $\left(x_1(t)\left[w'(t)\right]^{\alpha}\right)' > 0$ for $t \ge t_3$. Now, for $v \ge u \ge t_3$, we obtain

$$\begin{aligned} w(u) > w(u) - w(v) &= -\int_u^v -x_1^{-1/\alpha}(\tau)(x_1(\tau)(w'(\tau))^{\alpha})^{1/\alpha}d\tau \\ &\ge x_1^{1/\alpha}(v)(-w'(v)))\left(\int_u^v x_1^{-1/\alpha}(\tau)d\tau\right) \\ &= x_1^{1/\alpha}(v)(-w'(v))A_1(u,v). \end{aligned}$$

Taking $u = \xi(t)$ and $v = g(t,d)$, we obtain

$$w(g(t,d)) > A_1(g(t,d),\xi(t))\,(x_1^{1/\alpha}(\xi(t))(-w'(\xi(t)))) = A_1(g(t,d),\xi(t))\;y(\xi(t)),$$

where $y(t) = x_1^{1/\alpha}(\xi(t))(-w'(\xi(t))) > 0$ for $t \ge t_3$. From Equation (1), we have that $y(t)$ is decreasing and $g(t,d) \le \xi(t) \le \delta(t) \le t$; thus, we obtain

$$(x_2(t)z'(t))' + \frac{p(t)}{x_1(\delta(t))}z(\delta(t)) \ge k_1\,(\theta g(t,d))^{\beta}\,\widetilde{q}(t,\varrho)A_1(g(t,d),\xi(t))z^{\frac{\beta}{\alpha}-1}(\xi(t))z(\xi(t)).$$

Since z is decreasing and $\alpha \ge \beta$, there exists a constant ℓ such that $z^{\frac{\beta}{\alpha}-1}(t) \ge \ell$ for $t \ge t_3$. Thus, we obtain

$$(x_2(t)z'(t))' \ge \left(\ell\, k_1\,(\theta\, g(t,d))^{\beta}\,\widetilde{q}(t,\varrho)A_1(g(t,d),\xi(t)) - \frac{p(t)}{x_1(\delta(t))}\right)z(\xi(t)).$$

Proceeding the rest of the proof in Lemma (3), we arrive at the required conclusion, and so it is omitted. □

4. Oscillation—Riccati Method

This section deals with some oscillation criteria for Equation (1) using the Ricatti Method.

Theorem 2. *Assume $\alpha \ge \beta$ and the conditions in Equation (2) hold, Equation (3) is non-oscillatory. Suppose there exists η, $\xi \in C^1(I,\mathbb{R})$ such that $g(t,\varrho) \le \xi(t) \le \delta(t) \le t$, $\xi'(t) \ge 0$ and $\eta > 0$ for $t \ge t_1$ with*

$$\limsup_{t\to\infty} \int_{t_5}^{t} \left(k_1\,\eta(s)\,\widetilde{q}(s,\varrho) - \frac{A^2(s)}{4B(s)}\right)ds = \infty \text{ for all } t_1 \in I, \tag{17}$$

where, for $t \geq t_1$,

$$A(t) = \frac{\eta'(t)}{\eta(t)} - \frac{p(t)}{x_1(\delta(t))} A_2(t_1, \delta(t)) \tag{18}$$

and

$$B(t) = \frac{\beta \ell_2^{\beta-\alpha} g'(t,d)}{\eta(t)} \Big(A_4(t_1, g(t,d)) \Big)^{\beta-1} \Big(A_3(t_1, g(t,d)) \Big)^{1/\alpha}, \tag{19}$$

also Equations (8) *or* (9) *hold with* $\Theta(t)$ *as in Theorem 1. Then every solution of Equation* (1) *is oscillatory.*

Proof. Suppose that Equation (1) has a non-oscillatory solution $u(t)$. Assume that, there exists a $t \geq t_1$ such that $u(t) > 0$ and $u(g(t,\varrho)) > 0$ for some $t \geq t_0$. From Lemma 1, $u(t)$ has the conditions either $L^{[2]}u(t) > 0$ or $L^{[2]}u(t) < 0$ for $t \geq t_1$. If condition $L^{[2]}u(t) < 0$ holds, the proof follows from Theorem 1.

Next, if condition $L^{[2]}u(t) > 0$ holds, define

$$\omega(t) = \eta(t) \frac{L^{[3]}u(t)}{u^{\beta}(g(t,d))}, \quad t \in I, \tag{20}$$

then $\omega(t) > 0$ for $t \geq t_1$. From Equation (6) and $L^{[4]}u(t) < 0$, we have

$$\omega(t) = \eta(t) \frac{L^{[3]}u(t)}{u^{\beta}(g(t,d))} \leq \eta(t) \frac{L^{[3]}u(g(t,d))}{u^{\beta}(g(t,d))} \leq \eta(t)(A_4(t_1, g(t,d)))^{-\alpha} u^{\alpha-\beta}(g(t,d)), \tag{21}$$

for $t \geq t_1$. From Equation (5) and definition $L^{[2]}u(t)$, we find

$$u'(g(t,d)) = L^{[1]}u(g(t,d)) \geq A_3(t_1, g(t,d))(L^{[3]}u(\delta(t)))^{1/\alpha} \geq A_3(t_1, g(t,d))(L^{[3]}u(g(t,d)))^{1/\alpha}.$$

Then,

$$\begin{aligned} \frac{u'(g(t,d))}{u(g(t,d))} &\geq \left(\frac{A_3(t_1, g(t,d))}{\eta(\delta(t))} \right)^{1/\alpha} \frac{\eta^{1/\alpha}(\delta(t))(L^{[3]}u(t))^{1/\alpha}}{u^{\beta/\alpha}(g(\delta(t),d))} u^{\beta/\alpha-1}(g(\delta(t),d)) \\ &= \left(\frac{A_3(t_1, g(t,d))}{\eta(t)} \right)^{1/\alpha} \omega^{1/\alpha}(t) u^{\beta/\alpha-1}(g(\delta(t),d)). \end{aligned} \tag{22}$$

Furthermore, since there exists a constant ℓ_1 and $t_2 \geq t_1$ such that for $L^{[3]}u(t) \leq L^{[3]}u(t_2) = \ell_1$. Therefore,

$$\begin{aligned} L^{[2]}u(t) &= L^{[2]}u(t_2) + \int_{t_2}^{t} (L^{[2]}u(s))' ds \leq L^{[2]}u(t_2) + \ell_1 \int_{t_2}^{t} \frac{ds}{x_2(s)} \\ &= L^{[2]}u(t_2) + \ell_1 A_2(t_2, t) = \left[\frac{L^{[2]}u(t_2)}{A_2(t_2, t)} + \ell_1 \right] A_2(t_2, t) \\ &\leq \left[\frac{L^{[2]}u(t_2)}{A_2(t_2, t_3)} + \ell_1 \right] A_2(t_2, t) = \ell_1^* A_2(t_2, t), \end{aligned} \tag{23}$$

holds for all $t \geq t_2$, where $\ell_1^* = \ell_1 + \frac{L^{[2]}u(t_1)}{A_2(t_2,t_3)}$, which implies that

$$u'(t) = u'(t_3) + \int_{t_3}^{t} u''(s)ds \leq u'(t_3) + \int_{t_3}^{t} \left(\frac{\ell_1^* A_2(t_2,s)}{x_1(s)} \right)^{1/\alpha} ds$$
$$= u(t_3) + \left(\ell_1^*\right)^{1/\alpha} A_3(t_3,t) = \ell_2 \, A_3(t_3,t),$$

holds for all $t \geq t_3(\geq t_2)$, where $\ell_2 = \frac{u(t_2)}{A_3(t_3,t_4)} + (\ell_1^*)^{1/\alpha}$. Then,

$$u(t) = u(t_4) + \int_{t_4}^{t} u'(s)ds \leq u(t_4) + \int_{t_4}^{t} \Big(\ell_2 \, A_3(t_3,s) \Big) ds$$
$$= u(t_4) + \ell_2 \, A_4(t_4,t) = \ell_2^* \, A_4(t_4,t), \tag{24}$$

holds for all $t \geq t_4(\geq t_3)$, where $\ell_2^* = \frac{u(t_4)}{A_4(t_4,t_1)} + \ell_2$. Further,

$$u^{\beta/\alpha-1}(g(t,d)) \geq \left(\ell_2^*\right)^{\beta/\alpha-1} \left(A_4(t_4, g(t,d))\right)^{\beta/\alpha-1}, \quad t \geq t_4. \tag{25}$$

By using Equation (24) in Equation (21), we obtain

$$\omega(t) \leq \left(\ell_2^*\right)^{\alpha-\beta} \eta(t) \left(A_4(t_1, g(t,d))\right)^{-\beta}, \tag{26}$$

and hence

$$\omega^{\frac{1}{\alpha}-1}(t) \leq \left(\ell_2^*\right)^{(\alpha-\beta)(\frac{1}{\alpha}-1)} \eta^{\frac{1}{\alpha}-1}(t) \left(A_4(t_1, g(t,d))\right)^{-\beta(\frac{1}{\alpha}-1)}. \tag{27}$$

Now differentiating Equation (20), we obtain

$$\omega'(t) = \frac{\eta'(t)}{\eta(t)}\omega(t) + \frac{L^{[4]}u(t)}{L^{[3]}u(t)}\omega(t) - \beta g'(t,d)\frac{u'(g(t,d))}{u(g(t,d))}\omega(t). \tag{28}$$

Using Equations (1) and (4) in Equation (28), we have

$$\begin{aligned} \omega'(t) &\leq \left[\frac{\eta'(t)}{\eta(t)} - \frac{p(t)}{x_1(g(t,d))} A_2(t_4, g(t,d))\right]\omega(t) - k_1\eta(t)\widetilde{q}(t,\varrho) - \beta g'(t)\frac{u'(g(t,d))}{u(g(t,d))}\omega(t) \\ &\leq A(t)\omega(t) - k_1\eta(t)\widetilde{q}(t,\varrho) - \beta g'(t)\frac{u'(g(t,d))}{u(g(t,d))}\omega(t). \end{aligned} \tag{29}$$

By using Equations (22), (25) and (28) in Equation (29), we have

$$\begin{aligned} \omega'(t) &\leq A(t)\omega(t) - k_1\eta(t)\widetilde{q}(t,\varrho) - \frac{\beta\, \ell_2^{\beta-\alpha} g'(t)}{\eta(t)}\Big(A_4(t_1, g(t,d))\Big)^{\beta-1}\Big(A_3(t_1, g(t,d))\Big)^{1/\alpha}\omega^2(t) \\ &= A(t)\omega(t) - k_1\eta(t)\widetilde{q}(t,\varrho) + B(t)\omega^2(t) \end{aligned} \tag{30}$$

$$\begin{aligned} &= -k_1\eta(t)\widetilde{q}(t,\varrho) + \left[\sqrt{B(t)}\,\omega(t) - \frac{1}{2}\frac{A(t)}{\sqrt{B(t)}}\right]^2 + \frac{1}{4}\frac{A^2(t)}{B(t)} \\ &\leq -k_1\eta(t)\widetilde{q}(t,\varrho) + \frac{1}{4}\frac{A^2(t)}{B(t)}. \end{aligned} \tag{31}$$

Integrating Equation (31) from $t_5(> t_4)$ to t gives

$$\int_{t_5}^{t} \left(k_1 \, \eta(s) \, \widetilde{q}(s,\varrho) - \frac{1}{4}\frac{A^2(s)}{B(s)} \right) ds \leq \omega(t_5), \tag{32}$$

which contradicts Equation (17). $\square$

Corollary 1. *Assume $\alpha \geq \beta$ and the conditions in Equation (2) hold, Equation (3) is non-oscillatory. Suppose there exists $\eta, \xi \in C^1(I, \mathbb{R})$ such that $g(t, \varrho) \leq \xi(t) \leq \delta(t) \leq t$, $\xi'(t) \geq 0$ and $\eta > 0$ for $t \geq t_1$ such that the function $A(t) \leq 0$,*

$$\limsup_{t \to \infty} \int_{t_5}^{t} \Big(\eta(s)\, \tilde{q}(s, \varrho)\Big) ds = \infty \text{ for all } t_1 \in I, \tag{33}$$

where $A(t)$ is defined in Equation (18), and Equations (8) or (9) holds with $\Theta(t)$ as in Theorem 1. Then, every solution of Equation (1) is oscillatory.

Next, we examine the oscillation results of solutions to Equation (1) by Philos-type. Let $\mathbb{D}_0 = \{(t, s) : a \leq s < t < +\infty\}$, $\mathbb{D} = \{(t, s) : a \leq s \leq t < +\infty\}$, the continuous function $H(t, s)$, $H : \mathbb{D} \to \mathbb{R}$ belongs to the class function $\mathbb{R}$:

(i) $H(t, t) = 0$ for $t \geq t_0$ and $H(t, s) > 0$ for $(t, s) \in \mathbb{D}_0$;
(ii) H has a continuous and non-positive partial derivative on $\mathbb{D}_0$ with respect to the second variable such that

$$-\frac{\partial H(t, s)}{\partial s} = h(t, s)[H(t, s)]^{1/2},$$

for all $(t, s) \in \mathbb{D}_0$.

Theorem 3. *Assume $\alpha \geq 1$ and the conditions in Equation (2) hold, and Equation (3) is non-oscillatory. Suppose there exists $\eta, \xi \in C^1(I, \mathbb{R})$ such that $g(t, \varrho) \leq \xi(t) \leq \delta(t) \leq t$, $\xi'(t) \geq 0$, $\eta > 0$ and $H(t, s) \in \mathbb{R}$ for $t \geq t_1$ with*

$$\limsup_{t \to \infty} \frac{1}{H(t, t_5)} \int_{t_5}^{t} \left(k_1\, \eta(s)\, \tilde{q}(s, \varrho) H(t, s) - \frac{P^2(t, s)}{4B(s)} \right) ds = \infty \quad \text{for all} \quad t_1 \in I, \tag{34}$$

where $P(t, s) = h(t, s) - A(s)\sqrt{H(t, s)}$ and $A(t)$, $B(t)$ are defined in Theorem 2, and Equations (8) or (9) holds with $\Theta(t)$ as in Theorem 1. Then, every solution of Equation (1) is oscillatory.

Proof. Suppose that Equation (1) has a non-oscillatory solution $u(t)$. Assume that there exists a $t \geq t_1$ such that $u(t) > 0$ and $u(g(t, \varrho)) > 0$ for some $t \geq t_0$. Proceeding as in the proof of Theorem 2, we obtain the inequality from Equation (30), i.e.,

$$\omega'(t) \;\leq\; A(t)\omega(t) - k_1 \eta(t) \tilde{q}(t, \varrho) + B(t)\omega^2(t),$$

and so,

$$\begin{aligned}
\int_{t_5}^{t} H(t, s)\eta(s)\tilde{q}(s, \varrho) ds \;&\leq\; \int_{t_5}^{t} H(t, s)[-\omega'(s) + A(s)\omega(s) - B(s)\omega^2(s)] ds \\
&= -H(t, s)\Big[\omega(s)\Big]_{t_5}^{t} + \int_{t_5}^{t} \Big[\frac{\partial H(t, s)}{\partial s}\omega(s) \\
&\quad + H(t, s)\Big[A(s)\omega(s) - B(s)\omega^2(s)\Big]\Big] ds \\
&= H(t, t_5)\omega(t_5) - \int_{t_5}^{t} \Big[\omega^2(s)B(s)H(t, s) \\
&\quad + \omega(s)\Big(h(t, s)\sqrt{H(t, s)} - H(t, s)A(s)\Big)\Big] ds \\
&\leq H(t, t_5)\omega(t_5) + \int_{t_5}^{t} \frac{P^2(t, s)}{4B(s)} ds,
\end{aligned}$$

which contradicts Equation (34). The rest of the proof is similar to that of Theorem 2 and hence is omitted. □

5. Examples

Below, we present an example to show the application of the main results. This example is given to demonstrate Theorem 2.

Example 1. *For $t \geq 1$, consider the fourth-order differential equation*

$$\left(1/2t\left(9e^{-t}(t)\left(u''(t)\right)\right)'\right)' + 36e^{-s/2}u''(t/2) + \int_1^2 \frac{t}{3}u(\varrho, 36e^{t/3})d\varrho = 0. \tag{35}$$

Here, $x_1 = 9e^{-t}$, $x_2 = 1/2t$, $\alpha = \beta = 1$, $p(t) = 36e^{-s/2}$, $q(t,\varrho) = t/3$ and $\delta(t) = t/2$, $g(t,\varrho) = t/3$. Now, pick $\eta(t) = 36e^{t/3}$, so we obtain

$$A_1(t_1,t) = \int_1^t (9e^s)^{-1}ds = 9(e^t - e),$$
$$A_2(t_1,t) = \int_1^t 2s\,ds = t^2 - 1 = (t+1)(t-1),$$
$$A_3(t_1,t/3) = \int_1^{t/3} (9e^s)^{-1}(s^2-1)ds = e^{t/3}(t-3)^2,$$
$$\widetilde{q}(s,\varrho) = \frac{s}{3}\int_1^2 d\varrho = s/3,$$

$A^2(s) = \frac{(3t^2-5)^2}{9}$ and $B(s) = \frac{(s-3)^2}{36}$. Now,

$$\limsup_{t\to\infty}\int_2^t \left(k_1\,\eta(s)\,\widetilde{q}(s,\varrho) - \frac{A^2(s)}{4B(s)}\right)ds = \limsup_{t\to\infty}\int_2^t \left(12k_1\,s\,e^{s/3} - \left(\frac{3s^2-5}{s-3}\right)^2\right)ds \to \infty \text{ as } t\to\infty,$$

and all hypotheses of Theorem 2 are satisfied, so every solution of Equation (35) is oscillatory.

6. Conclusions

The form in Equation (1) is clearly more generic than all of the problems covered in the literature. In this paper, we provided some oscillatory properties using the appropriate Riccati-type transformation, integral averaging condition, and comparison method, ensuring that any solution of Equation (1) oscillates under the assumption of $A_1(t_1,t) \to \infty$, $A_2(t_1,t) \to \infty$ as $t \to \infty$. Furthermore, based on the condition of $A_1(t_1,t) < \infty$, $A_2(t_1,t) < \infty$ as $t \to \infty$, it would be desirable to expand the oscillation criteria of Equation (1).

Author Contributions: Formal analysis, B.A., A.H.A. and A.M.L.; Data curation, O.B. and A.M.L.; Funding acquisition, A.M.L.; Methodology, B.A., A.H.A. and O.B.; Project administration, A.M.L.; Resources, A.H.A. and O.B.; Software, A.H.A. and O.B.; Supervision, A.M.L. and O.B.; Validation, A.M.L. and O.B.; Visualization, A.H.A.; Writing—review and editing, A.H.A. and O.B. All authors have read and agreed to the published version of the manuscript.

Funding: Princess Nourah bint Abdulrahman University Researchers Supporting Project number (PNURSP2022R216), Princess Nourah bint Abdulrahman University, Riyadh, Saudi Arabia.

Institutional Review Board Statement: Not applicable.

Informed Consent Statement: Not applicable.

Data Availability Statement: Not applicable.

Conflicts of Interest: The authors declare no conflict of interest.

References

1. Althobati, S.; Bazighifan, O.; Yavuz, M. Some Important Criteria for Oscillation of Non-Linear Differential Equations with Middle Term. *Mathematics* **2021**, *9*, 346. [CrossRef]
2. Agarwal, R.P.; Bazighifan, O.; Ragusa, M.A. Nonlinear Neutral Delay Differential Equations of Fourth-Order: Oscillation of Solutions. *Entropy* **2021**, *23*, 129. [CrossRef] [PubMed]
3. Agarwal, R.P.; Grace, S.R. The oscillation of higher-order differential equations with deviating arguments. *Comput. Math. Appl.* **1999**, *38*, 185–199. [CrossRef]
4. Agarwal, R.P.; Grace, S.R.; O'Regan, D. Oscillation of certain fourth order functional differential equations. *Ukrain. Mat. Zh.* **2007**, *59*, 291–313. [CrossRef]
5. Bartušek, M.; Cecchi, M.; Došlá, Z.; Marini, M. Fourth-order differential equation with deviating argument. *Abstr. Appl. Anal.* **2012**, *2012*, 185242. [CrossRef]
6. Agarwal, R.P.; Grace, S.R.; Wong, P.J. On the bounded oscillation of certain fourth order functional differential equations. *Nonlinear Dyn. Syst. Theory* **2005**, *5*, 215–227.
7. Bartušek, M.; Došlá, Z.: Oscillatory solutions of nonlinear fourth order differential equations with a middle term. *Electron. J. Qual. Theory Differ. Equ.* **2014**, *55*, 1–9. [CrossRef]
8. Grace, S.R.; Sun, S.; Akın, E. Oscillation Criteria for Certain Fourth-Order Nonlinear Delay Differential Equations. *Mediterr. J. Math.* **2016**, *13*, 2383–2396. [CrossRef]
9. Grace, S.R. Oscillation theorems for nth-order differential equations with deviating arguments. *J. Math. Anal. Appl.* **1984**, *101*, 268–296. [CrossRef]
10. Grace, S.R. Oscillation criteria for third-order nonlinear delay differential equations with damping. *Opuscula Math.* **2015**, *35*, 485–497. [CrossRef]
11. Grace, S.R.; Džurina, J.; Jadlovská, I.; Li, T. On the oscillation of fourth-order delay differential equations. *Adv. Differ. Equ.* **2019**, *1*, 1–15. [CrossRef]
12. Wei, M.; Jiang, C.; Li. T. Oscillation of third-order neutral differential equations with damping and distributed delay. *Adv. Differ. Equ.* **2019**, *1*, 1–11. [CrossRef]
13. Kumar, M.S.; Bazighifan, O.; Al-Shaqsi, K.; Wannalookkhee, F.; Nonlaopon, K. Symmetry and Its Role in Oscillation of Solutions of Third-Order Differential Equations. *Symmetry* **2021**, *13*, 1485. [CrossRef]
14. Hou, C.; Cheng, S.S. Asymptotic dichotomy in a class of fourth-order nonlinear delay differential equations with damping. *Abstr. Appl. Anal.* **2009**, *2009*, 7. [CrossRef]
15. Philos, C.G. On the existence of nonoscillatory solutions tending to zero at ∞ for differential equations with positive delays. *Arch. Math.* **1981**, *36*, 168–178. [CrossRef]
16. Saker, S.; Sethki, A. Riccati Technique and Oscillation of Second Order Nonlinear Neutral Delay Dynamic Equations. *J. Comput. Anal. Appl.* **2021**, *2021*, 266.
17. Jadlovská, I.; Chatzarakis, G.E.; Džurina, J.; Grace, S.R. On Sharp Oscillation Criteria for General Third-Order Delay Differential Equations. *Mathematics* **2021**, *9*, 1675. [CrossRef]
18. Tiryaki, A.; Aktas, M.F. Oscillation criteria of a certain class of third order nonlinear delay differential equations with damping. *J. Math. Anal. Appl.* **2007**, *325*, 54–68. [CrossRef]
19. Bazighifan, O.; Minhos, F.; Moaaz, O. Sufficient Conditions for Oscillation of Fourth-Order Neutral Differential Equations with Distributed Deviating Arguments. *Axioms* **2020**, *9*, 39. [CrossRef]
20. Dzurina, J.; Baculíková, B.; Jadlovská, I. New oscillation results to fourth order delay differential equations with damping. *Electron. J. Qual. Theory Differ. Equ.* **2016**, 1–15. [CrossRef]
21. Agarwal, R.P.; Grace, S.R.; O'Regan, D. Oscillation theory for second order dynamic equations. In *Series Mathematical Analysis Applications*; Taylor and Francis: London, UK, 2003; Volume 5.

mathematics

Article

Controllability Results for First Order Linear Fuzzy Differential Systems

Rui Liu [1], Michal Fečkan [2,3,*], Donal O'Regan [4] and Jinrong Wang [1]

1 Department of Mathematics, Guizhou University, Guiyang 550025, China; gzliuruiha@163.com (R.L.); jrwang@gzu.edu.cn (J.W.)
2 Department of Mathematical Analysis and Numerical Mathematics, Faculty of Mathematics, Physics and Informatics, Comenius University in Bratislava, Mlynská Dolina, 842 48 Bratislava, Slovakia
3 Mathematical Institute, Slovak Academy of Sciences, Štefánikova 49, 814 73 Bratislava, Slovakia
4 School of Mathematical and Statistical Sciences, National University of Ireland, H91 TK33 Galway, Ireland; donal.oregan@nuigalway.ie
* Correspondence: Michal.Feckan@fmph.uniba.sk

Abstract: In this paper, we investigate the controllability of first order linear fuzzy differential systems. We use the direct construction method to derive the controllability results for three types of first order linear fuzzy controlled systems via $(c1)$-solution and $(c2)$-solution, respectively. An example is presented to illustrate our theoretical results.

Keywords: linear fuzzy differential systems; controllability; $(c1)$-solution; $(c2)$-solution

MSC: 34A37; 93C05

check for updates

Citation: Liu, R.; Fečkan, M.; O'Regan, D.; Wang, J. Controllability Results for First Order Linear Fuzzy Differential Systems. *Mathematics* **2022**, *10*, 1193. https://doi.org/10.3390/math10071193

Academic Editor: Clemente Cesarano

Received: 5 March 2022
Accepted: 2 April 2022
Published: 6 April 2022

1. Introduction

The study of controllability is an important area of research and classical differential controlled systems have been discussed in many papers. In [1], controllability of impulsive differential equations was studied for the first time. The authors in [2] established some sufficient conditions for controllability by applying the measure of noncompactness and Mönch's fixed point theorem. In [3], the authors presented complete controllability of the considered impulsive control system and in [4], the author gave local and global controllability properties for nonlinear systems. In [5], the controllability and the observability of continuous linear time-varying systems with norm-bounded parameter perturbations was considered. When using classical mathematical models alone, many equations need to be solved due to the uncertainty of the measurement parameters and this can be time-consuming and can be financially burdensome to compute. The disadvantages of uncertainty can be overcome by fuzzy differential systems. The concept of controllability was introduced by Kalman [6] in 1963 and has played an important role in control theory since then. We have done some research on fuzzy differential systems [7,8], in this paper we will go further investigate the controllability of first-order linear fuzzy differential systems and it can resolve the controllability problem with uncertainty more accurately.

The main contributions are as follows:

Comparing the relevant literature above, we use the direct construction method to present the controllability of three types of first order linear fuzzy differential systems in the case of $a > 0$, $(c1)$-solution; $a > 0$, $(c2)$-solution; $a < 0$, $(c1)$-solution; $a < 0$, $(c2)$-solution, respectively.

While the method used is standard in some sense the constructed approach is technical and the control function given for our problem is a function with hyperbolic sine and cosine.

The controllability of first order fuzzy differential systems in iterative feedback tuning in fuzzy control system, fuzzy control design for non-holonomic wheeled mobile and fuzzy optimal control for complex nonlinear systems is useful and for more details we refer the reader to [9–11].

In this paper, we consider the controllability of three types of first order linear fuzzy differential systems

$$\begin{cases} y'(t) = a(t)y(t) + b(t) + \tilde{d}u(t), \ t \in J = [0,T], \\ y(0) = y_0, \end{cases} \tag{1}$$

and

$$\begin{cases} y'(t) + (-a(t))y(t) = b(t) + \tilde{d}u(t), \ t \in J, \\ y(0) = y_0, \end{cases} \tag{2}$$

and

$$\begin{cases} y'(t) + (-(b(t) + \tilde{d}u(t))) = a(t)y(t), \ t \in J, \\ y(0) = y_0, \end{cases} \tag{3}$$

where $a : J \to \mathbb{R}$, $y_0 \in \mathbb{R}_F$ and $b : J \to \mathbb{R}_F$, $u : J \to \mathbb{R}_F$, $\tilde{d} \in \mathbb{R}_+$.

In Section 2, we present notations, concepts, and lemmas needed in this paper. In Section 3, we establish some theorems concerning the controllability of first order linear fuzzy differential systems. Finally, in the last section, we give an example to illustrate our main results.

2. Preliminaries

We collect some concepts which will be used throughout the paper; for more details, see for [12,13].

Denote by $\mathbb{R}_F := \{v \mid v : \mathbb{R} \to [0,1]\}$ the class of the fuzzy subsets of the real axis satisfying the following properties:

(X_1) v is normal (i.e., $\exists\, x_0 \in \mathbb{R}$ s.t. $v(x_0) = 1$).

(X_2) v is convex fuzzy set (i.e., $v(\xi s_0 + (1-\xi)s_1) \geq \min\{v(s_0), v(s_1)\}$) for all s_0, $s_1 \in \mathbb{R}$ and $\xi \in [0,1]$.

(X_3) v is upper semicontinuous on $\mathbb{R}$.

(X_4) $[v]^0 = \overline{\{x \in \mathbb{R} : v(x) > 0\}}$ is compact.

Let $\alpha \in (0,1]$. Consider the α-level set of $v \in \mathbb{R}_F$ by $[v]^\alpha = \{s \in \mathbb{R} \mid v(s) \geq \alpha\}$, which is a nonempty compact interval for all $\alpha \in (0,1]$. We use $[v]^\alpha = [\underline{v}_\alpha, \overline{v}_\alpha]$ to denote explicitly the α-level set of v. We call $\underline{v}_\alpha$ and $\overline{v}_\alpha$ the lower and upper branches of v, respectively. We use the notation $diam([v]^\alpha) = \overline{v}_\alpha - \underline{v}_\alpha$ to denote the length of v.

Now $\forall \alpha \in [0,1]$, u, $v \in \mathbb{R}_F$ and $\xi \in \mathbb{R}$, we define the sum $u + v$ and the produce ξu as $[u+v]^\alpha = [u]^\alpha + [v]^\alpha = [\underline{u}_\alpha + \underline{v}_\alpha, \overline{u}_\alpha + \overline{v}_\alpha]$ and $[\xi u]^\alpha = \xi[u]^\alpha$.

Consider the Hausdorff distance $D : \mathbb{R}_F \times \mathbb{R}_F \to \mathbb{R}_+ \cup \{0\}$ where $D(u,v) = \sup_{0 \leq \alpha \leq 1} d_H([u]^\alpha, [v]^\alpha) = \sup_{0 \leq \alpha \leq 1} \max\{|\underline{u}_\alpha - \underline{v}_\alpha|, |\overline{u}_\alpha - \overline{v}_\alpha|\}$ (see [14]). Then $(\mathbb{R}_F, D)$ is a complete metric space (see [15]) and (i) $D(u+e, v+e) = D(u,v)$, $\forall\, u, v, e \in \mathbb{R}_F$, (ii) $D(\varsigma u, \varsigma v) = |\varsigma| D(u,v)$, $\forall\, \varsigma \in \mathbb{R}$, $u, v \in \mathbb{R}_F$, (iii) $D(u+e, v+\varrho) \leq D(u,v) + D(e,\varrho)$, $\forall\, u, v, e, \varrho \in \mathbb{R}_F$ are satisfied.

Definition 1 (see [13] (Definition 2.1)). *Let $f : [a,b] \to \mathbb{R}_F$ be measurable and integrably bounded. The integral of f over $[a,b]$, denoted by $\int_a^b f(t)dt$, is defined levelwise by the expression*

$$\begin{aligned}\left[\int_a^b f(t)dt\right]^\alpha &:= \int_a^b [f(t)]^\alpha dt \\ &= \left\{\int_a^b \tilde{f}(t)dt \mid \tilde{f} : [a,b] \to \mathbb{R}_F \text{ is a measurable selection for } [f(\cdot)]^\alpha\right\},\end{aligned}$$

for every $\alpha \in [0,1]$.

Throughout this paper, we use the symbol $\ominus$ to represent the H-difference. Note that $a_1 \ominus a_2 \neq a_1 + (-1)a_2 := a_1 - a_2$. In the sequel, we fix $J = [0,T]$, for $T > 0$.

Here we simplify the classes of strongly generalized differentiable by considering case (i) and case (ii) as in paper [16].

Definition 2 (see [13] (Definition 2.2)). *Let $Q : J \to \mathbb{R}_F$ and fix $n_0 \in J$. We say Q is differentiable at n_0, if we have an element $Q'(n_0) \in \mathbb{R}_F$ such that either*

(c1) for all $p > 0$ sufficiently close to 0, the H-differences $Q(n_0 + p) \ominus Q(n_0)$, $Q(n_0) \ominus Q(n_0 - p)$ exist and the limits (in the metric D)

$$\lim_{p\to 0^+} \frac{Q(n_0+p) \ominus Q(n_0)}{p} = \lim_{p\to 0^+} \frac{Q(n_0) \ominus Q(n_0-p)}{p} = Q'(n_0),$$

or

(c2) for all $p > 0$ sufficiently close to 0, the H-differences $Q(n_0) \ominus Q(n_0 + p)$, $Q(n_0 - p) \ominus Q(n_0)$ exist and the limits (in the metric D)

$$\lim_{p\to 0^+} \frac{Q(n_0) \ominus Q(n_0+p)}{-p} = \lim_{p\to 0^+} \frac{Q(n_0-p) \ominus Q(n_0)}{-p} = Q'(n_0).$$

Definition 3 (See [13] (Definition 2.5)). *Let $Q : J \to \mathbb{R}_F$. We say Q is $(c1)$-differentiable on J if Q is differentiable in the sense (c1) in Definition 2 and its derivative is denoted D_1Q. Similarly we can define $(c2)$-differentiable and denote it by D_2Q.*

Theorem 1 (see [13] (Theorem 2.6)). *Let $Q : J \to \mathbb{R}_F$ and put $[Q(t)]^\alpha = [p_\alpha(t), q_\alpha(t)]$ for each $\alpha \in [0,1]$.*

(i) If Q is $(c1)$-differentiable then p_α and q_α are differentiable functions and $[D_1Q(t)]^\alpha = [p'_\alpha(t), q'_\alpha(t)]$.

(ii) If Q is $(c2)$-differentiable then p_α and q_α are differentiable functions and we have $[D_2Q(t)]^\alpha = [q'_\alpha(t), p'_\alpha(t)]$.

Theorem 2 (see [17] (Theorem 2.2)). *Let $K : J \to \mathbb{R}_F$ be a differentiable fuzzy number-valued mapping and we suppose that the derivative K' is integrable over J. Then for each $t \in J$, we have*

(a) if K is $(c1)$-differentiable, then $K(t) = K(b) + \int_b^t K'(s)ds$;

(b) if K is $(c2)$-differentiable, then $K(t) = K(b) \ominus \int_b^t -K'(s)ds$.

Theorem 3 (see [13] (Theorem 2.7)). *Let Q be (c2)-differentiable on J and assume that the derivative Q' is integrable over J. Then for each $t \in J$ we have*

$$Q(t) = Q(a) \ominus \int_a^t -Q'(\tau)d\tau.$$

Theorem 4 (see [18] (Theorem 2.4)). *Let $K : J \to \mathbb{R}_F$ be continuous. Define the integral $G(t) := \sigma \ominus \int_v^t -K(s)ds$, $t \in J$, where $\sigma \in \mathbb{R}_F$ is such that the preceding H-difference exist on J. Then $G(t)$ is $(c2)$-differentiable and $G'(t) = K(t)$.*

Consider the following conditions (here $p : \mathbb{R} \to \mathbb{R}_F$):

Hypothesis 1. *For a given $t \in J$, $p(t+h) \ominus p(t)$ and $p(t) \ominus p(t-h)$ exist for $h \to 0^+$;*

Hypothesis 2. *For a given $t \in J$, $p(t) \ominus p(t+h)$ and $p(t-h) \ominus p(t)$ exist for $h \to 0^+$.*

3. Controllability of First Order Linear Fuzzy Differential Systems

Definition 4. *Fuzzy system (1)–(3) is called controllable on $[t_0, T]$ $(T > t_0)$, if for an arbitrary initial state $y_0 \in \mathbb{R}_F$ at t_0, final state $y_1 \in \mathbb{R}_F$ at time T_1(here $t_0 = 0$ and $T_1 = T$), there exists a control $u : J \to \mathbb{R}_F$ such that the system (1)–(3) has a solution y that satisfies $y(T_1) = y_1$(i.e., $[y(T_1)]^\alpha = [y_1]^\alpha$).*

Consider the following system (see [17]):

$$\begin{cases} y'(t) = a(t)y(t) + b(t) + \tilde{d}u(t), \ t \in J, \\ y(0) = y_0, \end{cases}$$

where $a : J \to \mathbb{R}$, $y_0 \in \mathbb{R}_F$ and $b : J \to \mathbb{R}_F$, $u : J \to \mathbb{R}_F$, $\tilde{d} \in \mathbb{R}_+$.

From [17] (Theorem 3.1, $a < 0$), the $(c1)$-solution of (1) can be written in the form:

$$\begin{aligned} y(t) &= \cosh\left(\int_0^t a(v)dv\right)\left(y_0 + \int_0^t \left[(b(s) + \tilde{d}u(s))\cosh\left(\int_0^s a(v)dv\right)\right.\right. \\ &\quad \left.\left.\ominus (b(s) + \tilde{d}u(s))\sinh\left(\int_0^s a(v)dv\right)\right]ds\right) \\ &\quad + \sinh\left(\int_0^t a(v)dv\right)\left(y_0 + \int_0^t \left[(b(s) + \tilde{d}u(s))\cosh\left(\int_0^s a(v)dv\right)\right.\right. \\ &\quad \left.\left.\ominus (b(s) + \tilde{d}u(s))\sinh\left(\int_0^s a(v)dv\right)\right]ds\right). \end{aligned}$$

From [17] (Theorem 3.1, $a < 0$), the $(c2)$-solution of (1) can be written in the form:

$$y(t) = e^{\int_0^t a(v)dv} y_0 \ominus e^{\int_0^t a(v)dv} \int_0^t (-b(s)) e^{-\int_0^s a(v)dv} ds \ominus e^{\int_0^t a(v)dv} \int_0^t (-\tilde{d}u(s)) e^{-\int_0^s a(v)dv} ds.$$

Thus, we consider two cases to study the controllability of (1): The $(c1)$-solution and the $(c2)$-solution.

Case 1.1 Consider $a < 0$ via $(c1)$-solution.

Theorem 5. *In Case 1.1, system (1) is controllable, if the control function $u(t)$ is given by*

$$\begin{aligned} u(t) &= \frac{1}{T_1 \tilde{d}}\left[y_1 \cosh\left(\int_0^{T_1} a(v)dv - \int_0^t a(v)dv\right) \ominus y_1 \sinh\left(\int_0^{T_1} a(v)dv - \int_0^t a(v)dv\right)\right. \\ &\quad \left.\ominus\left(\cosh\left(\int_0^t a(v)dv\right)y_0 + \sinh\left(\int_0^t a(v)dv\right)y_0\right)\right] \ominus \frac{1}{\tilde{d}}b(t), \ t \in J, \end{aligned}$$

where the H-differences exist.

Proof. Since the H-differences exist in $u(t)$, for $T_1 > 0$, we obtain

$$\begin{aligned} y(T_1) &= \cosh\left(\int_0^{T_1} a(v)dv\right)y_0 + \sinh\left(\int_0^{T_1} a(v)dv\right)y_0 \\ &+\frac{1}{T_1}\int_0^{T_1} \cosh\left(\int_0^{s} a(v)dv\right)\cosh\left(\int_0^{s} a(v)dv\right)y_1 ds \\ &+\frac{1}{T_1}\int_0^{T_1} \sinh\left(\int_0^{s} a(v)dv\right)\cosh\left(\int_0^{s} a(v)dv\right)y_1 ds \\ &\ominus\frac{1}{T_1}\int_0^{T_1} \cosh\left(\int_0^{s} a(v)dv\right)\sinh\left(\int_0^{s} a(v)dv\right)y_1 ds \\ &\ominus\frac{1}{T_1}\int_0^{T_1} \sinh\left(\int_0^{s} a(v)dv\right)\sinh\left(\int_0^{s} a(v)dv\right)y_1 ds \\ &\ominus\frac{1}{T_1}\int_0^{T_1} \cosh\left(\int_0^{T_1} a(v)dv + \int_0^{s} a(v)dv\right)\cosh\left(\int_0^{s} a(v)dv\right)y_0 ds \\ &\ominus\frac{1}{T_1}\int_0^{T_1} \sinh\left(\int_0^{T_1} a(v)dv + \int_0^{s} a(v)dv\right)\cosh\left(\int_0^{s} a(v)dv\right)y_0 ds \\ &+\frac{1}{T_1}\int_0^{T_1} \cosh\left(\int_0^{T_1} a(v)dv + \int_0^{s} a(v)dv\right)\sinh\left(\int_0^{s} a(v)dv\right)y_0 ds \\ &+\frac{1}{T_1}\int_0^{T_1} \sinh\left(\int_0^{T_1} a(v)dv + \int_0^{s} a(v)dv\right)\sinh\left(\int_0^{s} a(v)dv\right)y_0 ds \\ &= \left(\cosh\left(\int_0^{T_1} a(v)dv\right)y_0 \ominus \cosh\left(\int_0^{T_1} a(v)dv\right)y_0\right) \\ &+\left(\sinh\left(\int_0^{T_1} a(v)dv\right)y_0 \ominus \sinh\left(\int_0^{T_1} a(v)dv\right)y_0\right) + y_1 \\ &= y_1. \end{aligned}$$

Thus, the system (1) is controllable in this case. □

Case 1.2 Consider $a < 0$ via the $(c2)$-solution.

Theorem 6. *In Case 1.2, system* (1) *is controllable, if* $W_0^{-1}[0, T_1]$ *exists; here*

$$W_0[0, T_1] = \int_0^{T_1} e^{-\int_0^s a(v)dv}\tilde{d}\tilde{d}e^{-\int_0^s a(v)dv}ds. \tag{4}$$

Proof. From $W_0[0, T_1] = \int_0^{T_1} e^{-\int_0^s a(v)dv}\tilde{d}\tilde{d}e^{-\int_0^s a(v)dv}ds$, for $T_1 > 0$, for any final state $y_1 \in \mathbb{R}_F$ we choose a control function as follows:

$$u_1(t) = -\tilde{d}e^{-\int_0^t a(v)dv}W_0^{-1}[0, T_1]\left(y_0 \ominus \int_0^{T_1}(-b(s))e^{-\int_0^s a(v)dv}ds \ominus e^{-\int_0^{T_1} a(v)dv}y_1\right),\ t \in J,$$

where the H-differences exist.

Then we get

$$\begin{aligned} y(T_1) &= e^{\int_0^{T_1} a(v)dv} y_0 \ominus e^{\int_0^{T_1} a(v)dv} \int_0^{T_1} (-b(s))e^{-\int_0^s a(v)dv} ds \\ &\ominus e^{\int_0^{T_1} a(v)dv} \int_0^{T_1} (-\tilde{d}u_1(s))e^{-\int_0^s a(v)dv} ds \\ &= e^{\int_0^{T_1} a(v)dv} y_0 \ominus e^{\int_0^{T_1} a(v)dv} \int_0^{T_1} (-b(s))e^{\int_0^s a(v)dv} ds \\ &\ominus e^{\int_0^{T_1} a(v)dv} \int_0^{T_1} e^{-\int_0^s a(v)dv} \tilde{d}\tilde{d}e^{-\int_0^s a(v)dv} W_0^{-1}[0, T_1] \\ &\times \left(y_0 \ominus \int_0^{T_1} (-b(s))e^{-\int_0^s a(v)dv} ds \ominus e^{-\int_0^{T_1} a(v)dv} y_1 \right) ds \\ &= e^{\int_0^{T_1} a(v)dv} y_0 \ominus e^{\int_0^{T_1} a(v)dv} \int_0^{T_1} (-b(s))e^{\int_0^s a(v)dv} ds \\ &\ominus e^{\int_0^{T_1} a(v)dv} y_0 + e^{\int_0^{T_1} a(v)dv} \int_0^{T_1} (-b(s))e^{\int_0^s a(v)dv} ds \\ &+ e^{\int_0^{T_1} a(v)dv} e^{-\int_0^{T_1} a(v)dv} y_1 \\ &= y_1. \end{aligned}$$

That means system (1) is controllable in this case. □

From [17] (Theorem 3.2, $a > 0$), the $(c1)$-solution of (1) can be written in the form:

$$y(t) = e^{\int_0^t a(v)dv} \left(y_0 + \int_0^t b(s)e^{-\int_0^s a(v)dv} ds + \int_0^t \tilde{d}u(s)e^{-\int_0^s a(v)dv} ds \right).$$

From [17] (Theorem 3.2, $a > 0$), the $(c2)$-solution of (1) can be written in the form:

$$\begin{aligned} y(t) &= \cosh\left(\int_0^t a(v)dv \right) \left(y_0 \ominus \int_0^t \left[(b(s) + \tilde{d}u(s)) \sinh\left(\int_0^s a(v)dv \right) \right. \right. \\ &\left. \left. -(b(s) + \tilde{d}u(s)) \cosh\left(\int_0^s a(v)dv \right) \right] ds \right) \\ &\ominus - \sinh\left(\int_0^t a(v)dv \right) \left(y_0 \ominus \int_0^t \left[(b(s) + \tilde{d}u(s)) \sinh\left(\int_0^s a(v)dv \right) \right. \right. \\ &\left. \left. -(b(s) + \tilde{d}u(s)) \cosh\left(\int_0^s a(v)dv \right) \right] ds \right). \end{aligned}$$

Thus, we consider two cases to study the controllability of (1): The $(c1)$-solution and the $(c2)$-solution.

Case 2.1 Consider $a > 0$ via the $(c1)$-solution.

Theorem 7. *In Case 2.1, system* (1) *is controllable, if* $V_\varepsilon^{-1}[0, T_1]$ *exists; here*

$$V_\varepsilon[0, T_1] = \int_0^{T_1} e^{-\int_0^s a(v)dv} \tilde{d}\tilde{d}e^{-\int_0^s a(v)dv} ds. \tag{5}$$

Proof. We set

$$u_0(t) = -\tilde{d}e^{-\int_0^t a(v)dv} V_\varepsilon^{-1}[0, T_1] \left(y_0 \ominus \int_0^{T_1} (-b(s))e^{-\int_0^s a(v)dv} ds \ominus e^{-\int_0^{T_1} a(v)dv} y_1 \right), \; t \in J,$$

where the H-differences exist. The rest of proof is the same as in Case 1.2. □

Case 2.2 Consider $a > 0$ via the $(c2)$-solution.

Theorem 8. *In Case 2.2, system* (1) *is controllable, if the control function* $u_2(t)$ *is given by*

$$\begin{aligned} u_2(t) &= \frac{1}{T_1\tilde{d}}\Bigg[\Bigg(-\Bigg(\cosh\Big(\int_0^t a(v)dv\Big)y_0 \ominus -\sinh\Big(\int_0^t a(v)dv\Big)y_0\Bigg)\Bigg) \\ &\quad \ominus -y_1\cosh\Big(\int_0^{T_1} a(v)dv - \int_0^t a(v)dv\Big) \\ &\quad \ominus y_1\sinh\Big(\int_0^{T_1} a(v)dv - \int_0^t a(v)dv\Big)\Bigg] \ominus \frac{1}{\tilde{d}}b(t),\ t \in J, \end{aligned}$$

where the H-differences exist.

Proof. Since the H-differences exist in $u_2(t)$, for $T_1 > 0$, we get

$$\begin{aligned} &y(T_1) \\ = &\Bigg[\cosh\Big(\int_0^{T_1} a(v)dv\Big)y_0 \ominus \Big(\frac{1}{T_1}\int_0^{T_1}\cosh\Big(\int_0^{T_1} a(v)dv + \int_0^s a(v)dv\Big)\cosh\Big(\int_0^s a(v)dv\Big)y_0 ds \\ &\ominus \frac{1}{T_1}\int_0^{T_1}\sinh\Big(\int_0^{T_1} a(v)dv + \int_0^s a(v)dv\Big)\sinh\Big(\int_0^s a(v)dv\Big)y_0 ds\Big)\Bigg] \\ &+\Bigg[-\Big(\frac{1}{T_1}\int_0^{T_1}\sinh\Big(\int_0^{T_1} a(v)dv + \int_0^s a(v)dv\Big)\cosh\Big(\int_0^s a(v)dv\Big)y_0 ds \\ &\ominus \frac{1}{T_1}\int_0^{T_1}\cosh\Big(\int_0^{T_1} a(v)dv + \int_0^s a(v)dv\Big)\sinh\Big(\int_0^s a(v)dv\Big)y_0 ds\Big) \\ &\ominus -\sinh\Big(\int_0^{T_1} a(v)dv\Big)y_0\Bigg] + \frac{1}{T_1}\int_0^{T_1}\cosh\Big(\int_0^s a(v)dv\Big)\cosh\Big(\int_0^s a(v)dv\Big)y_1 ds \\ &\ominus \frac{1}{T_1}\int_0^{T_1}\sinh\Big(\int_0^s a(v)dv\Big)\sinh\Big(\int_0^s a(v)dv\Big)y_1 ds \\ &\ominus -\Big(\frac{1}{T_1}\int_0^{T_1}\sinh\Big(\int_0^s a(v)dv\Big)\cosh\Big(\int_0^s a(v)dv\Big)y_1 ds \\ &\ominus \frac{1}{T_1}\int_0^{T_1}\cosh\Big(\int_0^s a(v)dv\Big)\sinh\Big(\int_0^s a(v)dv\Big)y_1 ds\Big) \\ = &\ y_1. \end{aligned}$$

Thus, system (1) is controllable in this case. □

Consider system (2) (see [17]):

$$\begin{cases} y'(t) + (-a(t))y(t) = b(t) + \tilde{d}u(t),\ t \in J, \\ y(0) = y_0, \end{cases}$$

where $a : J \to \mathbb{R}$, $y_0 \in \mathbb{R}_F$ and $b : J \to \mathbb{R}_F$, $u : J \to \mathbb{R}_F$, $\tilde{d} \in \mathbb{R}_+$.

From [17] (Theorem 3.3, $a < 0$), the $(c1)$-solution of (2) can be written in the form:

$$y(t) = e^{\int_0^t a(v)dv}\Big(y_0 + \int_0^t (b(s) + \tilde{d}u(s))e^{-\int_0^s a(v)dv} ds\Big).$$

From [17] (Theorem 3.3, $a < 0$), the $(c2)$-solution of (2) can be written in the form:

$$\begin{aligned} y(t) &= \cosh\left(\int_0^t a(v)dv\right)\left(y_0 \ominus (-1)\int_0^t \left[(b(s)+\tilde{d}u(s))\cosh\left(\int_0^s a(v)dv\right)\right.\right. \\ &\left.\left.\ominus (b(s)+\tilde{d}u(s))\sinh\left(\int_0^s a(v)dv\right)\right]ds\right) \\ &+\sinh\left(\int_0^t a(v)dv\right)\left(y_0 \ominus (-1)\int_0^t \left[(b(s)+\tilde{d}u(s))\cosh\left(\int_0^s a(v)dv\right)\right.\right. \\ &\left.\left.\ominus (b(s)+\tilde{d}u(s))\sinh\left(\int_0^s a(v)dv\right)\right]ds\right). \end{aligned}$$

Thus, we consider two cases to study the controllability of (1): The $(c1)$-solution and the $(c2)$-solution.

Case 3.1 Consider $a < 0$ via the $(c1)$-solution.

Theorem 9. *In Case 3.1, system* (2) *is controllable, if* $W_1^{-1}[0, T_1]$ *exists; here*

$$W_1[0, T_1] = \int_0^{T_1} e^{-\int_0^s a(v)dv}\tilde{d}\tilde{d}e^{-\int_0^s a(v)dv}ds. \tag{6}$$

Proof. Let

$$u_3(t) = \tilde{d}e^{-\int_0^t a(v)dv}W_1^{-1}[0, T_1]\left(e^{-\int_0^{T_1} a(v)dv}y_1 \ominus y_0 \ominus \int_0^{T_1}(-b(s))e^{-\int_0^s a(v)dv}ds\right),\ t \in J,$$

where the H-differences exist. The method of proof is the same as in Case 1.2, so we omit it here. □

Case 3.2 Consider $a < 0$ via the $(c2)$-solution.

Theorem 10. *In Case 3.2, system* (2) *is controllable, if the control function* $u_4(t)$ *is given by*

$$\begin{aligned} u_4(t) &= \frac{1}{T_1\tilde{d}}\left[\left(-\cosh\left(\int_0^t a(v)dv\right)y_0 - \sinh\left(\int_0^t a(v)dv\right)y_0\right)\right. \\ &\ominus -y_1\cosh\left(\int_0^{T_1} a(v)dv - \int_0^t a(v)dv\right) \\ &\left.-y_1\sinh\left(\int_0^{T_1} a(v)dv - \int_0^t a(v)dv\right)\right] \ominus \frac{1}{\tilde{d}}b(t),\ t \in J, \end{aligned}$$

where the H-differences exist.

Proof. The method of proof is the same as in Case 1.1, so we omit it here. □

From [17] (Theorem 3.5, $a > 0$), the $(c1)$-solution of (2) can be written in the form:

$$\begin{aligned} y(t) &= \cosh\left(\int_0^t a(v)dv\right)\left(y_0 + \int_0^t \left[(b(s)+\tilde{d}u(s))\sinh\left(\int_0^s a(v)dv\right)\right.\right. \\ &\left.\left.-(b(s)+\tilde{d}u(s))\cosh\left(\int_0^s a(v)dv\right)\right]ds\right) \\ &\ominus -\sinh\left(\int_0^t a(v)dv\right)\left(y_0 + \int_0^t \left[(b(s)+\tilde{d}u(s))\sinh\left(\int_0^s a(v)dv\right)\right.\right. \\ &\left.\left.-(b(s)+\tilde{d}u(s))\cosh\left(\int_0^s a(v)dv\right)\right]ds\right). \end{aligned}$$

From [17] (Theorem 3.5, $a > 0$), the $(c2)$-solution of (2) can be written in the form:

$$y(t) = e^{\int_0^t a(v)dv}\left(y_0 \ominus \int_0^t (-b(s) - \tilde{d}u(s))e^{-\int_0^s a(v)dv}ds\right).$$

Thus, we consider two cases to study the controllability of (1): The $(c1)$-solution and the $(c2)$-solution.

Case 4.1 Consider $a > 0$ via the $(c1)$-solution.

Theorem 11. *In Case 4.1, system* (2) *is controllable, if the control function $u_5(t)$ is given by*

$$\begin{aligned} u_5(t) &= \frac{1}{T_1\tilde{d}}\Big[-y_1\cosh\left(\int_0^{T_1} a(v)dv - \int_0^t a(v)dv\right) + y_1\sinh\left(\int_0^{T_1} a(v)dv - \int_0^t a(v)dv\right) \\ &\ominus - \left(\cosh\left(\int_0^t a(v)dv\right)y_0 \ominus - \sinh\left(\int_0^t a(v)dv\right)y_0\right)\Big] \ominus \frac{1}{\tilde{d}}b(t),\ t \in J, \end{aligned}$$

where the H-differences exist.

Proof. The method of proof is the same as in Case 1.1, so we omit it here. □

Case 4.2 Consider $a > 0$ via the $(c2)$-solution.

Theorem 12. *In Case 4.2, system* (2) *is controllable, if $W_2^{-1}[0, T_1]$ exists; here*

$$W_2[0, T_1] = \int_0^{T_1} e^{-\int_0^s a(v)dv}\tilde{d}\tilde{d}e^{-\int_0^s a(v)dv}ds. \tag{7}$$

Proof. Let

$$u_6(t) = -\tilde{d}e^{-\int_0^t a(v)dv}W_2^{-1}[0, T_1]\left(y_0 \ominus \int_0^{T_1} (-b(s))e^{-\int_0^s a(v)dv}ds \ominus e^{-\int_0^{T_1} a(v)dv}y_1\right),\ t \in J,$$

where the H-differences exist. The method of proof is the same as in Case 1.2, so we omit it here. □

Next, we consider the following system (3):

$$\begin{cases} y'(t) + (-(b(t) + \tilde{d}u(t))) = a(t)y(t),\ t \in J, \\ y(0) = y_0, \end{cases}$$

where $a : J \to \mathbb{R}$, $y_0 \in \mathbb{R}_F$ and $b : J \to \mathbb{R}_F$, $u : J \to \mathbb{R}_F$, $\tilde{d} \in \mathbb{R}_+$.

The method of proof is same as in systems (1) and (2), so we omit it here.

4. An Example

In this section we give an example to prove our theorems.

Example 1. *Consider linear fuzzy differential equations:*

$$\begin{cases} y'(t) = -y(t) + \tilde{d}u(t),\ t \in [0, 0.3], \\ y(0) = \gamma, \end{cases} \tag{8}$$

where $a = -1$, $[\gamma]^\alpha = [\alpha - 1, 1 - \alpha]$, $y_1 = 0.6\gamma$, $\tilde{d} = 3$.

According to Theorem 6,

$$W_0[0, T_1] \triangleq \int_0^{0.3} 9(e^{\int_0^s dv})^2 ds = 9\int_0^{0.3} e^{2s}ds = \frac{9}{2}(e^{0.6} - 1) = 3.6995,$$

then $W_0^{-1}[0,0.3] = 0.1827$. *Note*

$$\begin{aligned} u_1(t) &= -\tilde{d}e^{-\int_0^t a(v)dv}W_0^{-1}[0,0.3]\left(y_0 \ominus \int_0^{0.3}(-b(s))e^{-\int_0^s a(v)dv}ds \ominus e^{-\int_0^{0.3} a(v)dv}y_1\right) \\ &= -0.4439e^{-t}\gamma. \end{aligned}$$

To sum up, $W_0^{-1}[0,T_1]$ *exists, so the system is controllable.*

5. Conclusions

In this paper we mainly study the controllability of first order linear fuzzy differential equations with the direct construction method. In future work, we shall study the controllability of first order linear and non-linear impulsive fuzzy differential equations.

Author Contributions: The contributions of all authors (R.L., M.F., D.O. and J.W.) are equal. All the main results were developed together. All authors have read and agreed to the published version of the manuscript.

Funding: This work was supported by Postgraduate Education Innovation Program in Guizhou Province [YJSCXJH(2020)082], the Slovak Research and Development Agency under the contract No. APVV-18-0308, and the Slovak Grant Agency VEGA No. 1/0358/20 and No. 2/0127/20.

Institutional Review Board Statement: Not applicable.

Informed Consent Statement: Not applicable.

Data Availability Statement: Not applicable.

Conflicts of Interest: The authors declare no conflict of interest.

References

1. Leela, S.; McRae, F.A.; Sivaundaram, S. Controllability of impulsive differential equations. *J. Math. Anal. Appl.* **1993**, *177*, 24–30. [CrossRef]
2. Ji, S.; Li, G.; Wang, M. Controllability of impulsive differential systems with nonlocal conditions. *Appl. Math. Comput.* **2011**, *217*, 6981–6989. [CrossRef]
3. George, R.K.; Nandakumaran, A.K.; Arapostathis, A. A note on controllability of impulsive systems. *J. Math. Anal. Appl.* **2000**, *241*, 276–283. [CrossRef]
4. Aeyels, D. Local and global controllability for nonlinear systems. *Syst. Control. Lett.* **1984**, *5*, 19–26. [CrossRef]
5. Chung, D.; Park, G.G.; Lee, J.G. Robustness of controllability and observability of continuous linear time-varying systems with parameters perturbations. *IEEE Trans. Autom. Control* **1999**, *44*, 1919–1923. [CrossRef]
6. Kalman, R.E. Mathematical description of linear dynamical systems. *J. Soc. Ind. Appl. Math. Ser. Control* **1963**, *2*, 152–192. [CrossRef]
7. Liu, R.; Wang, J. R.; O'Regan, D. On the solution of first-order linear impulsive fuzzy differential equations. *Fuzzy Sets Syst.* **2020**, *400*, 1–33. [CrossRef]
8. Liu, R.; Wang, J. R.; O'Regan, D. Ulam type stability of first-order linear impulsive fuzzy differential equations. *Fuzzy Sets Syst.* **2020**, *400*, 34–89. [CrossRef]
9. Preitl, S.; Precup, R.E.; Fodor, J.; Bede, B. Iterative feedback tuning in fuzzy control systems. Theory and application. *Acta Polytech. Hung.* **2006**, *3*, 81–96. [CrossRef]
10. Guechi, E.H.; Lauber, J.; Dambrine, M. PDC control design for non-holonomic wheeled mobile robots with delayed outputs. *J. Intell. Robot. Syst.* **2010**, *60*, 395–414. [CrossRef]
11. Chen, T.; Babanin, A.; Muhammad, A. Modified evolved bat algorithm of fuzzy optimal control for complex nonlinear systems. *Rom. J. Inf. Sci. Technol.* **2020**, *23*, T28–T40. [CrossRef]
12. Lakshmikantham, V.; Mohapatra, R.N. *Theory of Fuzzy Differential Equation and Inclusions*; Taylor & Francis: London, UK, 2003. [CrossRef]
13. Khastan, A.; Nieto, J.J.; Rodríguez-López, R. Variation of constant formula for first order fuzzy differential equations. *Fuzzy Sets Syst.* **2011**, *177*, 20–33.
14. Diamond, P. Brief note on the variation of constants formula for fuzzy differential equations. *Fuzzy Sets Syst.* **2002**, *129*, 65–71. [CrossRef]
15. Nieto, J.J.; Rodríguez-López, R.; Franco, D. Linear first order fuzzy differential equations. *Int. J. Uncertain. Fuzziness -Knowl.-Based Syst.* **2006**, *14*, 687–709.

16. Bede, B.; Gal, S.G. Generalizations of the differentiability of fuzzy-number-valued functions with applications to fuzzy differential equations. *Fuzzy Sets Syst.* **2005**, *151*, 581–599.
17. Khastan, A.; Rodríguez-López, R. On the solutions to first order linear fuzzy differential equations. *Fuzzy Sets Syst.* **2016**, *295*, 114–135. [CrossRef]
18. Shen, Y. On the Ulam stability of first order linear fuzzy differential equations under generalized differentiability. *Fuzzy Sets Syst.* **2015**, *280*, 27–57. [CrossRef]

mathematics

MDPI

Article

Regional Controllability and Minimum Energy Control of Delayed Caputo Fractional-Order Linear Systems

Touria Karite [1,†], Adil Khazari [2,†] and Delfim F. M. Torres [3,*,†]

1 Laboratory of Engineering, Systems and Applications, Department of Electrical Engineering & Computer Science, National School of Applied Sciences, Sidi Mohamed Ben Abdellah University, Avenue My Abdallah Km 5 Route d'Imouzzer, Fez BP 72, Morocco
2 Laboratory of Analysis, Mathematics and Applications, National School of Commerce & Management, Sidi Mohamed Ben Abdellah University, Fez BP 1796, Morocco
3 Center for Research and Development in Mathematics and Applications (CIDMA), Department of Mathematics, University of Aveiro, 3810-193 Aveiro, Portugal
* Correspondence: delfim@ua.pt; Tel.: +351-234-370-668
† These authors contributed equally to this work.

Abstract: We study the regional controllability problem for delayed fractional control systems through the use of the standard Caputo derivative. First, we recall several fundamental results and introduce the family of fractional-order systems under consideration. Afterward, we formulate the notion of regional controllability for fractional systems with control delays and give some of their important properties. Our main method consists of defining an attainable set, which allows us to prove exact and weak controllability. Moreover, the main results include not only those of controllability but also a powerful Hilbert uniqueness method, which allows us to solve the minimum energy optimal control problem. More precisely, an explicit control is obtained that drives the system from an initial given state to a desired regional state with minimum energy. Two examples are given to illustrate the obtained theoretical results.

Keywords: regional controllability; fractional-order systems; Caputo derivatives; control delays; optimal control; minimum energy

MSC: 26A33; 49J20; 93B05

check for updates

Citation: Karite, T.; Khazari, A.; Torres, D.F.M. Regional Controllability and Minimum Energy Control of Delayed Caputo Fractional-Order Linear Systems. *Mathematics* **2022**, *10*, 4813. https://doi.org/10.3390/math10244813

Academic Editors: Alexandra M.S.F. Galhano, António Lopes and Carlo Cattani

Received: 11 November 2022
Accepted: 14 December 2022
Published: 18 December 2022

1. Introduction

The celebrated letter addressed by L'Hopital to Leibniz about the possibilities that can be obtained when the order n of the derivative is a fraction $1/2$, revolutionized calculus and marked the birth of fractional calculus [1]. Since its beginnings, fractional calculus has attracted many great mathematicians, who directly or indirectly contributed to its development [2]. Today, many researchers consider fractional calculus an important tool for solving different problems in various fields, e.g., physics, thermodynamics, chemistry, biology, classical and quantum mechanics, viscoelasticity, finance, engineering, signal and image processing, and automatics and control [3–5].

Let Ω be a bounded subset of $\mathbb{R}^n$ with a regular boundary $\partial\Omega$, the final time be $\tau > 0$, $Q = \Omega \times [0,\tau]$, and $\Sigma = \partial\Omega \times [0,\tau]$. We then consider the system

$$\begin{cases} {}_0^C\mathbb{D}_t^r z(x,t) = \mathcal{A}z(x,t) + \mathfrak{B}\mathrm{u}(t-h), & t \geq 0, \\ z(x,0) = z_0(x), \\ \mathrm{u}(t) = \varphi(t), & -h \leq t \leq 0, \end{cases} \tag{1}$$

where ${}_0^C\mathbb{D}_t^r$ denotes the left-sided Caputo fractional order derivative of order $r \in (0,1)$ [6,7]. Note that z is a function of two parameters but the derivative is an operator that acts on t. The linear operator $\mathcal{A}$ is an infinitesimal generator of a C_0-semi-group $(\mathcal{T}(t))_{t\geq 0}$ on the Hilbert state space $L^2(\Omega)$ [8,9]. Here, $h > 0$ is the time control delay and $\varphi(t)$ is the initial control function. In the sequel, we have $z(x,t) \in \mathcal{Z} = L^2(0,\tau;L^2(\Omega))$ and the control $u \in \mathfrak{U} = L^2(0,\tau;\mathbb{R}^p)$. The initial state $z_0 \in L^2(\Omega)$ and the linear control operator $\mathfrak{B} : \mathbb{R}^p \to L^2(\Omega)$, which might be unbounded, depend on the number p and the structure of the actuators.

The notion of controllability seeks to find a command or control that brings the system under study from an initial state to a desired final state. This is generally difficult to achieve, in particular, for fractional order diffusion systems. This explains why a large number of scholars have been investigating control problems using the notion of "regional controllability". This concept was first introduced by El Jai et al. in 1995 [10] for parabolic systems and was then extended to the case of hyperbolic systems [11]. The concept is widely used to investigate problems where the target of interest is not fully specified as a state and relates only to a smaller internal region ω of the system domain Ω. It is especially crucial when it comes to real-world problems since the transfer costs are lower in a regional case, for instance, in the case of wildfires, where the main purpose is to control it in a smaller region and one tries to minimize the costs [12–14].

In various processes, future states are dependent on the current and previous states of the system, which implies that the models describing these processes should include delays, either in the state or control variables or both. If the delays are in the inputs, we are faced with systems with delayed commands. Due to the number of mathematical models describing dynamical systems with delays in the controls, solving controllability problems for such systems is of great importance. In particular, controllability problems for linear continuous-time fractional systems with a delayed control have been the subject of several works [15–18]. However, it should be noted that the majority of research in this area deals with the global case, that is, controllability is treated on the whole evolution domain. Here, we are interested in studying the concept in a specific region $\omega \in \Omega$.

Fractional delayed differential equations are equations involving fractional derivatives and delays. Unlike ordinary derivatives, they are nonlocal derivatives by nature and are able to model memory effects. Indeed, time delays express the history of a past state [19]. Many real-world problems can be modeled more accurately by including fractional derivatives and delays in a specific subregion ω of the whole evolution domain of the system Ω. For instance, when it comes to modeling several epidemiological problems, regional controllability of fractional delayed differential systems can be more plausible. In the case of monitoring glucose rates, fractional-order models provide a reasonable rate of movement of glucose from the blood into the environment [20,21].

Over the years, numerous mathematicians, utilizing their own notations and approaches, have defined different types of fractional derivatives and integrals. In this paper, we treat the controllability problem of a fractional diffusion equation in the sense of Caputo with a delay in the control. Recent works were expanded to solve optimal control problems with delays by combining conformable and Caputo–Fabrizio fractional derivatives via artificial neural networks [22]. Here, we define the regional controllability in the exact and weak senses; we give the necessary and sufficient conditions under which the system is controllable and we obtain the control that minimizes the energy cost functional.

The rest of this paper is structured as follows. Some definitions and fundamentals of fractional calculus are given in Section 2. In Section 3, a definition of regional fractional controllability for delayed systems is given and a necessary and sufficient condition to verify it is proved. Our main findings on controllability and optimal control are then formulated and proved in Section 4. In Section 5, we provide illustrative examples for cases of both a bounded and an unbounded control operator. We conclude with Section 6, providing a summary of the main conclusions and some insightful open questions that still deserve in-depth investigations.

2. Preliminary Results

We begin with some definitions, properties, and known results of fractional calculus that are used to study System (1). In particular, we recall the two more standard notions for fractional derivatives: the concept of the solution to System (1) and the fractional Green formula. In what follows, $g : \mathbb{R}^+ \to \mathbb{R}$ is a given function.

Definition 1 (Caputo fractional derivatives; see, e.g., [6]). *The Caputo fractional derivative of order $r > 0$ of a function $g : [0,\infty) \to \mathbb{R}$ is defined as*

$$ {}_0^C\mathbb{D}_t^r g(t) := {}_0I_t^{n-r} g^{(n)}(t) := \frac{1}{\Gamma(n-r)} \int_0^t (t-\sigma)^{n-r-1} g^{(n)}(\sigma) d\sigma, \tag{2}$$

where $n = -[-r]$, provided the right side is pointwise, is defined on $\mathbb{R}^+$ and ${}_0I_t^{n-r}$ is the left-sided Riemann–Liouville fractional integral of order $n-r > 0$ defined by

$$ {}_0I_t^{n-r} g(t) := \frac{1}{\Gamma(n-r)} \int_0^t (t-\sigma)^{n-r-1} g(\sigma) d\sigma, \quad t > 0. \tag{3}$$

Definition 2 (Riemann–Liouville fractional derivatives; see, e.g., [23–25]). *The left- and right-hand sides of the Riemann–Liouville fractional derivatives of order r of function g are expressed by*

$$ {}_0\mathbb{D}_t^r g(t) := \left(\frac{d}{dt}\right)^n {}_0I_t^{n-r} g(t) := \frac{1}{\Gamma(n-r)} \frac{d^n}{dt^n} \int_0^t (t-\sigma)^{n-r-1} g(\sigma) d\sigma, \quad t > 0, \tag{4}$$

and

$$ {}_t\mathbb{D}_\tau^r g(t) := \frac{1}{\Gamma(n-r)} \left(-\frac{d}{dt}\right)^n \int_t^\tau (\sigma - t)^{n-r-1} g(\sigma) d\sigma, \quad t < \tau, \tag{5}$$

respectively, where $r \in (n-1, n)$, $n \in \mathbb{N}$.

Definition 3 (Mittag–Leffler function; see, e.g., [26]). *The generalized Mittag–Leffler function is defined by*

$$ E_{r,s}(y) := \sum_{i=1}^{\infty} \frac{y^i}{\Gamma(ri+s)}, \quad \mathrm{Re}(r) > 0, \quad s, y \in \mathbb{C}. \tag{6}$$

Definition 4 (Three-parameter Mittag–Leffler function [7]). *The Prabhakar generalized Mittag–Leffler function is given by*

$$ E_{\alpha,\beta}^{\gamma}(y) := \frac{1}{\Gamma(\gamma)} \sum_{n=0}^{\infty} \frac{\Gamma(\gamma+n) y^n}{n!\Gamma(\alpha n + \beta)}, \quad \mathrm{Re}(\alpha) > 0, \quad \alpha, \beta, \gamma \in \mathbb{C}. \tag{7}$$

Definition 5 (See, e.g., [27]). *For any given $F(x,t) \in \mathcal{Z}$, $0 < r < 1$, a function $z(x,t) \in \mathcal{Z}$ is said to be the general solution of*

$$ \begin{cases} {}_0^C\mathbb{D}_t^r z(x,t) = \mathcal{A}z(x,t) + F(x,t), & t \geq 0, \\ z(x,0) = z_0(x), \end{cases} $$

and is expressed by

$$ z(x,t) = \mathfrak{R}_r(t) z_0(x) + \int_0^t (t-\sigma)^{r-1} \mathcal{S}_r(t-\sigma) F(\sigma) d\sigma, $$

where

$$ \mathfrak{R}_r(t) = \int_0^\infty \Phi_r(\alpha) \mathcal{T}(t^r \alpha) d\alpha, \tag{8}$$

and

$$\mathcal{S}_r(t) = r\int_0^\infty \alpha\Phi_r(\alpha)\mathcal{T}(t^r\alpha)d\alpha. \quad (9)$$

Here, $\{\mathcal{T}(t)\}_{t\geq 0}$ is the strongly continuous semigroup generated by operator $\mathcal{A}$,

$$\Phi_r(\alpha) = \frac{1}{r}\alpha^{-1-1/r}\psi_r(\alpha^{-1/r}),$$

and ψ_r is the probability density function defined by

$$\psi_r(\alpha) = \frac{1}{\pi}\sum_{n=1}^{\infty}(-1)^{n-1}\alpha^{-rn-1}\frac{\Gamma(nr+1)}{n!}\sin(n\pi r), \quad \alpha\in(0,\infty). \quad (10)$$

Remark 1. *The density function given by* (10) *satisfies the properties*

$$\int_0^\infty e^{-v\alpha}\psi_r(\alpha)d\alpha = e^{-v^r}, \qquad \int_0^\infty \psi_r(\alpha)d\alpha = 1, \quad r\in(0,1), \quad (11)$$

and

$$\int_0^\infty \alpha^{\nu}\Phi_r(\alpha)d\alpha = \frac{\Gamma(1+\nu)}{\Gamma(1+r\nu)}, \quad \nu\geq 0. \quad (12)$$

The following hypotheses are used in our results:

(H_1) The control operator $\mathfrak{B}$ is dense and $\mathfrak{B}^*$ exists;

(H_2) $(\mathfrak{B}\mathcal{S}_r(t))^*$ exists and $(\mathfrak{B}\mathcal{S}_r(t))^* = \mathcal{S}_r^*(t)\mathfrak{B}^*$.

Note that (H_1) and (H_2) hold when $\mathfrak{B}$ is bounded and linear. Throughout this paper, we use $z(x,t)$ for the state of the system. Next, we introduce the notion of a mild solution of System (1), using for it the notation $z_{\mathrm{u}}(x,t)$.

Definition 6 (Mild solution of System (1) [28]). *We say that a function $z_{\mathrm{u}}(x,t)\in\mathcal{Z}$ is a mild solution of System* (1) *if it satisfies*

$$z_{\mathrm{u}}(x,t) = \mathfrak{R}_r(t)z_0(x) + \int_0^{t-h}(t-\sigma-h)^{r-1}\mathcal{S}_r(t-\sigma-h)\mathfrak{B}\mathrm{u}(\sigma)d\sigma + \int_{-h}^{0}(t-\sigma-h)^{r-1}\mathcal{S}_r(t-\sigma-h)\mathfrak{B}\varphi(\sigma)d\sigma. \quad (13)$$

We define $\mathcal{H}: L^2(0,\tau-h;\mathbb{R}^p)\to L^2(\Omega)$ by

$$\mathcal{H}\mathrm{u} = \int_0^{\tau-h}(\tau-\sigma-h)^{r-1}\mathcal{S}_r(\tau-\sigma-h)\mathfrak{B}\mathrm{u}(\sigma)d\sigma, \quad \text{for all} \quad \mathrm{u}\in L^2(0,\tau-h;\mathbb{R}^p). \quad (14)$$

Assume that (H_1)–(H_2) hold and $(\mathcal{T}^*(t))_{t\geq 0}$ is a semigroup generated by $\mathcal{A}^*$ on $L^2(\Omega)$, which is strong and continuous. For $\nu\in L^2(\Omega)$, we have

$$\begin{aligned}\langle\mathcal{H}\mathrm{u},\nu\rangle &= \left\langle\int_0^{\tau-h}(\tau-\sigma-h)^{r-1}\mathcal{S}_r(\tau-\sigma-h)\mathfrak{B}\mathrm{u}(\sigma)d\sigma,\nu\right\rangle_{L^2(\Omega)}\\ &= \int_0^{\tau-h}\left\langle(\tau-\sigma-h)^{r-1}\mathcal{S}_r(\tau-\sigma-h)\mathfrak{B}\mathrm{u}(\sigma),\nu\right\rangle_{L^2(\Omega)}d\sigma\\ &= \int_0^{\tau-h}\left\langle\mathrm{u}(\sigma),\mathfrak{B}^*(\tau-\sigma-h)^{r-1}\mathcal{S}_r^*(\tau-\sigma-h)\nu\right\rangle_{\mathfrak{U}}d\sigma\\ &= \langle\mathrm{u},\mathcal{H}^*\nu\rangle,\end{aligned} \quad (15)$$

where $\langle\cdot,\cdot\rangle$ is the inner product on the vector space and

$$\mathcal{S}_r^*(t) = r\int_0^\infty \alpha\Phi_r(\alpha)\mathcal{T}^*(t^r\alpha)d\alpha.$$

Then, one has

$$\mathcal{H}^* v = \mathfrak{B}^* (\tau - \sigma - h)^{r-1} \mathcal{S}_r^* (\tau - \sigma - h) v, \quad \text{for all} \quad v \in L^2(\Omega). \tag{16}$$

Let $\mathcal{H}$ be defined as in (14) and let us define the operator $\mathfrak{L}_\varphi$ in such a way that

$$\begin{aligned} \mathfrak{L}_\varphi : L^2(-h, 0; L^2(\Omega)) &\longrightarrow L^2(\Omega) \\ \varphi &\longmapsto \int_{-h}^{0} (\tau - \sigma - h)^{r-1} \mathcal{S}_r (\tau - \sigma - h) \mathfrak{B} \varphi(\sigma) d\sigma. \end{aligned} \tag{17}$$

Following the same steps in the computation of $\mathcal{H}^*$, we obtain

$$\mathfrak{L}_\varphi^* v = \mathfrak{B}^* (\tau - \sigma - h)^{r-1} \mathcal{S}_r^* (\tau - \sigma - h) v, \quad \text{for all} \quad v \in L^2(\Omega).$$

Remark 2. *The solutions of* (1) *are considered in the weak sense.*

The subsequent lemmas are necessary to demonstrate our main results: Lemma 1 is used in the proof of Theorem 1, whereas Lemma 3 is useful for proving Theorem 2.

Lemma 1 (See [29,30])**.** *The operators $\mathfrak{R}_r(t)$ and $\mathcal{S}_r(t)$ are bounded and linear. Moreover, for every $z \in L^2(\Omega)$, we have*

$$\|\mathfrak{R}_r(t) z\| \le M \|z\|, \qquad \text{and} \qquad \|\mathcal{S}_r(t) z\| \le \frac{rM}{\Gamma(1+r)} \|z\|. \tag{18}$$

Lemma 2 (See, e.g., [31])**.** *If the reflection operator $\mathcal{Q}$ on $[0, \tau]$ is defined for a differentiable and integrable function g by*

$$\mathcal{Q} g(t) := g(\tau - t), \tag{19}$$

then it satisfies the properties

$$\begin{aligned} \mathcal{Q}_0 I_t^r g(t) &= {}_t I_\tau^r \mathcal{Q} g(t), \qquad & \mathcal{Q}_0 \mathbb{D}_t^r g(t) &= {}_t \mathbb{D}_\tau^r \mathcal{Q} g(t), \\ {}_0 I_t^r \mathcal{Q} g(t) &= \mathcal{Q}_t I_\tau^r g(t), \qquad & {}_0 \mathbb{D}_t^r \mathcal{Q} g(t) &= \mathcal{Q}_t \mathbb{D}_\tau^r g(t). \end{aligned} \tag{20}$$

In the following lemmas, we recall the integration by parts and the fractional Green formulas that relate the left-sided Caputo derivatives with the right-sided Riemann–Liouville derivatives.

Lemma 3 (Fractional integration by parts formula; see, e.g., [32])**.** *For $t \in [0, \tau]$ and $r \in (n-1, n)$, $n \in \mathbb{N}$, the integration by parts relation,*

$$\int_0^\tau f(t) {}_0^C \mathbb{D}_t^r g(t) dt = \sum_{i=0}^{n-1} (-1)^{n-1-i} \left[g^i(t) {}_t \mathbb{D}_\tau^{r-1-i} f(t) \right]_0^\tau + (-1)^n \int_0^\tau g(t) {}_t \mathbb{D}_\tau^r f(t) dt \tag{21}$$

holds.

Remark 3. *If $0 < r < 1$, we obtain from Lemma 3*

$$\int_0^\tau f(t) {}_0^C \mathbb{D}_t^r g(t) dt = \left[g(t) {}_t I_\tau^{1-r} f(t) \right]_0^\tau - \int_0^\tau g(t) {}_t \mathbb{D}_\tau^r f(t) dt. \tag{22}$$

Lemma 4 (Fractional Green formula; see, e.g., [24,33]). *Let $0 < r \leq 1$ and $t \in [0, \tau]$. Then,*

$$\begin{aligned}\int_0^\tau \int_\Omega & ({}_0^C\mathbb{D}_t^r z(x,t) - \mathcal{A}z(x,t))\phi(x,t)dxdt \\ &= \int_0^\tau \int_\Omega z(x,t)(-{}_0\mathbb{D}_\tau^r \phi(x,t) - \mathcal{A}^*\phi(x,t))dxdt \\ &+ \int_{\partial\Omega} z(x,\tau){}_t I_\tau^{1-r}\phi(x,\tau)d\Gamma - \int_{\partial\Omega} z(x,0){}_t I_\tau^{1-r}\phi(x,0)d\Gamma \\ &+ \int_0^\tau \int_{\partial\Omega} z(x,t)\frac{\partial\phi(x,t)}{\partial\nu_\mathcal{A}}d\Gamma dt - \int_0^\tau \int_{\partial\Omega} \frac{\partial z(x,t)}{\partial\nu_\mathcal{A}}\phi(x,t)d\Gamma,\end{aligned} \tag{23}$$

for any $\phi \in C^\infty(\overline{Q})$.

As a corollary of Lemma 4, the following result can be derived.

Corollary 1. *Let $0 < r < 1$. Then, for any $\phi \in C^\infty(\overline{Q})$ such that $\phi(x,\tau) = 0$ in Ω and $\phi = 0$ on Σ, we obtain*

$$\begin{aligned}\int_0^\tau \int_\Omega & ({}_0^C\mathbb{D}_t^r z(x,t) - \mathcal{A}z(x,t))\phi(x,t)\mathrm{d}x\mathrm{d}t \\ &= -\int_\Omega z(x,0){}_t I_\tau^{1-r}\phi(x,0)\mathrm{d}x + \int_0^\tau \int_{\partial\Omega} z(x,t)\frac{\partial\phi(x,t)}{\partial\nu_\mathcal{A}}\mathrm{d}\Gamma\mathrm{d}t \\ &+ \int_0^\tau \int_\Omega z(x,t)(-{}_0\mathbb{D}_\tau^r\phi(x,t) - \mathcal{A}^*\phi(x,t))\mathrm{d}x\,\mathrm{d}t.\end{aligned} \tag{24}$$

3. Regional Fractional Controllability

Let ω be a given region and a subset of Ω with a positive Lebesgue measure. The projection operator on ω is denoted by the restriction mapping

$$\begin{aligned} P_\omega : L^2(\Omega) &\longrightarrow L^2(\omega) \\ y &\longmapsto P_\omega y = y|_\omega. \end{aligned} \tag{25}$$

Definition 7 (Regional exact controllability at time τ). *We say that System* (1) *is ω-exactly controllable at time τ if, for any $z_d \in L^2(\omega)$, there exists a control $\mathrm{u} \in \mathfrak{U}$ such that*

$$P_\omega z_\mathrm{u}(x,\tau) = z_d. \tag{26}$$

Definition 8 (Regional weak controllability at time τ). *We say that System* (1) *is ω-weakly controllable at time τ if, for every $z_d \in L^2(\omega)$, given $\varepsilon > 0$, there is a control $\mathrm{u} \in \mathfrak{U}$ such that*

$$\|P_\omega z_\mathrm{u}(x,\tau) - z_d\|_{L^2(\omega)} \leq \varepsilon. \tag{27}$$

Remark 4. *It is equivalent saying that System* (1) *is regionally exactly (resp. regionally weakly) controllable or that System* (1) *is ω-exactly (resp. ω-weakly) controllable.*

Taking into account that System (1) is linear, for $\mathrm{u} \in \mathfrak{U}$, let us consider the attainable set $\mathbb{A}(t)$ in $L^2(\Omega)$ defined by

$$\mathbb{A}(t) = \left\{a(\cdot,t) \in L^2(\Omega) \,|\, a(x,t) = \mathcal{H}\mathrm{u} + \mathfrak{L}_\varphi\varphi\right\}, \tag{28}$$

where operators $\mathcal{H}$ and $\mathfrak{L}_\varphi$ are defined by Equations (14) and (17), respectively. The following result holds.

Theorem 1 (Necessary and sufficient conditions for regional controllability). *For any given $\tau > 0$, System* (1) *is regionally exactly (resp. regionally approximately) controllable if, and only if,*

$$P_\omega\mathbb{A}(\tau) = L^2(\omega) \qquad (resp.\ \overline{P_\omega\mathbb{A}(\tau)} = L^2(\omega)).$$

Proof. We prove the approximate controllability case. Using the same proof as in [34,35] and assuming that $u \equiv 0$ for all t, System (1) admits a unique solution

$$\mathcal{X}(z_0)(x,t) \in \mathcal{Z}, \qquad \text{such that} \qquad \mathcal{X}(z_0)(x,t) = \mathfrak{R}_r(t)z_0(x). \tag{29}$$

Then, using Lemma 1, $\exists c > 0$ that satisfies $\|\mathcal{X}(z_0)\|_{L^2(\Omega)} \leq c\|z_0\|_{L^2(\Omega)}$. Hence, (29) is well defined. For every $\nu \in L^2(\omega)$, since $P_\omega \mathcal{X}(z_0)(\cdot,\tau) \in L^2(\omega)$, we obtain

$$(\nu - P_\omega \mathcal{X}(z_0))(\cdot,\tau) \in L^2(\omega),$$

and

$$\begin{aligned}\|P_\omega z(x,\tau) - \nu\|_{L^2(\omega)} &= \|P_\omega z(x,\tau) + P_\omega \mathcal{X}(z_0)(x,\tau) - (P_\omega \mathcal{X}(z_0) - \nu)(x,\tau)\|_{L^2(\omega)} \\ &= \|(P_\omega z - P_\omega \mathcal{X}(z_0))(x,\tau) - (\nu - P_\omega \mathcal{X}(z_0))(x,\tau)\|_{L^2(\omega)} \\ &= \|P_\omega a(x,\tau) - (\nu - P_\omega \mathcal{X}(z_0))(x,\tau)\|_{L^2(\omega)}.\end{aligned}$$

If $\overline{P_\omega \mathbb{A}(\tau)} = L^2(\omega)$, for any $\varepsilon > 0$, we can find that $u \in \mathfrak{U}$ satisfies

$$\|P_\omega a(\cdot,\tau) - (\nu - P_\omega \mathcal{X}(z_0))(\cdot,\tau)\| \leq \varepsilon,$$

where $a(\cdot,\cdot)$ is an element of the attainable set (28). This implies that $\|P_\omega z_u(\cdot,\tau) - \nu\| \leq \varepsilon$, where $z_u(\cdot,\tau) = \mathcal{X}(z_0)(\cdot,\tau) + a(\cdot,\tau)$ is the mild solution of System (1). Then, System (1) is ω-weakly controllable at time τ.

On the other hand, for a given $\tau > 0$, System (1) is ω-weakly controllable if for any $z_d \in L^2(\omega)$, given $\varepsilon > 0$, there is a control $u \in \mathfrak{U}$ such that

$$\begin{aligned}\|P_\omega z_u(x,\tau) - z_d\|_{L^2(\omega)} &= \|P_\omega z_u(\cdot,\tau) + P_\omega \mathcal{X}(z_0)(\cdot,\tau) - P_\omega \mathcal{X}(z_0)(\cdot,\tau) - z_d\|_{L^2(\omega)} \\ &= \|P_\omega z_u(\cdot,\tau) - P_\omega \mathcal{X}(z_0)(\cdot,\tau) - (z_d - P_\omega \mathcal{X}(z_0)(\cdot,\tau))\|_{L^2(\omega)} \\ &= \|P_\omega a(\cdot,\tau) - (z_d - P_\omega \mathcal{X}(z_0)(\cdot,\tau))\|_{L^2(\omega)} \\ &\leq \varepsilon.\end{aligned}$$

One has $(P_\omega z_u(\cdot,\tau) - P_\omega \mathcal{X}(z_0)(\cdot,\tau)) \in P_\omega \mathbb{A}(\tau)$. Then, $(z_d - P_\omega \mathcal{X}(z_0)(\cdot,\tau)) \in L^2(\omega)$. Thus $\overline{P_\omega \mathbb{A}(\tau)} = L^2(\omega)$. □

Proposition 1. *The following properties are equivalent:*

(1) System (1) *is ω-exactly controllable;*
(2) $\mathrm{Im}(P_\omega(\mathcal{H} + \mathfrak{L}_\varphi)) = L^2(\omega)$;
(3) $\mathrm{Ker}(P_\omega) + \mathrm{Im}(\mathcal{H} + \mathfrak{L}_\varphi) = L^2(\Omega)$.

Proof. (1) $\Leftrightarrow$ (2). Suppose that System (1) is ω-exactly controllable. Then, there exists $z_d \in L^2(\omega)$ such that $P_\omega z_u(x,\tau) = z_d$, which is equivalent to

$$P_\omega \mathfrak{R}_r z_0 + P_\omega \mathcal{H} u + P_\omega \mathfrak{L}_\varphi \varphi = z_d.$$

For $z_0 = 0$, we have $P_\omega \mathcal{H} u + P_\omega \mathfrak{L}_\varphi \varphi = z_d \Leftrightarrow \mathrm{Im}(P_\omega \mathcal{H}) + \mathrm{Im}(P_\omega \mathfrak{L}_\varphi) = L^2(\omega)$.

(2) $\Rightarrow$ (3). For every $z \in L^2(\omega)$, we designate by $\tilde{z}$ the prolongation of z to $L^2(\Omega)$. Given $\mathrm{Im}(P_\omega \mathcal{H}) + \mathrm{Im}(P_\omega \mathfrak{L}_\varphi) = L^2(\omega)$, there is a control $u \in \mathfrak{U}$, $\varphi \in L^2(-h,0;L^2(\Omega))$, and $z_1 \in \mathrm{Ker}(P_\omega)$, such that $\tilde{z} = z_1 + \mathcal{H} u + \mathfrak{L}_\varphi \varphi$.

(3) $\Rightarrow$ (2). For every $\tilde{z} \in L^2(\Omega)$, it follows from (3) that $\tilde{z} = z_1 + z_2 + z_3$, where $z_1 \in \mathrm{Ker}(P_\omega)$, $z_2 \in \mathrm{Im}\,\mathcal{H}$, and $z_3 \in \mathrm{Im}\,\mathfrak{L}$. Then, there exists $u \in \mathfrak{U}$ such that $\mathcal{H} u = z_2$ and $\varphi \in L^2(-h,0;L^2(\Omega))$ such that $\mathfrak{L}_\varphi \varphi = z_3$. Hence, it follows from (25) that

$$\mathrm{Im}(P_\omega \mathcal{H}) + \mathrm{Im}(P_\omega \mathfrak{L}_\varphi) = L^2(\omega).$$

The proof is complete. □

Proposition 2. *The following properties are equivalent:*

(1) *System* (1) *is ω-weakly controllable;*

(2) $\overline{\mathrm{Im}(P_\omega(\mathcal{H}+\mathfrak{L}_\varphi))} = L^2(\omega)$;

(3) $\mathrm{Ker}(P_\omega) + \overline{\mathrm{Im}(\mathcal{H}+\mathfrak{L}_\varphi)} = L^2(\Omega)$.

Proof. The proof that (1)⇔ (2)⇔ (3) is similar to the proof of Proposition 1 and is left to the reader. □

We give an illustrative example of the application of our results.

Example 1. *Consider the time fractional order differential system with a zonal actuator governed by the state equation*

$$\begin{cases} {}_0^C\mathbb{D}_t^r z(x,t) = \Delta z(x,t) + P_{[\beta_1,\beta_2]}\mathrm{u}(t-h) & \text{in } [0,1]\times[0,\tau-h], \\ z(0) = z_0, \\ \mathrm{u}(t) = \varphi(t), \end{cases} \tag{30}$$

where $0 < r < 1$, $\mathfrak{B}\mathrm{u} = P_{[\beta_1,\beta_2]}\mathrm{u}$ and $0 \le \beta_1 \le \beta_2 \le 1$. Moreover, since $\mathcal{A} = \dfrac{\partial^2 \cdot}{\partial x^2}$ is a self-adjoint operator, we find that the eigenvalues of the operator $\mathcal{A}$ are given by $v_i = -i^2\pi^2$ and its eigenfunctions by $\zeta_i(x) = \sqrt{2}\sin(i\pi x)$. The uniformly continuous semigroup generated by $\mathcal{A}$ is

$$\Xi(t)z(x,t) = \sum_{i=1}^{\infty} e^{(v_i t)}(z,\zeta_i)_{L^2(0,1)}\zeta_i(x).$$

It implies

$$\mathcal{S}_r(t)z(x,t) = r\int_0^\infty \theta\phi_r(\theta)\Xi(t^r\theta)z(x,t)d\theta = r\sum_{i=1}^{\infty} E^2_{r,r+1}(v_i t^r)(z,\zeta_i)_{L^2(0,1)}\zeta_i(x),$$

and we obtain that

$$\begin{aligned} (\mathcal{H}+\mathfrak{L}_\varphi)^* z(x,t) &= 2\Big[\mathfrak{B}^*(\tau-\sigma-h)^{r-1}\mathcal{S}_r^*(\tau-\sigma-h)z\Big](x,t) \\ &= 2r\mathfrak{B}^*(\tau-\sigma-h)^{r-1}\sum_{i=1}^{\infty} E^2_{r,r+1}(v_i(\tau-\sigma-h)^r)(z,\zeta_i)_{L^2(0,1)}\zeta_i(x) \\ &= 2r(\tau-\sigma-h)^{r-1}\sum_{i=1}^{\infty} E^2_{r,r+1}(v_i(\tau-\sigma-h)^r)(z,\zeta_i)_{L^2(0,1)}\int_{\beta_1}^{\beta_2}\zeta_i(x)dx, \end{aligned} \tag{31}$$

whereas from $\int_{\beta_1}^{\beta_2}\zeta_i(x)dx = \dfrac{\sqrt{2}}{i\pi}\sin\dfrac{i\pi(\beta_1+\beta_2)}{2}\sin\dfrac{i\pi(\beta_1-\beta_2)}{2}$ we obtain $\mathrm{Ker}(\mathcal{H}+\mathfrak{L}_\varphi)^* \neq \{0\}$ i.e., $\mathrm{Im}(\mathcal{H}+\mathfrak{L}_\varphi) \neq L^2(\omega)$, which means that System (30) *is not controllable on $\Omega = [0,1]$. Let $[\beta_1 = 0, \beta_2 = 1/3]$. We then have*

$$\begin{aligned} \int_0^{1/3}\zeta_i(x)dx &= \int_0^{1/3}\sqrt{2}\sin(i\pi x)dx \\ &= \sqrt{2}\Big[-\frac{\cos(i\pi x)}{i\pi}\Big]_0^{1/3} \\ &= \frac{\sqrt{2}}{i\pi}(1-\cos(i\pi/3)). \end{aligned}$$

If $\omega = [1/3, 2/3] \subset [0,1]$, then

$$\begin{aligned}(\mathcal{H}+\mathfrak{L}_{\varphi})^{*}P_{\omega}^{*}(P_{\omega}\zeta_j) &= 2r(\tau-\sigma-h)^{r-1}\sum_{k=1}^{\infty}E_{r,r+1}^{2}(v_k(\tau-\sigma-h)^{r})\langle\zeta_i,\zeta_j\rangle_{L^2(\omega)}\int_0^{1/3}\zeta_i(x)dx \\ &= \sum_{k=1}^{\infty}\frac{rE_{r,r+1}^{2}(v_k(\tau-\sigma-h)^{r})}{2(\tau-\sigma-h)^{1-r}}\langle\zeta_i,\zeta_j\rangle_{L^2(\omega)}\int_0^{1/3}\zeta_i(x)dx \\ &= \sum_{k=1}^{\infty}\frac{rE_{r,r+1}^{2}(v_k(\tau-\sigma-h)^{r})}{2(\tau-\sigma-h)^{1-r}}\int_{1/3}^{2/3}\zeta_i(x)\zeta_j(x)dx\frac{\sqrt{2}}{i\pi}(1-\cos(i\pi/2)) \\ &= \sum_{k=1}^{\infty}\frac{rE_{r,r+1}^{2}(v_k(\tau-\sigma-h)^{r})}{\sqrt{2}i\pi(\tau-\sigma-h)^{1-r}}\int_{1/3}^{2/3}\zeta_i(x)\zeta_j(x)dx[1-\cos(i\pi/2)] \\ &\neq 0.\end{aligned} \tag{32}$$

We conclude that the state ζ_j is reachable on ω.

4. Optimal Control with a Regional Target

Fractional optimal control is a rapidly developing topic (see, for instance, [36–39]). This section is motivated by the results of [10,40–44] and is devoted to the proof that the steering control is a minimizer of a suitable optimal control problem. For this, we use an extended version of the Hilbert uniqueness method (HUM) first introduced by Lions [45,46].

Let F be a closed subspace of $L^2(\Omega)$. Our extended optimal control problem consists of seeking a minimum-norm control that drives the system to F at time τ. More precisely, we consider

$$\inf_{\mathrm{u}}\mathcal{J}(\mathrm{u}) = \inf_{\mathrm{u}}\left\{\int_0^{\tau}\frac{1}{2}\|\mathrm{u}(t)\|^2dt \quad : \quad \mathrm{u}\in\mathfrak{U}_{ad}\right\}, \tag{33}$$

where $\mathfrak{U}_{ad} = \{\mathrm{u}\in\mathfrak{U} \mid P_{\omega}z_{\mathrm{u}}(\cdot,\tau)-z_d\in F\}$, and the set

$$F^{\circ} = \left\{f\in L^2(\Omega) \mid f=0 \quad \text{in } \Omega\setminus\omega\right\}.$$

For $\psi_0\in F^{\circ}$, we consider the system

$$\begin{cases} {}_t^C\mathbb{D}_{\tau}^{r}\mathcal{Q}\psi(t) = -\mathcal{A}^{*}\mathcal{Q}\psi(t), \\ \lim\limits_{t\to\tau^-}{}_tI_{\tau}^{1-r}\mathcal{Q}\psi(t) = \psi_0\subset L^2(\Omega), \end{cases} \tag{34}$$

in $L^2(\Omega)$ and let

$$\|\psi_0\|_{F^{\circ}}^2 = \int_0^{\tau}\|\mathfrak{B}^{*}(\tau-\sigma-h)^{r-1}\mathcal{S}_r^{*}(\tau-\sigma-h)P_{\omega}^{*}\psi(\sigma)\|^2d\sigma, \tag{35}$$

which is a semi-norm on F°.

Using Lemma 2, we can rewrite (34) as

$$\begin{cases} {}_0^C\mathbb{D}_{t}^{r}\psi(t) = -\mathcal{A}^{*}\psi(t), \\ \lim\limits_{t\to 0^+}{}_0I_{t}^{1-r}\psi(t) = \psi_0\subset L^2(\Omega), \end{cases} \tag{36}$$

with the solution given by $\psi(t) = -t^{r-1}K_r^{*}\psi_0$.

Theorem 2. *If* u *spans* $\mathfrak{U}$, *then* $\overline{z_{\mathrm{u}}(x,\tau)} = L^2(\Omega)$.

Proof. Take $z(x,0)=z_0(x)=0$ and suppose that $z_{\mathrm{u}}(x,\tau)$ is not dense in $L^2(\Omega)$. Consequently, there is $\psi_0\in L^2(\Omega)$, $\psi_0\neq 0$, such that

$$\langle z_{\mathrm{u}}(x,\tau),\psi_0\rangle = 0, \qquad \forall\mathrm{u}\in\mathfrak{U}. \tag{37}$$

By multiplying both sides of (34) by $z(t)$ and then integrating it over Q, we obtain

$$\begin{aligned}\int_{\Omega}\int_0^{\tau} z(x,t){}_0^C\mathbb{D}_t^r \mathcal{Q}\psi(t)dtdx &= \int_0^{\tau}\langle z(x,t), -\mathcal{A}^*\mathcal{Q}\psi(t)\rangle_{\Omega}dt \\ &= -\int_0^{\tau}\langle \mathcal{A}z(x,t), \mathcal{Q}\psi(t)\rangle_{\Omega}dt.\end{aligned} \tag{38}$$

From Lemma 3, we have

$$\begin{aligned}\int_{\Omega}\int_0^{\tau} z(x,t){}_0^C\mathbb{D}_t^r \mathcal{Q}\psi(t)dtdx &= \int_{\Omega}\Big[z(x,t){}_tI_{\tau}^{1-r}\mathcal{Q}\psi(t)\Big]_0^{\tau} - \int_{\Omega}\int_0^{\tau}\mathcal{Q}\psi(t){}_0^C\mathbb{D}_t^r z(x,t)dtdx \\ &= \Big\langle z(x,\tau), \lim_{t\to\tau}{}_tI_{\tau}^{1-r}\mathcal{Q}\psi(t)\Big\rangle_{\Omega} - \Big\langle z(x,0), \lim_{t\to 0}{}_tI_{\tau}^{1-r}\mathcal{Q}\psi(t)\Big\rangle_{\Omega} \\ &\quad - \int_0^{\tau}\Big\langle \mathcal{Q}\psi(t), {}_0^C\mathbb{D}_t^r z(x,t)\Big\rangle_{\Omega}dt \\ &= \langle z(x,\tau), \psi_0\rangle - \int_0^{\tau}\Big\langle \mathcal{Q}\psi(t), {}_0^C\mathbb{D}_t^r z(x,t)\Big\rangle_{\Omega}dt \\ &= \langle z(x,\tau), \psi_0\rangle - \int_0^{\tau}\langle \mathcal{Q}\psi(t), \mathcal{A}z(x,t) + \mathfrak{B}u(t-h)\rangle dt.\end{aligned} \tag{39}$$

From Equations (38) and (39), one has

$$\langle z(x,\tau), \psi_0\rangle = \int_0^{\tau}\langle \mathcal{Q}\psi(t), \mathfrak{B}u(t-h)\rangle dt. \tag{40}$$

Using (37), we have

$$\int_0^{\tau}\langle \mathcal{Q}\psi(t), \mathfrak{B}u(t-h)\rangle dt = 0 \Leftrightarrow \mathcal{Q}\psi(t) = \psi(\tau - t) \equiv 0, \quad \text{in } L^2(\Omega),\ \forall t\in[0,\tau]. \tag{41}$$

Then $\psi_0 = 0$, which is a contradiction. The proof is complete. □

To proceed with the HUM approach, we first need to prove that the semi-norm $||\cdot||$ on F° in (35) is a norm. We prove the next results.

Lemma 5. *Assuming that (H_1)–(H_2) hold, (35) defines a norm of F° when System (1) is ω-weakly controllable.*

Proof. Suppose that system (1) is ω-weakly controllable. Then, $\mathrm{Ker}((\mathcal{H}+\mathfrak{L}_{\varphi})^*P_{\omega}^*) = \{0\}$, that is,

$$\mathfrak{B}^*(\tau-\sigma-h)^{r-1}\mathcal{S}_r^*(\tau-\sigma-h)P_{\omega}^*\psi = 0 \Longrightarrow \psi = 0.$$

Therefore, for every $\psi_0 \in F^\circ$, it follows that

$$\begin{aligned}||\psi_0||_{F^\circ} = \int_0^{\tau}||\mathfrak{B}^*(\tau-t-h)^{r-1}\mathcal{S}_r^*(\tau-t-h)P_{\omega}^*\psi(t)||^2dt = 0 \\ \Leftrightarrow \mathfrak{B}^*(\tau-t-h)^{r-1}\mathcal{S}_r^*(\tau-t-h)P_{\omega}^*\psi(t) = 0.\end{aligned}$$

Then, (35) is a norm. □

Furthermore, let us define an operator $\mathcal{M}: F^{\circ *} \to F^\circ$ by

$$\mathcal{M}f = \mathcal{P}(\phi(\tau)), \tag{42}$$

where $\mathcal{P} = P_{\omega}^*P_{\omega}$ and ϕ is defined by

$$\begin{cases}{}_0^C\mathbb{D}_t^r\phi(t) = \mathcal{A}\phi(t) + \mathfrak{B}\mathfrak{B}^*(\tau-t-h)^{r-1}\mathcal{S}_r^*(\tau-t-h)\psi(t), \\ \phi(0) = z_0.\end{cases} \tag{43}$$

We then decompose $\mathcal{M}$ as

$$\mathcal{M}f = \mathcal{P}(\phi_1(\tau) + \phi_2(\tau)),$$

where

$$\begin{cases} {}_0^C\mathbb{D}_t^r \phi_1(t) = \mathcal{A}\phi_1(t), \\ \phi_1(0) = z_0, \end{cases} \tag{44}$$

and

$$\begin{cases} {}_0^C\mathbb{D}_t^r \phi_2(t) = \mathcal{A}\phi_2(t) + \mathfrak{B}\mathfrak{B}^*(\tau - t - h)^{r-1}\mathcal{S}_r^*(\tau - t - h)\psi(t), \quad t \in [0, \tau - h], \\ \phi_2(0) = 0. \end{cases} \tag{45}$$

Let

$$\Lambda\psi_0 = P_\omega \mathcal{P}(\phi_2(\tau)). \tag{46}$$

Then, the regional controllability problem leads to the resolution of the equation

$$\Lambda\psi_0 = z_d - P_\omega \mathcal{P}(\phi_1(\tau)). \tag{47}$$

For any $f \in (F^\circ)^*$ and $g \in F$, by Holder's inequality we have that

$$\begin{aligned} \langle \Lambda f, g \rangle &= \int_\Omega P_\omega \int_0^\tau (\tau - \sigma - h)^{r-1}\mathcal{S}_r(\tau - \sigma - h)\mathfrak{B}\mathfrak{B}^*(\tau - \sigma - h)^{r-1} \\ &\quad \times \mathcal{S}_r^*(\tau - \sigma - h)P_\omega^* f(\sigma)d\sigma g(x)dx \\ &\leq \int_0^\tau \| \mathfrak{B}^*(\tau - \sigma - h)^{r-1}\mathcal{S}_r^*(\tau - \sigma - h)P_\omega^* f(\sigma) \|^2 d\sigma \| g \| \\ &\leq \| f \| \| g \|, \end{aligned}$$

and $\| \Lambda f \| \leq \| f \|$. Moreover, for any $f \in (F^\circ)^*$, we obtain

$$\begin{aligned} &\langle \Lambda\psi_0, \psi_0 \rangle_{F^{\circ *}, F^\circ} = \langle P_\omega \mathcal{P}(\phi_2(\tau)), \psi_0 \rangle \\ &= \left\langle \int_0^\tau P_\omega(\tau - \sigma - h)^{r-1}\mathcal{S}_r(\tau - \sigma - h)\mathfrak{B}\mathfrak{B}^*(\tau - \sigma - h)^{r-1}\mathcal{S}_r^*(\tau - \sigma - h)P_\omega^*\psi_0(\sigma)d\sigma, \psi_0(\sigma) \right\rangle \\ &= \int_0^\tau \left\langle \mathfrak{B}^*(\tau - \sigma - h)^{r-1}\mathcal{S}_r^*(\tau - \sigma - h)P_\omega^*\psi_0(\sigma), \mathfrak{B}^*(\tau - \sigma - h)^{r-1}\mathcal{S}_r(\tau - \sigma - h)P_\omega^*\psi_0(\sigma) \right\rangle d\sigma \\ &= \int_0^\tau \|\mathfrak{B}^*(\tau - \sigma - h)^{r-1}\mathcal{S}_r^*(\tau - \sigma - h)P_\omega^*\psi_0(\sigma)\|^2 d\sigma \\ &= \|\psi_0\|_{F^\circ}^2. \end{aligned}$$

Consequently, if System (1) is ω-weakly controllable at τ, we find that $\psi_0 = 0$. From the uniqueness of $\mathcal{M}$, we find that Λ defined in (46) is an isomorphism.

Theorem 3. *If System* (1) *is ω-weakly controllable, for any $z_d \in L^2(\omega)$,* (47) *has a unique solution $\psi_0 \in F^\circ$ and the control*

$$\mathrm{u}^*(t) = \begin{cases} \mathfrak{B}^*(\tau - t - h)^{r-1}\mathcal{S}_r^*(\tau - t - h)P_\omega^*\psi(t), & 0 \leq t \leq \tau - h, \\ \varphi(t), & \tau - h \leq t \leq \tau, \end{cases} \tag{48}$$

steers the system to z_d in ω. Moreover, u^* solves the minimization optimal control problem in (33).

Proof. If System (1) is ω-weakly controllable, (35) is a norm. Let us consider the completion of F° regarding the norm in (35) and let us denote it again by F°. Now, we prove that (47) admits a unique solution in F°. For any $\psi_0 \in F^\circ$, one has

$$\langle \Lambda\psi_0, \psi_0 \rangle_{F^{\circ *}, F^\circ} = \langle P_\omega \mathcal{P}(\phi_1(\tau)), \psi_0 \rangle = \|\psi_0\|_{F^\circ}^2.$$

Using Theorems 1.1 and 2.1 in [45], one can see that (47) has a unique solution. Moreover, by setting $\mathfrak{u} = \mathfrak{u}^*$ in System (1), we have $P_\omega z_{\mathfrak{u}^*}(x,\tau) = z_d$. For $\mathfrak{u}$ and v in $\mathfrak{U}_{ad}$, one has $P_\omega z_{\mathfrak{u}}(x,\tau) = P_\omega z_v(x,\tau) = z_d$ so $P_\omega[z_{\mathfrak{u}}(x,\tau) - z_v(x,\tau)] = 0$. We can easily find, for any $\psi_0 \in F^\circ$, that

$$\begin{aligned}
&\langle \psi_0, P_\omega[z_{\mathfrak{u}}(x,\tau) - z_v(x,\tau)]\rangle = 0 \\
&\Leftrightarrow \langle \psi_0, P_\omega \int_0^{\tau-h} (\tau-\sigma-h)^{r-1}\mathcal{S}_r(\tau-\sigma-h)\mathfrak{B}[\mathfrak{u}(\sigma) - v(\sigma)]d\sigma\rangle = 0 \\
&\Leftrightarrow \int_0^{\tau-h} \langle \mathfrak{B}^*(\tau-\sigma-h)^{r-1}\mathcal{S}_r^*(\tau-\sigma-h)P_\omega^*\psi_0, \mathfrak{u}(\sigma) - v(\sigma)\rangle d\sigma = 0,
\end{aligned}$$

and

$$\begin{aligned}
\mathcal{J}'(\mathfrak{u})(\mathfrak{u} - v) &= \int_0^{\tau-h} \langle \mathfrak{u}(\sigma), \mathfrak{u}(\sigma) - v(\sigma)\rangle d\sigma \\
&= \int_0^{\tau-h} \langle \mathfrak{B}^*(\tau-\sigma-h)^{r-1}\mathcal{S}_r^*(\tau-\sigma-h)P_\omega^*\psi_0, \mathfrak{u}(\sigma) - v(\sigma)\rangle d\sigma \\
&= 0.
\end{aligned}$$

Because $\mathfrak{U}_{ad}$ is convex, by using Theorem 1.3 in [45], we establish the optimality of $\mathfrak{u}^*$. □

5. Examples

We provide two illustrative examples for cases of both a bounded (Section 5.1) and an unbounded control operator (Section 5.2).

5.1. Example 1: Zonal Actuator

We consider the system

$$\begin{cases}
{}_0^C\mathbb{D}_t^{0.3} z(x,t) = \Delta z(x,t) + P_{[\beta_1,\beta_2]}\mathfrak{u}(t-h), & [0,1]\times[0,\tau-h], \\
z(x,0) = 0, \\
\mathfrak{u}(t) = \varphi(t), \quad -h \le t \le 0,
\end{cases} \tag{49}$$

with a fractional order $r = 0.3$. Here, the control operator $\mathfrak{B}$ is bounded, $[\beta_1, \beta_2] = [0, 1/2]$, $\mathcal{A} = \Delta = \frac{\partial^2 \cdot}{\partial x^2}$, and the semigroup $(\mathcal{T}(t))_{t\ge 0}$ is given by

$$\mathcal{T}(t)z(x,t) = \sum_{i=1}^{\infty} e^{v_i t}(z,\zeta_i)\zeta_i(x), \qquad x \in [0,1],$$

where $v_i = -i^2\pi^2$ and $\zeta_i(x) = \sqrt{2}\sin(i\pi x)$, $i = 1,2,\ldots$ Then, $(\mathcal{T}(t))_{t\ge 0}$ is uniformly bounded. Moreover, one has

$$\begin{aligned}
\mathcal{S}_{0.3}(t)z(x,t) &= 0.3\int_0^\infty \theta\Phi_{0.3}(\theta)\mathcal{T}(t^{0.3}\theta)z(x,t)d\theta \\
&= 0.3\int_0^\infty \theta\Phi_{0.3}(\theta)\sum_{i=1}^{\infty} e^{v_i t^{0.3}\theta}(z,\zeta_i)\zeta_i(x)d\theta \\
&= 0.3\sum_{i=1}^{\infty}\sum_{n=0}^{\infty}\int_0^\infty \frac{(v_i t^{0.3})^n}{n!}\theta^{n+1}\Phi_{0.3}(\theta)d\theta(z,\zeta_i)\zeta_i(x) \\
&= 0.3\sum_{i=1}^{\infty}\sum_{n=0}^{\infty}\frac{(v_i t^{0.3})^n}{n!}\frac{\Gamma(n+2)}{\Gamma(1+0.3n+0.3)}(z,\zeta_i)\zeta_i(x) \\
&= 0.3\sum_{i=1}^{\infty} E^2_{0.3,1.3}\left(v_i t^{0.3}\right)(z,\zeta_i)\zeta_i(x).
\end{aligned} \tag{50}$$

Similarly, we have

$$\mathfrak{R}_{0.3}(t)z(x,t) = \sum_{i=1}^{\infty}(z,\zeta_i)E_{0.3,1}\left(v_i t^{0.3}\right)\zeta_i(x). \tag{51}$$

Let $\omega = [0.25, 0.75]$. We can easily verify that System (49) is $[0.25, 0.75]$-controllable using the same arguments in (32) and, by Theorem 3, we obtain

$$\|f\|_{F^\circ} = \int_0^{\tau-h} \|0.3(\tau-\sigma-h)^{-0.7}\sum_{i=1}^{\infty} E^2_{0.3,1.3}\left(v_i(\tau-\sigma-h)^{0.3}\right)(z,\zeta_i)P^*_\omega \int_0^{1/2} f(x)dx\|^2 d\sigma,$$

which defines a norm on F°. Moreover, $\Lambda f = P_\omega \mathcal{P}(\phi_2(\tau))$ is an isomorphism and we obtain the control

$$u^*(t) = 0.3(\tau-\sigma-h)^{-0.7}\sum_{i=1}^{\infty} E^2_{0.3,1.3}\left(v_i(\tau-\sigma-h)^{0.3}\right)(z,\zeta_i)P^*_\omega \int_0^{1/2} f(x)dx,$$

steering system in (49) from $z_0(x)$ to z_d with minimum energy.

5.2. Example 2: Pointwise Actuator

Let us now consider the same system as in Example 1 with the control operator $\mathfrak{B} = \delta(x-b)$, where $0 < b < 1$ is the control action point. The system is given by

$$\begin{cases} {}^C_0\mathbb{D}^{0.3}_t z(x,t) = \Delta z(x,t) + \delta(x-b)u(t-h), & [0,1]\times[0,\tau-h], \\ z(x,0) = 0, \\ u(t) = \varphi(t), \quad -h \le t \le 0, \end{cases} \tag{52}$$

with δ the impulse function defined by

$$\begin{cases} \delta(t) = 0, \quad \text{for} \quad t \neq 0, \\ \int_{-\infty}^{+\infty} \delta(t)dt = 1. \end{cases} \tag{53}$$

Here the operator $\mathfrak{B}$ is unbounded. Using Equations (50) and (51), one has

$$\begin{aligned} (\mathcal{H}+\mathfrak{L}_\varphi)^* z(x,t) &= 2\left[\mathfrak{B}^*(\tau-\sigma-h)^{r-1}\mathcal{S}^*_r(\tau-\sigma-h)z\right](x,t) \\ &= 0.6\sum_{i=1}^{\infty}\frac{E^2_{0.3,1.3}(v_i(\tau-\sigma-h)^{0.3})(z,\zeta_i)_{L^2(0,1)}\zeta_i(b)}{(\tau-\sigma-h)^{0.7}}. \end{aligned} \tag{54}$$

If $b \in \mathbb{Q}$, System (52) is not Ω-controllable. Considering $\omega = [1/3, 3/4]$ and $b = 1/2$, System (52) is ω-controllable. Indeed,

$$\begin{aligned} (\mathcal{H}+\mathfrak{L}_\varphi)^* z(x,t) &= 0.6\sum_{i=1}^{\infty}\frac{E^2_{0.3,1.3}(v_i(\tau-\sigma-h)^{0.3})(z,\zeta_i)_{L^2(0,1)}\sqrt{2}\sin\left(i\frac{\pi}{2}\right)}{(\tau-\sigma-h)^{0.7}} \\ &= 0.6\sqrt{2}\sum_{i=1}^{\infty}\frac{E^2_{0.3,1.3}(v_i(\tau-\sigma-h)^{0.3})(z,\zeta_i)_{L^2(0,1)}\sin\left(i\frac{\pi}{2}\right)}{(\tau-\sigma-h)^{0.7}}. \end{aligned}$$

Then,

$$
\begin{aligned}
(\mathcal{H}+\mathfrak{L}_{\varphi})^{*}P_{\omega}^{*}(P_{\omega}\zeta_j) &= 0.6\sqrt{2}\sum_{k=1}^{\infty}\frac{E_{0.3,1.3}^{2}\left(v_k(\tau-\sigma-h)^{0.3}\right)}{2(\tau-\sigma-h)^{0.7}}\langle\zeta_i,\zeta_j\rangle_{L^2(\omega)}\sin\left(i\frac{\pi}{2}\right)\\
&= 0.6\sqrt{2}\sum_{k=1}^{\infty}\frac{E_{0.3,1.3}^{2}\left(v_k(\tau-\sigma-h)^{0.3}\right)}{2(\tau-\sigma-h)^{0.7}}\sin\left(i\frac{\pi}{2}\right)\int_{1/3}^{2/3}\zeta_i(x)\zeta_j(x)dx\\
&\neq 0.
\end{aligned}
$$

Moreover,

$$
\begin{aligned}
\|\psi_0\|_{F^\circ} &= \int_0^{\tau-h}\left\|(\tau-\sigma-h)^{-0.7}\mathcal{S}_{0.3}^{*}(\tau-\sigma-h)P_{[1/3,3/4]}^{*}\psi(b)\right\|^2 d\sigma\\
&= \int_0^{\tau-h}\left\|0.3\sum_{i=1}^{\infty}\frac{E_{0.3,1.3}^{2}\left(v_i(\tau-\sigma-h)^{0.3}\right)(z,\zeta_i)P_{\omega}^{*}\psi(1/2)}{(\tau-\sigma-h)^{0.7}}\right\|^2 d\sigma.
\end{aligned}
\tag{55}
$$

From Lemma 5, (55) defines a norm on F° and (46) is an isomorphism from $F^{\circ *}$ to F°, where $\phi_2(\tau)$ is the solution of the system

$$
\begin{cases}
{}_0^C\mathbb{D}_t^{0.3}\phi_2(t) = \Delta\phi_2(t) + (\tau-t-h)^{-0.7}\mathcal{S}_{0.3}^{*}(\tau-t-h)\psi(b), & t\in[0,\tau-h],\\
\phi_2(0) = 0.
\end{cases}
\tag{56}
$$

By Theorem 3, we obtain the control

$$
u^{*}(t) = 0.3\sum_{i=1}^{\infty}\frac{E_{0.3,1.3}^{2}\left(v_i(\tau-\sigma-h)^{0.3}\right)(z,\zeta_i)P_{\omega}^{*}\psi(1/2)}{(\tau-\sigma-h)^{0.7}}d\sigma,
\tag{57}
$$

steering system in (52) to z_d, which is simultaneously the solution of the minimization problem in (33), where ψ is a solution of (47) and $\phi_1(\tau)$ is a solution of

$$
\begin{cases}
{}_0^C\mathbb{D}_t^{0.3}\phi_1(t) = \Delta\phi_1(t),\\
\phi_1(0) = z_0.
\end{cases}
\tag{58}
$$

6. Conclusions

In this paper, we dealt with a fractional Caputo diffusion equation defined in (1). We studied regional controllability with a delay in the control. By defining an attainable set, we proved the exact and weak controllability of such a system. We also formulated a minimum optimal energy control problem subject to System (1) and computed its optimal control. The solution of the optimal control problem was obtained via an extension of the Hilbert uniqueness method.

In future work, we intend to extend the obtained results (i) to the case of fractional semi-linear systems with delays in either the control, state variables, or both; (ii) to the case of neutral evolution systems [47,48] by extending the notion of regional controllability to such systems; (iii) to the case of the complete controllability of nonlinear fractional neutral functional differential equations [49] with delays; and (iv) to the case of the regional stability of fractional delay systems [50]. Another line of research consists of developing the numerical part and providing suitable numerical simulations for real problems. This is under investigation and will be addressed elsewhere.

Author Contributions: Conceptualization, T.K., A.K. and D.F.M.T.; methodology, T.K., A.K. and D.F.M.T.; formal analysis, T.K., A.K. and D.F.M.T.; investigation, T.K., A.K. and D.F.M.T.; writing—original draft preparation, T.K., A.K. and D.F.M.T.; writing—review and editing, T.K., A.K. and D.F.M.T. All authors have read and agreed to the published version of the manuscript.

Funding: This research was funded by The Portuguese Foundation for Science and Technology (FCT—Fundação para a Ciência e a Tecnologia), grant number UIDB/04106/2020 (CIDMA).

Institutional Review Board Statement: Not applicable.

Informed Consent Statement: Not applicable.

Data Availability Statement: Not applicable.

Acknowledgments: The authors are grateful to the anonymous reviewers for their comments and suggestions.

Conflicts of Interest: The authors declare no conflicts of interest. The funders had no role in the design of the study; in the collection, analyses, or interpretation of data; in the writing of the manuscript; or in the decision to publish the results.

References

1. Miller, K.; Ross, B. *An Introduction to the Fractional Calculus and Fractional Differential Equations*; John Wiley & Sons, Inc.: New York, NY, USA, 1993.
2. Machado, J.A.T.; Kiryakova, V.; Mainardi, F. A poster about the old history of fractional calculus. *Fract. Calc. Appl. Anal.* **2010**, *13*, 447–454.
3. Machado, J.A.T.; Kiryakova, V.; Mainardi, F. A poster about the recent history of fractional calculus. *Fract. Calc. Appl. Anal.* **2010**, *13*, 329–334.
4. Sene, N. Fractional input stability for electrical circuits described by the Riemann–Liouville and the Caputo fractional derivatives. *AIMS Math.* **2019**, *1*, 147–165. [CrossRef]
5. Abuaisha, T.; Kertzscher, J. Fractional-order modelling and parameter identification of electrical coils. *Fract. Calc. Appl. Anal.* **2019**, *22*, 193–216. [CrossRef]
6. Podlubny, I. *Fractional Differential Equations*; Academic Press, Inc.: San Diego, CA, USA, 1999.
7. Kilbas, A.A.; Srivastava, H.M.; Trujillo, J.J. *Theory and Applications of Fractional Differential Equations*; North-Holland Mathematics Studies, 204; Elsevier Science B.V.: Amsterdam, The Netherlands, 2006.
8. Engel, K.J.; Nagel, R. *A Short Course on Operator Semigroups*; Springer: New York, NY, USA, 2006.
9. Renardy, M.; Rogers, R.C. *An Introduction to Partial Differential Equations*; Springer: New York, NY, USA, 2004.
10. Jai, A.E.; Pritchard, A.J.; Simon, M.C.; Zerrik, E. Regional controllability of distributed systems. *Int. J. Control* **1995**, *62*, 1351–1365. [CrossRef]
11. Aadi, A.A.; Zerrik, E.H. Constrained regional control problem of a bilinear plate equation. *Int. J. Control* **2022**, *95*, 996–1002. [CrossRef]
12. Jai, A.E.; Pritchard, A.J. *Sensors and Controls in the Analysis of Distributed Systems*; Halsted Press: New York, NY, USA, 1988.
13. Ge, F.; Chen, Y.Q.; Kou, C. Regional gradient controllability of sub-diffusion processes. *J. Math. Anal. Appl.* **2016**, *440*, 865–884. [CrossRef]
14. Ge, F.; Chen, Y.Q.; Kou, C. Regional boundary controllability of time fractional diffusion processes. *IMA J. Math. Control Inf.* **2016**, *34*, 871–888. [CrossRef]
15. Kaczorek, T. *Selected Problems of Fractional Systems Theory*; Lecture Notes in Control and Information Sciences; Springer: Berlin/Heidelberg, Germany, 2011.
16. Kaczorek, T.; Rogowski, K. *Fractional Linear Systems and Electrical Circuits*; Studies in Systems, Decision and Control; Springer: Cham, Switzerland, 2015.
17. Ghasemi, M.; Nassiri, K. Controllability of linear fractional systems with delay in control. *J. Funct. Spaces* **2022**, *2022*, 5539770 . [CrossRef]
18. Jerzy, K. Controllability of fractional linear systems with delays. In Proceedings of the 25th International Conference on Methods and Models in Automation and Robotics (MMAR), Miedzyzdroje, Poland, 23–26 August 2021; pp. 331–336.
19. Singh, B.K.; Agrawal, S. Study of time fractional proportional delayed multi-pantograph system and integro-differential equations. *Math. Methods Appl. Sci.* **2022**, *45*, 8305–8328. [CrossRef]
20. Sakulrang, S.; Moore, E.J.; Sungnul, S.; Gaetano, A.D. A fractional differential equation model for continuous glucose monitoring data. *Adv. Differ. Equ.* **2017**, *2017*, 150. [CrossRef]
21. Gaetano, A.D.; Sakulrang, S.; Borri, A.; Pitocco, D.; Sungnul, S.; Moore, E.J. Modeling continuous glucose monitoring with fractional differential equations subject to shocks. *J. Theor. Biol.* **2021**, *526*, 110776. [CrossRef] [PubMed]
22. Kheyrinataj, F.; Nazemi, A. On delay optimal control problems with a combination of conformable and Caputo-Fabrizio fractional derivatives via a fractional power series neural network. *Netw. Comput. Neural Syst.* **2022**, *33*, 62–94. [CrossRef] [PubMed]
23. Agrawal, O.P. A general formulation and solution scheme for fractional optimal control problems. *Nonlinear Dyn.* **2004**, *38*, 323–337. [CrossRef]
24. Bahaa, G.M. Fractional optimal control problem for differential system with control constraints. *Filomat* **2016**, *30*, 2177–2189. [CrossRef]

25. Mophou, G.M. Optimal control of fractional diffusion equation. *Comput. Math. Appl.* **2011**, *61*, 68–78. [CrossRef]
26. Tomovski, Z.; Metzler, R.; Gerhold, S. Fractional characteristic functions, and a fractional calculus approach for moments of random variables. *Fract. Calc. Appl. Anal.* **2022**, *25*, 1307–1323. [CrossRef]
27. Mainardi, F.; Paradisi, P.; Gorenflo, R. Probability distributions generated by fractional diffusion equations. *arXiv* **2007**, arXiv:0704.0320v1
28. Wei, J. The controllability of fractional control systems with control delay. *Comput. Math. Appl.* **2012**, *64*, 3153–3159. [CrossRef]
29. Zhou, Y.; Jiao, F. Existence of mild solutions for fractional neutral evolution equations. *Comput. Math. Appl.* **2010**, *59*, 1063–1077. [CrossRef]
30. Ge, F.; Chen, Y.-Q.; Kou, C.; Podlubny, I. On the regional controllability of the sub-diffusion process with Caputo fractional derivative. *Fract. Calc. Appl. Anal.* **2016**, *19*, 1262–1281. [CrossRef]
31. Małgorzata, K. *On Solutions of Linear Fractional Differential Equations of a Variational Type*; Czestochowa University of Technology: Czestochowa, Poland, 2009.
32. Pudlubny, I.; Chen, Y.Q. Adjoint fractional differential expressions and operators. In Proceedings of the ASME 2007 International Design Engineering Technical Conferences & Computers and Information in Engineering Conference IDETC/CIE 2007, Las Vegas, NV, USA, 4–7 September 2007.
33. Bahaa, G.M.; Tang, Q. Optimality conditions of fractional diffusion equations with weak Caputo derivatives and variational formulation. *J. Fract. Calc. Appl.* **2018**, *9*, 100–119.
34. Sakamoto, K.; Yamamoto, M. Initial value/boundary value problems for fractional diffusion-wave equations and applications to some inverse problems. *J. Math. Anal. Appl.* **2011**, *382*, 426–447. [CrossRef]
35. Sakamoto, K.; Yamamoto, M. Approximate controllability for fractional diffusion equations by interior control. *Appl Anal.* **2014**, *93*, 1793–1810.
36. Mozyrska, D.; Torres, D.F.M. Minimal modified energy control for fractional linear control systems with the Caputo derivative. *Carpathian J. Math.* **2010**, *26*, 210–221.
37. Mozyrska, D.; Torres, D.F.M. Modified optimal energy and initial memory of fractional continuous-time linear systems. *Signal Process.* **2011**, *91*, 379–385. [CrossRef]
38. Eroğlu, B.B.; şkan, D.Y. Comparative analysis on fractional optimal control of an SLBS model. *J. Comput. Appl. Math.* **2023**, *421*, 114840. [CrossRef]
39. Mamehrashi, K. Ritz approximate method for solving delay fractional optimal control problems. *J. Comput. Appl. Math.* **2023**, *417*, 114606. [CrossRef]
40. Jai, A.E.; Pritchard, A.J. *Sensors and Actuators in Distributed Systems Analysis*; Wiley: New York, NY, USA, 1988.
41. Zerrik, E. Regional Analysis of Distributed Parameter Systems. PhD Thesis, University of Rabat, Rabat, Morocco, 1993.
42. Karite, T.; Boutoulout, A. Regional enlarged controllability for parabolic semilinear systems. *Int. J. Appl. Pure Math.* **2017**, *113*, 113–129. [CrossRef]
43. Karite, T.; Boutoulout, A. Regional boundary controllability of semilinear parabolic systems with state constraints. *Int. J. Dyn. Syst. Differ. Equ.* **2018**, *8*, 150–159.
44. Karite, T.; Boutoulout, A.; Torres, D.F.M. Enlarged controllability of Riemann–Liouville fractional differential equations. *J. Comput. Nonlinear Dyn.* **2018**, *13*, 6. [CrossRef]
45. Lions, J.L. *Optimal Control of Systems Governed by Partial Differential Equations*; Springer: New York, NY, USA, 1971.
46. Lions, J.-L. *Contrôlabilité Exacte, Perturbations et Stabilisation de Systèmes Distribués. Tome 1*; Recherches en Mathématiques Appliquées: Paris, France, 1988.
47. Huang, H.; Fu, X. Optimal control problems for a neutral integro-differential system with infinite delay. *Evol. Equ. Control Theory* **2022**, *11*, 177–197. [CrossRef]
48. Xi, X.X.; Hou, M.; Zhou, X.F.; Wen, Y. Approximate controllability of fractional neutral evolution systems of hyperbolic type. *Evol. Equ. Control Theory* **2022**, *11*, 1037–1069. [CrossRef]
49. Wen, Y.; Xi, X.X. Complete controllability of nonlinear fractional neutral functional differential equations. *Adv. Contin. Discret. Models* **2022**, *2022*, 1–11. [CrossRef]
50. Elshenhab, A.M.; Wang, X.; Cesarano, C.; Almarri, B.; Moaaz, O. Finite-time stability analysis of fractional delay system. *Mathematics* **2022**, *10*, 1883. [CrossRef]

mathematics

MDPI

Article

Asymptotic Regularity and Existence of Time-Dependent Attractors for Second-Order Undamped Evolution Equations with Memory

Xuan Wang [1], Didi Hu [2,*] and Chenghua Gao [1]

1 College of Mathematics and Statistics, Northwest Normal University, Lanzhou 730070, China; wangxuan@nwnu.edu.cn (X.W.); gaochenghua@nwnu.edu.cn (C.G.)
2 State Key Laboratory of Cryospheric Science, Northwest Institute of Eco-Environment and Resources, Chinese Academy of Sciences, Lanzhou 730070, China
* Correspondence: hudidi@lzb.ac.com

Abstract: Our purpose in this article is to study the asymptotic behavior of undamped evolution equations with fading memory on time-dependent spaces. By means of the theory of processes on time-dependent spaces, asymptotic a priori estimate and the technique of operator decomposition and the existence and asymptotic regularity of time-dependent attractors are, respectively, established in the critical case. At the same time, we also obtain the asymptotic regularity of the solution.

Keywords: asymptotic regularity; time-dependent attractor; undamped evolution equation; fading memory

MSC: 34Q35; 35B40; 35B41

check for updates

Citation: Wang, X.; Hu, D.; Gao, C. Asymptotic Regularity and Existence of Time-Dependent Attractors for Second-Order Undamped Evolution Equations with Memory. *Mathematics* **2022**, *10*, 2198. https://doi.org/10.3390/math10132198

Academic Editors: Alexandra M.S.F. Galhano, António M. Lopes and Carlo Cattani

Received: 4 June 2022
Accepted: 21 June 2022
Published: 23 June 2022

1. Introduction

In this paper, we study the following n-dimensional system of undamped abstract evolution equations with memory:

$$\begin{cases} \varepsilon(t)u_{tt}+k(0)A^{\theta}u+\int_0^{\infty}k'(s)A^{\theta}u(t-s)\mathrm{d}s+f(u)=g(x), & (x,t)\in\Omega\times[\tau,+\infty), \\ u(x,t)=0, & x\in\partial\Omega,\ t\in[\tau,+\infty), \\ u(x,t)=u_0(x,t),\ u_t(x,t)=u_1(x,t), & x\in\Omega,\ t\in(-\infty,\tau], \end{cases} \tag{1}$$

where Ω is a bounded domain of $\mathbb{R}^n$ $(n\geqslant 3)$ with a smooth boundary $\partial\Omega$, A is a Laplacian operator with the Dirichlet boundary condition, with domain $D(A)=H^2(\Omega)\cap H_0^1(\Omega)$, and

$$\theta\in(\frac{2n}{n+2},\frac{n}{2}),\ n\geqslant 3, \tag{2}$$

Through the linear time convolution of function $A^{\theta}u(\cdot)$ and memory kernel $k(\cdot)$, the fading memory term replaces the damping term and plays the role of energy dissipation in system (1). It follows that the solution semigroup (or the solution process) of undamped evolution equations with fading memory can generate a dissipative dynamical system.

Especially in recent years, one of the key problems in the study of abstract evolution equations with fading memory has been the asymptotic behavior of the solutions when time tends to infinity. Therefore, it has attracted the attention and research interest of many scholars (see, e.g., [1–4] and the references therein). Influenced by this, we also carried out a study of this issue. The problem in (1) we studied arises from isothermal viscoelasticity theory and describes the energy dissipation of an isotropic viscoelastic material (see, e.g., [5–10]). Therefore, it has a strong background in mathematical physics, and it can be naturally transformed into many concrete mathematical models such as the semilinear wave equation, Sine–Gordon equation, relative quantum mechanical equation, semilinear hyperbolic equation and the floating beam equation (see, e.g., [11–15]).

However, as far as we know, the undamped abstract evolution equations with fading memory is less considered. This is mainly because it is more difficult to verify the compactness of the solution semigroup (or solution process) and estimate the asymptotic regularity of solutions than in the damping case. Moreover, it is worth emphasizing that the energy dissipation of the system is only dependent on the fading memory term.

Regarding the abstract evolution equations with fading memory, known results are all in the case of $\varepsilon(t) \equiv 1$ and the asymptotic behavior of the solution can be studied by applying the usual dynamical system theory. Nevertheless, when $\varepsilon(\cdot)$ is a positive decreasing function, the standard theory fails to discuss the dissipative property involved in evolution equations. Therefore, the time-dependent terms are at a functional level; this can be found in (4).

To this end, let $\varepsilon(t)$ be a positive decreasing function which vanishes at infinity and satisfies:

$$\sup_{t\in\mathbb{R}}[|\varepsilon(t)| + |\varepsilon'(t)|] \leqslant L, \tag{3}$$

where $L > 0$. In this case, the natural energy functional associated with the system is defined in the standard way:

$$\mathfrak{E}(t) = \int_{\Omega} |A^{\frac{\theta}{2}}u|^2 dx + \varepsilon(t)\int_{\Omega} |u_t|^2 dx + \int_0^{\infty} \mu(s)\int_{\Omega} |A^{\frac{\theta}{2}}\eta^t|^2 dx ds, \tag{4}$$

which shows a structural dependence on time. Moreover, it is not hard to see that the vanishing property of $\varepsilon(\cdot)$ transforms the dissipative property and holds back the existence of absorbing sets in the usual sense, that is, the bounded sets of the phase space absorb all the trajectories after a certain period of time. In such a case, Conti et al. [16,17] put forward the notions and established theories of time-dependent attractors (the modified pullback attractors theory). The main idea is to obtain the existence of absorbing set and attractors by restricting the attraction domain of the compact pullback attracting family in the phase space.

By using the ideas in [16,17], some breakthrough progress was made in the research of the existence of time-dependent attractors and the regularity of solutions for the evolution equation problems. The semilinear wave equations have been treated in many papers, see, for example, [16,18–21]. Conti et al. [16] proved the existence and regularity of a time-dependent attractor, and they [18] obtained the asymptotic structure of a time-dependent attractor. Meng et al. [21,22] discussed and investigated the longtime dynamical behavior for the semilinear wave equation with nonlinear damping and the extensible Berger equation via a contractive function method, respectively. In addition, Meng et al. [20] gave some necessary and sufficient conditions to guarantee the existence of a time-dependent attractor. Liu et al. [23,24] considered the longtime dynamical behavior and achieved the existence of time-dependent attractor for the plate equation on a bounded domain or unbounded domain via an operator decomposition or a contractive function method, respectively. Furthermore, the time-dependent asymptotic behavior of the nonclassical reaction–diffusion equation was studied in [25,26].

Motivated by the ideas in [9,16,17], we were interested in analyzing the dynamical behavior of the undamped abstract evolution equations with fading memory, under the assumption that the nonlinear term satisfies critical growth. It is worth mentioning that the asymptotic regularity of the solutions and time-dependent attractors for the problem in (1) are discussed and investigated firstly in our paper.

The main goal of the present paper was to study the asymptotic behavior of the solutions of system (1). For the existence of time-dependent attractors, the compactness verification of the family of processes is a key ingredient. However, the critical nonlinearity, the memory space that lacks compactness and A^{θ} that is a fractional operator all contribute to the essential difficulties of the compactness verification. Furthermore, it seems hard to directly apply the methods of [9,16,17] to verify the asymptotic compactness in the time-dependent function space. Therefore, it is very important to study how to handle these natural difficulties brought by the critical nonlinearity, noncompact memory term and fractional operator in the undamped model when verifying the asymptotic compactness. At the same time, this is also a main problem in the research of the asymptotic behavior of nonlinear dynamical systems. By applying the process theory of time-dependent space, asymptotic a priori estimate and the technique of operator decomposition, we conquer the above difficulties, verify the compactness of the process and obtain our main results (see Theorem 5, Lemma 9 and Theorem 6).

The structure of the paper is as follows: in preliminary Section 2, we give a definition of some function sets, present the assumptions and recall some known abstract results; in Section 3, we state and prove our main results on the existence and regularity of a time-dependent attractor and the asymptotic regularity of solutions for system (1).

2. Preliminaries

In this section, we introduce some notations and abstract results about a time-dependent dynamical system.

Let $H = L^2(\Omega)$ and let $A = -\Delta$. A can be viewed as a self-adjoint and unbounded operator in H with domain $D(A)$.

We presume that $\{\lambda_j\}_{j\in\mathbb{N}}$ and $\{\omega_j\}_{j\in\mathbb{N}}$ are eigenvalues and eigenvectors of A, then $\{\omega_j\}_{j\in\mathbb{N}}$ can form a group of orthogonal basis of H, and

$$\begin{cases} A\omega_j = \lambda_j\omega_j, \\ 0 < \lambda_1 < \lambda_2 \leqslant \ldots \leqslant \lambda_j,\ \lambda_j \to \infty, \text{ as } j \to \infty. \end{cases}$$

Define the powers A^θ of A with domain $D(A^\theta) \subset H$ as follows:

$$D(A^\theta) = \{u \in H, \sum_{j=1}^{\infty} \lambda_j^{2\theta}(u,\omega_j)^2 < \infty\}, \tag{5}$$

and

$$\langle u,v\rangle_{2\theta} = \langle A^\theta\cdot, A^\theta\cdot\rangle, \quad \|u\|_{2\theta}^2 = \|A^\theta\cdot\|^2, \tag{6}$$

here, $\langle\cdot,\ \cdot\rangle_{2\theta}$ and $\|\cdot\|_{2\theta}$ are the inner product and norm in $D(A^\theta)$. Obviously, A^θ is also unbounded and self-adjoint.

Set $V_\theta = D(A^{\frac{\theta}{2}})$, for $\theta \in (\frac{2n}{n+2}, \frac{n}{2})$. Then, $V_0 = H = L^2(\Omega)$, $D(A^{\frac{\theta}{2}}) = V_\theta$, $D(A^{-\frac{\theta}{2}}) = V_{-\theta}$. In the paper, we assume that the forcing term $g(x)$ only belongs to $V_{-\theta}$. The spaces H and V_θ are endowed with the following inner products and norms, respectively:

$$\langle u,v\rangle = \int_\Omega u(x)v(x)\mathrm{d}x, \quad \|u\|^2 = \int_\Omega |u(x)|^2\mathrm{d}x, \quad \forall\, u,\, v \in H; \tag{7}$$

$$\langle u,v\rangle_\theta = \int_\Omega A^{\frac{\theta}{2}}u(x)A^{\frac{\theta}{2}}v(x)\mathrm{d}x, \quad \|u\|_\theta^2 = \int_\Omega |A^{\frac{\theta}{2}}u(x)|^2\mathrm{d}x, \quad \forall\, u,\, v \in V_\theta. \tag{8}$$

Therefore, we know that the compact embedding is

$$V_{\theta_1} \hookrightarrow V_{\theta_2}, \text{ as } \theta_1 > \theta_2, \tag{9}$$

the continuous embedding is

$$V_\theta \hookrightarrow L^{\frac{2n}{n-2\theta}}, \tag{10}$$

and the following Poincaré inequality holds:

$$\lambda_1^{\theta}\int_{\Omega}|v|^2\mathrm{d}x \leqslant \int_{\Omega}|A^{\frac{\theta}{2}}v|^2\mathrm{d}x, \quad \forall v\in V_{\theta}. \tag{11}$$

In addition, concerning the memory kernel function in system (1), we presume that $k(\infty)=1,\ k'(s)<0,\ \forall s\in\mathbb{R}^+$.

Suppose also that $\mu(s)=-k'(s)$ and that it satisfies:

$$\mu(s)\in C^1(\mathbb{R}^+)\cap L^1(\mathbb{R}^+),\ \mu(s)\geqslant 0,\ \mu'(s)\leqslant 0, \quad \forall s\in\mathbb{R}^+; \tag{12}$$

$$\int_0^{\infty}\mu(s)\mathrm{d}s = k_0; \tag{13}$$

$$\mu'(s)+\delta\mu(s)\leqslant 0, \quad \forall s\in\mathbb{R}^+, \tag{14}$$

where k_0, δ are two positive constants. Furthermore, consequently, the kernels $k(s)$ and $\mu(s)$ decay to zero with an exponential rate.

Hereafter, we introduce a new unknown function $\eta^t(x,s)$ and let it be equal to $u(x,t)-u(x,t-s)$, $(x,s)\in\Omega\times\mathbb{R}$, $t\in[\tau,+\infty)$. In virtue of the presumption $k(\infty)=1$, then the problem in (1) can be written in the form:

$$\begin{cases} \varepsilon(t)u_{tt}+A^{\theta}u+\int_0^{\infty}\mu(s)A^{\theta}\eta^t(s)\mathrm{d}s+f(u)=g(x),\ t\in[\tau,+\infty), \\ \eta_t^t=-\eta_s^t+u_t,\ t\in[\tau,+\infty), \end{cases} \tag{15}$$

with the initial-boundary conditions are:

$$\begin{cases} u(x,t)=0,\ \eta^t(x,s)=0, & x\in\partial\Omega, t\in[\tau,+\infty), \\ u(x,t)=u_0(x,t),\ u_t(x,t)=u_1(x,t), & x\in\Omega,\ t\in(-\infty,\tau], \\ \eta^{\tau}(x,s)=u_0(x,\tau)-u_0(x,\tau-s), & (x,s)\in\Omega\times\mathbb{R}^+, \end{cases} \tag{16}$$

where the unknown function $u(\cdot)$ satisfies the condition as follows: there exist two positive constants $\mathcal{R}$ and $\varrho=\min\{\frac{\delta}{2},\frac{\lambda_1}{2}\}$, such that

$$\int_0^{\infty}e^{-\varrho s}\|\nabla u(-s)\|^2\mathrm{d}s\leqslant\mathcal{R}, \tag{17}$$

were $\|\cdot\|$ denotes L^2-norm, and λ_1 is the first eigenvalue of the operator $-\Delta$ with a Dirichlet boundary condition.

We assume that $g\in V_{-\theta}$ and the nonlinear function $f(v)\in C^1(\mathbb{R})$ with $f(0)=0$ satisfies the following conditions.

Growth condition:

$$|f'(s)|\leqslant C(1+|s|^p),\forall s\in\mathbb{R}, \begin{cases} 0\leqslant p\leqslant\frac{4}{n-2}, & n\geqslant 3, \\ p\geqslant 0 \text{ and is arbitrary }, & n=1,2. \end{cases} \tag{18}$$

The assumption in (18) will be used to verify the compactness about the solution process.

Dissipation condition:

$$\liminf_{|s|\to\infty}\frac{f(s)}{s}>-\lambda_1^{\theta}, \tag{19}$$

and in view of (19), we obtain

$$2\langle F(s),1\rangle\geqslant-(1-\nu)\|s\|_{\theta}^2-C^*,$$

and we also presume that

$$2\langle f(s),s\rangle \geqslant 2\langle F(s),1\rangle-(1-\nu)\|s\|_{\theta}^{2}-C^{*}, \tag{20}$$

where $0<\nu<1,\ C^{*}>0, F(u)=\int_0^u f(r)\mathrm{d}r$ and $\|\cdot\|_{\theta}$ is the norm of V_{θ}.

Considering the assumption about memory kernel $\mu(\cdot)$, let $L_{\mu}^{2}(\mathbb{R}^{+};V_{\theta})$ be the family of Hilbert spaces of the V_{θ}-valued functions on $\mathbb{R}^{+}$. The scalar product and norm are defined by the formula:

$$\langle\varphi,\psi\rangle_{\mu,\theta}=\int_0^{\infty}\mu(s)\int_{\Omega}A^{\frac{\theta}{2}}\varphi A^{\frac{\theta}{2}}\psi\mathrm{d}x\mathrm{d}s,\quad \|\varphi\|_{\mu,\theta}^{2}=\int_0^{\infty}\mu(s)\int_{\Omega}|A^{\frac{\theta}{2}}\varphi|^{2}\mathrm{d}x\mathrm{d}s. \tag{21}$$

Then, we introduce the family of Hilbert spaces $\mathcal{H}_t^{\theta+\sigma}$

$$\mathcal{H}_t^{\theta+\sigma}=V_{\theta+\sigma}\times V_{\sigma}\times L_{\mu}^{2}(\mathbb{R}^{+};V_{\theta+\sigma}),$$

and endowed norm

$$\|z\|_{\mathcal{H}_t^{\theta+\sigma}}^{2}=\|(u,u_t,\eta^{t})\|_{\mathcal{H}_t^{\theta+\sigma}}^{2}=\|u\|_{\theta+\sigma}^{2}+\varepsilon(t)\|u_t\|_{\sigma}^{2}+\|\eta^{t}\|_{\mu,\theta+\sigma}^{2}. \tag{22}$$

Clearly, when $\sigma\equiv 0$, the family of Hilbert spaces $\mathcal{H}_t^{\theta}$ is defined by:

$$\mathcal{H}_t^{\theta}=V_{\theta}\times H\times L_{\mu}^{2}(\mathbb{R}^{+};V_{\theta}), \tag{23}$$

endowed with the norm:

$$\|z\|_{\mathcal{H}_t^{\theta}}^{2}=\|(u,u_t,\eta^{t})\|_{\mathcal{H}_t^{\theta}}^{2}=\|u\|_{\theta}^{2}+\varepsilon(t)\|u_t\|^{2}+\|\eta^{t}\|_{\mu,\theta}^{2}. \tag{24}$$

By use of assumptions (12)–(14), we can gain the preliminary result as follows ([27]).

Lemma 1. *If assumptions (12)–(14) about the memory kernel function $\mu(s)$ hold, then for any $\eta^{t}\in C^{1}([\tau,t];L_{\mu}^{2}(\mathbb{R}^{+};V_{r}))$, $0<r\leqslant 2\theta$, $\forall t\geqslant\tau$, $\theta\in(\frac{2n}{n+2},\frac{n}{2})$, there exists a positive constant δ, such that $\langle\eta^{t},\eta_s^{t}\rangle_{\mu,r}\geqslant\dfrac{\delta}{2}\|\eta^{t}\|_{\mu,r}^{2}$.*

We also need the following abstract results to prove the existence of time-dependent attractors.

Lemma 2 ([28]). *Let $(\mathcal{M},\mathrm{d})$ be a metric space and also let $U(t,\tau)$ be a Lipschitz continuous dynamical process in $\mathcal{M}$, i.e.,*

$$\mathrm{d}(U(t,\tau)m_1,U(t,\tau)m_2)\leqslant C\mathrm{e}^{K(t-\tau)}\mathrm{d}(m_1,m_2),$$

for appropriate constants C and K which are independent of m_i, t and τ. Assume further that there exist three subsets $M_1,M_2,M_3\subset\mathcal{M}$ such that

$$\mathrm{d}(U(t,\tau)M_1,U(t,\tau)M_2)\leqslant L_1\mathrm{e}^{-\nu_1(t-\tau)},$$

$$\mathrm{d}(U(t,\tau)M_2,U(t,\tau)M_3)\leqslant L_2\mathrm{e}^{-\nu_2(t-\tau)},$$

for some $\nu_1,\nu_2>0$ and $L_1,L_2>0$. Then, it follows that

$$\mathrm{d}(U(t,\tau)M_1,U(t,\tau)M_3)\leqslant L\mathrm{e}^{-\nu(t-\tau)},$$

where $\nu=\frac{\nu_1\nu_2}{K+\nu_1+\nu_2}$ and $L=CL_1+L_2$.

Lemma 3 ([13,29,30]). *Let $\mu(s) \in C^1(\mathbb{R}^+) \cap L^1(\mathbb{R}^+)$ be a nonnegative function that satisfies the following: if there exists $s_0 \in \mathbb{R}^+$ such that $\mu(s_0) = 0$, then $\mu(s) \equiv 0$, for all $s \geqslant s_0$. Moreover, let B_0, B_1 and B_2 be Banach spaces satisfying*

$$B_0 \hookrightarrow B_1 \hookrightarrow B_2,$$

where B_0 and B_1 are reflexive, and the embedding $B_0 \hookrightarrow B_1$ is compact. Assume that $\mathcal{C} \subset L^2_\mu(\mathbb{R}^+; B_1)$ and it satisfies

(i) $\mathcal{C} \subset L^2_\mu(\mathbb{R}^+; B_0) \cap H^1_\mu(\mathbb{R}^+; B_2)$;
(ii) $\sup_{\eta \in \mathcal{C}} \|\eta(s)\|^2_{B_1} \leqslant h(s)$, $\forall\, s \in \mathbb{R}^+$, $h(s) \in L^1_\mu(\mathbb{R}^+)$.

Then, $\mathcal{C}$ is relatively compact in $L^2_\mu(\mathbb{R}^+; B_1)$.

Subsequently, we review some basic concepts and abstract results about a process on a time-dependent system ([16,18,25]), which are used to study the long-time behavior of solutions.

Definition 1. *Let X_t be a family of normed spaces. A two-parameter family of operators $\{U(t,\tau) : X_\tau \to X_t, \tau \leqslant t, \tau \in \mathbb{R}\}$ is said to be a process, if for any $\tau \in \mathbb{R}$,*

(i) *$U(\tau,\tau) = \mathrm{Id}$ is the identity operator on X_τ;*
(ii) *$U(t,s)U(s,\tau) = U(t,\tau)$, $\forall\, \tau \leqslant s \leqslant t$.*

Assume that X_t is a family of normed spaces. For every $t \in \mathbb{R}$, we introduce the R-ball of X_t:

$$\mathbb{B}_t(R) = \{z \in X_t \,|\, \|z\|_{X_t} \leqslant R\}.$$

The Hausdorff semidistance of sets A, B of X_t is denoted by:

$$\delta_t(A,B) = \sup_{x \in A} \mathrm{dist}_{X_t}(x,B) = \sup_{x \in A} \inf_{y \in B} \|x - y\|_{X_t}.$$

Definition 2. *A family $\mathfrak{C} = \{C_t\}_{t\in\mathbb{R}}$ of bounded sets $C_t \subset X_t$ is called uniformly bounded, if there exists a constant $R > 0$ such that $C_t \subset \mathbb{B}_t(R)$, $\forall\, t \in \mathbb{R}$.*

Definition 3. *A uniformly bounded family $\mathfrak{B}_t = \{\mathbb{B}_t(R_0)\}_{t\in\mathbb{R}}$ is called a time-dependent absorbing set for the process $U(t,\tau)$, if for every $R > 0$, there exist a $t_0 = t_0(R) \leqslant t$ and $R_0 > 0$ such that*

$$\tau \leqslant t - t_0 \Rightarrow U(t,\tau)\mathbb{B}_\tau(R) \subset \mathbb{B}_t(R_0).$$

The process $U(t,\tau)$ is said to be dissipative as it possesses a time-dependent absorbing set.

Definition 4. *The smallest family $\mathfrak{A} = \{A_t\}_{t\in\mathbb{R}}$ is called a time-dependent attractor for the process $U(t,\tau)$, if $\mathfrak{A}$ satisfies the following properties:*

(i) *Each A_t is compact in X_t ;*
(ii) *$\mathfrak{A}$ is pullback attracting, that is, it is uniformly bounded, and the limit*

$$\lim_{\tau\to-\infty} \delta_t(U(t,\tau)C_\tau, A_t) = 0$$

holds for every uniformly bounded family $\mathfrak{C} = \{C_t\}_{t\in\mathbb{R}}$ and every $t \in \mathbb{R}$.

Theorem 1 ([16]). *If $U(t,\tau)$ is asymptotically compact, that is, the set*

$$\mathbb{K} = \{\mathfrak{K} = \{K_t\}_{t\in\mathbb{R}} |\ \text{Each } K_t \text{ is compact in } X_t,\ \mathfrak{K} \text{ is pullback attracting}\}$$

is not empty, then the time-dependent attractor $\mathfrak{A}$ *exists and coincides with* $\mathfrak{A} = \{A_t\}_{t\in\mathbb{R}}$. *In particular, it is unique.*

Definition 5. *A function* $t \to Z(t)$ *and* $Z(t) \in X_t$ *is a complete bounded trajectory* (CBT) *of the process* $U(t,\tau)$, *if and only if*

(i) $\sup_{t\in\mathbb{R}} \|Z(t)\|_{X_t} < \infty$;
(ii) $Z(t) = U(t,\tau)Z(\tau), \forall \tau \leqslant t,\ \tau \in \mathbb{R}$.

Definition 6. *A time-dependent attractor* $\mathfrak{A} = \{A_t\}_{t\in\mathbb{R}}$ *is invariant, if for all* $\tau \leqslant t$,

$$U(t,\tau)A_\tau = A_t.$$

Theorem 2. *If the time-dependent attractor* $\mathfrak{A} = \{A_t\}_{t\in\mathbb{R}}$ *of the process* $U(t,\tau)$ *is invariant, then it coincides with the set of all* CBTs *of the process* $U(t,\tau)$, *that is,*

$$\mathfrak{A} = \{Z | t \to Z(t) \in X_t \text{ and } Z(t) \text{ is a CBT of the process } U(t,\tau)\}.$$

3. Time-Dependent Global Attractor in $\mathcal{H}_t^\theta$

3.1. Well-Posedness

We start with the general existence and uniqueness of the solutions of the problem in (15) and (16). Based on the standard Faedo–Galerkin approximation method, see, e.g., [14], the following results can be easily obtained and the time-dependent function $\varepsilon(t)$ does not bring about any essential difficulties.

Theorem 3. *Let* Ω *be a bounded domain in* $\mathbb{R}^n$ *with smooth boundary* $\partial\Omega$. *If* (12)–(14) *and* (3) *hold,* $g \in V_{-\theta}$ *and* f *satisfies* (18)–(20), *then for any initial data* $z(\tau) = (u(\tau), u_t(\tau), \eta^\tau(s)) \in \mathcal{H}_\tau^\theta$, *there exists a unique solution* $z(t) = (u(t), u_t(t), \eta^t(s)) \in L^\infty([\tau,t];\mathcal{H}_t^\theta) \cap C([\tau,t];\mathcal{H}_t^\theta)$ *of the problem in* (15) *and* (16), *in the sense that*

$$u \in C([\tau,t];V_\theta), \quad u_t \in L^2([\tau,t];V_\theta), \quad \eta^t \in C([\tau,t];L^2_\mu(\mathbb{R}^+;V_\theta)),$$
$$\eta_t^t + \eta_s^t \in L^\infty([\tau,t];L^2_\mu(\mathbb{R}^+;H)) \cap L^2([\tau,t];L^2_\mu(\mathbb{R}^+;V_\theta))$$

and

$$\begin{cases} \langle \varepsilon(t)u_{tt}, v\rangle + \langle u, v\rangle_\theta + \langle \eta^t(s), v\rangle_{\mu,\theta} + \langle f(u), v\rangle = \langle g, v\rangle, \\ \langle \eta_t^t(s) + \eta_s^t(s), \varphi(s)\rangle_{\mu,\theta} = \langle u_t, \varphi(s)\rangle_{\mu,\theta}, \end{cases}$$

for all $t \geqslant \tau$ *and any* $v \in V_\theta$, $\varphi \in L^2_\mu(\mathbb{R}^+;V_\theta)$.

Moreover, for any $t \geqslant \tau$, the mapping $z(\tau) \mapsto z(t)$ is continuous from $\mathcal{H}_\tau^\theta$ to $\mathcal{H}_t^\theta$.
By Theorem 3, we can define a process $U(t,\tau)$ as follows:

$$z(t) = U(t,\tau)z(\tau) : \mathcal{H}_\tau^\theta \to \mathcal{H}_t^\theta,$$

which is continuous from $\mathcal{H}_\tau^\theta$ to $\mathcal{H}_t^\theta$.

In the next subsection, we prove that $U(t,\tau)$ satisfies the continuous dependence property on the initial data.

Lemma 4. *Let* $z_i(t)$, $i = 1,2$, *be the corresponding solutions of the problem in* (15) *and* (16) *with* $z_i(\tau) \in \mathcal{H}_\tau^\theta$ *satisfying* $\|z_i(\tau)\|_{\mathcal{H}_\tau^\theta} \leqslant R$, $i = 1,2$. *Suppose that* (12)–(14) *and* (3) *hold. If* $g \in V_{-\theta}$ *and* f *satisfies* (18)–(20), *then there exists a positive constant* K, *such that the following estimate holds:*

$$\|z_1(t) - z_2(t)\|^2_{\mathcal{H}^\theta_t} = \|U(t,\tau)z_1(\tau) - U(t,\tau)z_2(\tau)\|^2_{\mathcal{H}^\theta_t}$$
$$\leqslant Ce^{K(t-\tau)}\|z_1(\tau) - z_2(\tau)\|^2_{\mathcal{H}^\theta_\tau}, \quad \forall \tau \leqslant t.$$

3.2. Time-Dependent Absorbing Set in $\mathcal{H}^\theta_t$

In the subsequent estimates, we presume that $0 < \rho < 1$. Furthermore, we prove the following dissipative estimate.

Theorem 4. *Under the assumption of Lemma 4, for any initial data $z(\tau) \in \mathbb{B}_\tau(R) \subset \mathcal{H}^\theta_\tau$, then there exists $R_0 > 0$, such that the process $U(t,\tau)$ corresponding to the problem in* (15) *and* (16) *possesses a time-dependent absorbing set, namely, the family $\mathfrak{B}_t = \{\mathbb{B}_t(R_0)\}_{t\in\mathbb{R}}$.*

Proof. Multiplying (15) by $2(u_t + \rho u)$ and integrating over Ω, we obtain

$$\langle \varepsilon(t)u_{tt}, 2(u_t + \rho u)\rangle + \langle A^\theta u, 2(u_t + \rho u)\rangle + \langle \int_0^\infty \mu(s)A^\theta \eta^t(s)\mathrm{d}s, 2(u_t + \rho u)\rangle$$
$$+ \langle f(u), 2(u_t + \rho u)\rangle = \langle g, 2(u_t + \rho u)\rangle. \tag{25}$$

Thanks to *Lemma* 1, we have

$$\langle \int_0^\infty \mu(s)A^\theta \eta^t(s)\mathrm{d}s, 2u_t\rangle = \int_\Omega \int_0^\infty 2(\eta^t_t + \eta^t_s)\mu(s)A^\theta\eta^t(s)\mathrm{d}s\mathrm{d}x$$
$$\geqslant \frac{\mathrm{d}}{\mathrm{d}t}\|\eta^t\|^2_{\mu,\theta} + \delta\|\eta^t\|^2_{\mu,\theta}, \tag{26}$$

combining with the Hölder inequality, Cauchy inequality and (13), we obtain that

$$\langle \int_0^\infty \mu(s)A^\theta \eta^t(s)\mathrm{d}s, 2\rho u\rangle = 2\rho\int_\Omega u \int_0^\infty \mu(s)A^\theta\eta^t(s)\mathrm{d}s\mathrm{d}x$$
$$\geqslant -\frac{\rho\nu}{2}\int_\Omega |A^{\frac{\theta}{2}}u|^2\mathrm{d}x - \frac{2\rho}{\nu}\int_\Omega (\int_0^\infty \mu(s)|A^{\frac{\theta}{2}}\eta^t(s)|\mathrm{d}s)^2\mathrm{d}x$$
$$\geqslant -\frac{\rho\nu}{2}\|u\|^2_\theta - \frac{2\rho k_0}{\nu}\|\eta^t\|^2_{\mu,\theta}. \tag{27}$$

Therefore, we obtain from (20) that

$$\frac{\mathrm{d}}{\mathrm{d}t}(\|u\|^2_\theta + \varepsilon(t)\|u_t\|^2 + \|\eta^t\|^2_{\mu,\theta} + 2\rho\varepsilon(t)\langle u_t, u\rangle + 2\langle F(u), 1\rangle - 2\langle g, u\rangle + C)$$
$$+ \rho(\|u\|^2_\theta + \varepsilon(t)\|u_t\|^2 + \|\eta^t\|^2_{\mu,\theta} + 2\rho\varepsilon(t)\langle u_t, u\rangle + 2\langle F(u), 1\rangle - 2\langle g, u\rangle + C) \tag{28}$$
$$+ \frac{\rho\nu}{2}\|u\|^2_\theta - (\varepsilon'(t) + 3\rho\varepsilon(t))\|u_t\|^2 + (\delta - \frac{2k_0\rho}{\nu} - \rho)\|\eta^t\|^2_{\mu,\theta}$$
$$- 2\rho(\varepsilon'(t) + \rho\varepsilon(t))\langle u_t, u\rangle \leqslant \rho(C + C^*).$$

The functional is defined by the formula:

$$\mathcal{M}(t) = \|u\|^2_\theta + \varepsilon(t)\|u_t\|^2 + \|\eta^t\|^2_{\mu,\theta} + 2\rho\varepsilon(t)\langle u_t, u\rangle + 2\langle F(u), 1\rangle - 2\langle g, u\rangle + C, \tag{29}$$

where $C = \frac{2}{\nu\rho}\|g\|^2_{V_{-\theta}} + C^*$.

Then, we deduce from (3) and (11) that

$$2\rho\varepsilon(t)\langle u_t, u\rangle \leqslant 2\rho\varepsilon(t)|\langle u_t, u\rangle| \leqslant \frac{\rho\nu}{2}\|u\|^2_\theta + \frac{2\rho L}{\nu\lambda_1^\theta}\varepsilon(t)\|u_t\|^2,$$

and

$$\pm 2\langle g,u\rangle \leqslant 2|\langle g,u\rangle| \leqslant \frac{\rho\nu}{2}\|u\|_\theta^2 + \frac{2}{\nu\rho}\|g\|_{V_{-\theta}}^2.$$

Using (19), for some $0<\nu<1$, we have

$$2\langle F(u),1\rangle \geqslant -(1-\nu)\|u\|_\theta^2 - C^*.$$

Thus, choosing ρ small enough, we obtain that $\mathcal{M}(t)\geqslant 0$.

Namely,

$$\begin{aligned}&\frac{\mathrm{d}}{\mathrm{d}t}\mathcal{M}(t)+\rho\mathcal{M}(t)+\frac{\rho\nu}{2}\|u\|_\theta^2-(\varepsilon'(t)+3\rho\varepsilon(t))\|u_t\|^2+(\delta-\frac{2k_0\rho}{\nu}-\rho)\|\eta^t\|_{\mu,\theta}^2\\&\quad-2\rho(\varepsilon'(t)+\rho\varepsilon(t))\langle u_t,u\rangle\leqslant\rho(C+C^*).\end{aligned}\tag{30}$$

By (3) and (11), we find that

$$-2\rho(\varepsilon'(t)+\rho\varepsilon(t))\langle u_t,u\rangle \geqslant -\frac{\rho\nu}{2}\|u\|_\theta^2-\frac{2\rho}{\nu\lambda_1^\theta}L^2\|u_t\|^2.\tag{31}$$

Hence,

$$\frac{\mathrm{d}}{\mathrm{d}t}\mathcal{M}(t)+\rho\mathcal{M}(t)\leqslant\rho(C+C^*),$$

that is,

$$\mathcal{M}(t)\leqslant \mathrm{e}^{-\rho(t-\tau)}\mathcal{M}(\tau)+C+C^*.\tag{32}$$

For $\epsilon>0$, from (18), $\theta\in(\frac{2n}{n+2},\frac{n}{2})$ and the interpolation inequality, we have

$$\begin{aligned}\langle F(u),1\rangle&=\int_\Omega F(u)\mathrm{d}x\\&\leqslant C\int_\Omega(|u|^2+|u|^{\frac{2n}{n-2}})\mathrm{d}x\\&\leqslant C(\|u\|^2+\|\nabla u\|^{\frac{2n}{n-2}})\\&\leqslant C\|u\|^2+C(C_\epsilon\|u\|+\epsilon\|u\|_\theta)^{\frac{2n}{n-2}}\\&\leqslant C\|u\|_\theta^{\frac{2n}{n-2}}.\end{aligned}$$

Thus, for a small enough ρ there exist positive constants C, C_1 and C_2, such that

$$C\|z(t)\|_{\mathcal{H}_t^\theta}^2-C_1\leqslant\mathcal{M}(t)\leqslant C\|z(t)\|_{\mathcal{H}_t^\theta}^{\frac{2n}{n-2}}+C_2.\tag{33}$$

Combining with (32), there exists a constant $N_1>0$ such that

$$\|U(t,\tau)z(\tau)\|_{\mathcal{H}_t^\theta}^2\leqslant Q_1(\|z(\tau)\|_{\mathcal{H}_\tau^\theta})\mathrm{e}^{-\rho(t-\tau)}+N_1,$$

where $Q_1(\cdot)$ is an increasing positive function. Because $z(\tau)\in\mathbb{B}_\tau(R)$, the following inequality is valid

$$\|U(t,\tau)z(\tau)\|_{\mathcal{H}_t^\theta}^2\leqslant Q_1(R)\mathrm{e}^{-\rho(t-\tau)}+N_1\leqslant 2N_1=R_0^2,$$

provided that $\tau\leqslant t-t_0$, where $t_0=\max\{0,\frac{1}{\rho}\ln\frac{2Q_1(R)}{N_1}\}$.

This completes the proof. □

Proof. Proof of Lemma 4:

Assume that the initial data $z_i(\tau)$, $i = 1,2$, satisfy $\|z_i(\tau)\|_{\mathcal{H}_\tau^\theta} \leqslant R$. It follows from Theorem 4 that

$$\|z_i(t)\|_{\mathcal{H}_t^\theta} = \|U(t,\tau)z_i(\tau)\|_{\mathcal{H}_t^\theta} \leqslant R_0. \tag{34}$$

We substitute $\bar{z}(t) = (\bar{u}(t), \bar{u}_t(t), \bar{\eta}^t(s)) = z_1(t) - z_2(t)$ into (15). Then,

$$\varepsilon(t)\bar{u}_{tt} + A^\theta \bar{u} + \int_0^\infty \mu(s) A^\theta \bar{\eta}^t(s) ds + f(u_1) - f(u_2) = 0 \tag{35}$$

and $\bar{z}(\tau) = z_1(\tau) - z_2(\tau)$.

Taking the scalar product of (35) with $2\bar{u}_t(t)$, we have

$$\frac{\mathrm{d}}{\mathrm{d}t}\|\bar{z}\|_{\mathcal{H}_t^\theta}^2 - \varepsilon'(t)\|\bar{u}_t\|^2 + \delta\|\bar{\eta}^t\|_{\mu,\theta}^2 \leqslant -2\langle f(u_1) - f(u_2), \bar{u}_t\rangle. \tag{36}$$

From (18), (10) and (34), we have

$$\begin{aligned}
&-2\langle f(u_1) - f(u_2), \bar{u}_t\rangle \\
&\leqslant 2\int_\Omega |f'(\xi)||\bar{u}||\bar{u}_t| dx \\
&\leqslant 2C\Big(\int_\Omega (1 + |u_2|^{\frac{4}{n-2}} + |u_1|^{\frac{4}{n-2}})^{\frac{n}{\theta}} dx\Big)^{\frac{\theta}{n}} \Big(\int_\Omega |\bar{u}|^{\frac{2n}{n-2\theta}} dx\Big)^{\frac{n-2\theta}{2n}} \Big(\int_\Omega |\bar{u}_t|^2 dx\Big)^{\frac{1}{2}} \\
&\leqslant 2C(1 + \|u_2\|_\theta^{\frac{4}{n-2}} + \|u_1\|_\theta^{\frac{4}{n-2}})\|\bar{u}\|_\theta \|\bar{u}_t\| \\
&\leqslant C(\|\bar{u}\|_\theta^2 + \|\bar{u}_t\|^2),
\end{aligned} \tag{37}$$

where $\frac{2n}{n-2\theta} > \frac{4n}{(n-2)\theta}$. Substituting (37) into (36), we obtain

$$\begin{aligned}
\frac{\mathrm{d}}{\mathrm{d}t}\|\bar{z}(t)\|_{\mathcal{H}_t^\theta}^2 &\leqslant C(\|\bar{u}\|_\theta^2 + \|\bar{u}_t\|^2) \\
&= \frac{C}{\varepsilon(t)}(\varepsilon(t)\|\bar{u}\|_\theta^2 + \varepsilon(t)\|\bar{u}_t\|^2) \\
&\leqslant \frac{C}{\varepsilon(t)}(L\|\bar{u}\|_\theta^2 + \varepsilon(t)\|\bar{u}_t\|^2) \\
&\leqslant \frac{C(L+1)}{\varepsilon(t)}\|\bar{z}(t)\|_{\mathcal{H}_t^\theta}^2.
\end{aligned}$$

Applying the Gronwall lemma, we finally have

$$\begin{aligned}
\|z_1(t) - z_2(t)\|_{\mathcal{H}_t^\theta}^2 &\leqslant \|\bar{z}(\tau)\|_{\mathcal{H}_\tau^\theta}^2 \cdot \mathrm{e}^{C(L+1)\int_\tau^t \frac{1}{\varepsilon(s)} ds} \\
&\leqslant C\mathrm{e}^{K(t-\tau)}\|z_1(\tau) - z_2(\tau)\|_{\mathcal{H}_\tau^\theta}^2.
\end{aligned} \tag{38}$$

The proof is completed. □

3.3. *The Existence of a Time-Dependent Attractor in* $\mathcal{H}_t^\theta$

Devoted to the difficulties arising from the critical exponent and noncompact memory space, in this subsection, we use the method of asymptotic a priori estimates and the technique of operator decomposition to verify the necessary compactness.

Since $H \hookrightarrow V_{-\theta}$ is dense, for every $g \in V_{-\theta}$ and any $\varrho > 0$, there exists $g^\varrho \in H$ which depends on g and ϱ, such that

$$\|g - g^\varrho\|_{V_{-\theta}} \leqslant \varrho. \tag{39}$$

Assuming (18)–(20) hold, we write $f = f_0 + f_1$, where f_0, $f_1 \in C^2(\mathbb{R})$ fulfill

$$f_0(0) = f_1(0) = 0, \tag{40}$$

$$2\langle f_0(s), s\rangle \geqslant 2\langle F_0(s), 1\rangle - \frac{1-\nu}{2}\|s\|_\theta^2 - \varrho, \tag{41}$$

from which we obtain

$$2\langle f_1(s), s\rangle \geqslant 2\langle F_1(s), 1\rangle - \frac{1-\nu}{2}\|s\|_\theta^2 - C^* + \varrho, \tag{42}$$

$$2\langle F_0(s), 1\rangle \geqslant -\frac{1-\nu}{2}\|s\|_\theta^2 - \varrho, \tag{43}$$

$$2\langle F_1(s), 1\rangle \geqslant -\frac{1-\nu}{2}\|s\|_\theta^2 - C^* + \varrho, \tag{44}$$

where $0 < 1-\nu < \lambda_1^\theta$, $C^* > 0$, $F_i(u) = \int_0^u f_i(r)\mathrm{d}r, i = 1,2,$ and $\|\cdot\|_\theta$ is the norm of V_θ. Furthermore, in the space $\mathcal{H}_t^\theta$, we assume that

$$|f_0'(s)| \leqslant C(1+|s|^p), \forall s \in \mathbb{R},\ 0 \leqslant p \leqslant \frac{4}{n-2},\ n \geqslant 3, \tag{45}$$

$$|f_1'(s)| \leqslant C(1+|s|^p), \forall s \in \mathbb{R},\ 0 \leqslant p < \frac{4}{n-2},\ n \geqslant 3. \tag{46}$$

Let $\mathfrak{B}_t = \{\mathbb{B}_t(R_0)\}_{t\in\mathbb{R}}$ be a time-dependent absorbing set obtained in Theorem 4. For a fixed $\tau \in \mathbb{R}$ and any $\|z(\tau)\|_{\mathcal{H}_\tau^\theta} \leqslant R$, we decompose the solution $z(t) = (u(t), u_t(t), \eta^t)$ of the problem in (15) and (16) as follows:

$$z(t) = U(t,\tau)z(\tau) = V(t,\tau)z_1(\tau) + W(t,\tau)z_2(\tau) = z_1(t) + z_2(t),$$

where

$$z_1(t) = (v(t), v_t(t), \zeta^t(s)), \quad z_2(t) = (w(t), w_t(t), \xi^t(s))$$

satisfy

$$\begin{cases} \varepsilon(t)v_{tt} + A^\theta v + \int_0^\infty \mu(s)A^\theta\zeta^t(s)\mathrm{d}s + f_0(v) = g - g^\varrho, \\ \zeta_t^t = -\zeta_s^t + v_t, \\ v(x,t)|_{\partial\Omega} = 0,\ \zeta^t(x,t)|_{\partial\Omega} = 0, \\ v(x,\tau) = u_0(x,t),\ v_t(x,\tau) = u_1(x,t),\ x \in \Omega,\ t \leqslant \tau, \\ \zeta^\tau(x,s) = u_0(x,\tau) - u_0(x,\tau - s),\ (x,s) \in \Omega \times \mathbb{R}^+, \end{cases} \tag{47}$$

and

$$\begin{cases} \varepsilon(t)w_{tt} + A^\theta w + \int_0^\infty \mu(s)A^\theta\xi^t(s)\mathrm{d}s + f(u) - f_0(v) = g^\varrho, \\ \xi_t^t = -\xi_s^t + w_t, \\ w(x,t)|_{\partial\Omega} = 0,\ \xi^t(x,t)|_{\partial\Omega} = 0, \\ w(x,\tau) = 0,\ w_t(x,\tau) = 0,\ \xi^\tau(x,s) = 0. \end{cases} \tag{48}$$

By the Galerkin approximation method, the existence and uniqueness of the solution of (47) and (48) can be obtained.

Furthermore, akin to the proof of *Theorem* 4, for the solution $z_1(t)$ of (47), we get

Lemma 5. *Let $z_1(t)$ be the solution of problem* (47) *with initial data $z_1(\tau) = z(\tau)$ satisfying $\|z(\tau)\|^2_{\mathcal{H}^\theta_\tau} \leqslant R$. Suppose that $g \in V_{-\theta}$ and* (12)–(14), (3) *hold. If f_0 satisfies* (40), (41), (43) *and* (45), *for $\epsilon > 0$, then there exists a positive constant $\varrho = \varrho(\epsilon)$, such that the solution of problem* (47) *satisfies*

$$\|V(t,\tau)z(\tau)\|^2_{\mathcal{H}^\theta_t} \leqslant Q_2(\|z(\tau)\|_{\mathcal{H}^\theta_\tau})e^{-\rho_1(t-\tau)} \leqslant 2\epsilon, \text{ as } \tau \leqslant t - t_1. \tag{49}$$

where $Q_2(\cdot)$ is an increasing positive function, $\rho_1 = \rho_1(\|\mathfrak{B}_t\|_{\mathcal{H}^\theta_t})$ is small enough and $t_1 = t_1(\epsilon, Q_2(R), \rho_1)$.

Proof. Taking the inner product of (47) with $2(v_t(t) + \rho_1 v(t))$, we get

$$\begin{aligned}
&\frac{d}{dt}(\|v\|^2_\theta + \varepsilon(t)\|v_t\|^2 + \|\zeta^t\|^2_{\mu,\theta} + 2\rho_1\varepsilon(t)\langle v, v_t\rangle + 2\langle F_0(v), 1\rangle - 2\langle g - g^\varrho, v\rangle + C) \\
&+ \rho_1(\|v\|^2_\theta + \varepsilon(t)\|v_t\|^2 + \|\zeta^t\|^2_{\mu,\theta} + 2\rho_1\varepsilon(t)\langle v, v_t\rangle + 2\langle F_0(v), 1\rangle - 2\langle g - g^\varrho, v\rangle + C) \\
&+ \frac{\rho_1 \nu}{2}\|v\|^2_\theta - (\varepsilon'(t) + 3\rho_1\varepsilon(t))\|v_t\|^2 + (\delta - 2\rho_1 k_0 - \rho_1)\|\zeta^t\|^2_{\mu,\theta} \\
&- 2\rho_1(\varepsilon'(t) + \rho_1\varepsilon(t))\langle v, v_t\rangle \\
&\leqslant \rho_1(\varrho + C),
\end{aligned} \tag{50}$$

where $F_0(s) = \int_0^s f_0(r)dr$.

We define the functional as follows:

$$\mathcal{M}_1(t) = \|v\|^2_\theta + \varepsilon(t)\|v_t\|^2 + \|\zeta^t\|^2_{\mu,\theta} + 2\rho_1\varepsilon(t)\langle v, v_t\rangle + 2\langle F_0(v), 1\rangle - 2\langle g - g^\varrho, v\rangle + C, \tag{51}$$

where $C = 8\|g - g^\varrho\|^2_{V_{-\theta}} + \varrho = 8\varrho^2 + \varrho$.

In fact, by virtue of (41) and (45), we have $-\frac{1-\nu}{2}\|v\|^2_\theta - \varrho \leqslant 2\langle F_0(v), 1\rangle \leqslant C\|v\|^{\frac{2n}{n-2}}_\theta$. Similarly, we can deduce that

$$2\rho_1\varepsilon(t)\langle v, v_t\rangle \leqslant 2\rho_1\varepsilon(t)|\langle v, v_t\rangle| \leqslant \frac{1}{8}\|v\|^2_\theta + \frac{8\rho_1^2 L}{\lambda_1^\theta}\varepsilon(t)\|v_t\|^2,$$

$$2\langle g - g^\varrho, v\rangle \leqslant 2|\langle g - g^\varrho, v\rangle| \leqslant \frac{1}{8}\|v\|^2_\theta + 8\|g - g^\varrho\|^2_{V_{-\theta}}.$$

Then,

$$\frac{1}{4}\|V(t,\tau)z(\tau)\|^2_{\mathcal{H}^\theta_t} \leqslant \mathcal{M}_1(t) \leqslant C\|V(t,\tau)z(\tau)\|^{\frac{2n}{n-2}}_{\mathcal{H}^\theta_t} + 16\varrho^2 + \varrho. \tag{52}$$

From (3) and (11), we have

$$-2\rho_1(\varepsilon'(t) + \rho_1\varepsilon(t))\langle v, v_t\rangle \geqslant -\frac{\rho_1\nu}{4}\|v\|^2_\theta - \frac{4\rho_1 L^2}{\lambda_1^\theta \nu}\|v_t\|^2.$$

Choosing ρ_1 small enough, we have

$$-\varepsilon'(t) - 3\rho_1\varepsilon(t) - \frac{4\rho_1 L^2}{\lambda_1^\theta \nu} \geqslant 0, \quad \delta - 2\rho_1 k_0 - \rho_1 \geqslant 0.$$

Combining with the above estimates, we get

$$\frac{\mathrm{d}}{\mathrm{d}t}\mathcal{M}_1(t) + \rho_1 \mathcal{M}_1(t) \leqslant \rho_1(8\varrho^2 + 2\varrho). \tag{53}$$

Taking $\epsilon = 32\varrho^2 + 2\varrho + C(64\varrho^2 + 4\varrho)$, we obtain from (52) and (53)

$$\|V(t,\tau)z_1(\tau)\|^2_{\mathcal{H}^\theta_t} \leqslant Q_2(\|z(\tau)\|_{\mathcal{H}^\theta_\tau})e^{-\rho_1(t-\tau)} + \epsilon.$$

Due to $z(\tau) \in \mathbb{B}_\tau(R)$, we find the estimate

$$\|V(t,\tau)z(\tau)\|^2_{\mathcal{H}^\theta_t} \leqslant 2\epsilon,$$

provided that $\tau \leqslant t - t_1$, where $t_1 = \max\{0, \frac{1}{\rho_1}\ln\frac{Q_2(R)}{\epsilon}\}$. This completes the proof of Lemma 6. □

All in all, the following uniformly bound estimate holds:

$$\sup_{\tau \leqslant t-t^*}\{\|U(t,\tau)z(\tau)\|_{\mathcal{H}^\theta_t} + \|V(t,\tau)z(\tau)\|_{\mathcal{H}^\theta_t} + \|W(t,\tau)z(\tau)\|_{\mathcal{H}^\theta_t}\} \leqslant 2R_0, \tag{54}$$

where $t^* = \max\{0, \frac{1}{\rho_1}\ln\frac{Q_2(R)}{2\epsilon}, \frac{1}{\rho}\ln\frac{2Q_1(R)}{R_0^2}\}$.

Lemma 6. *Let $z_2(t)$ be the solution of (48) with initial data $z_2(\tau)$ satisfying $\|z_2(\tau)\|_{\mathcal{H}^\theta_\tau} = 0$. If the assumptions (12)–(14), (18), (3) and (40)–(46) hold, then there exists $N_2 = N_2(\mathfrak{B}_t) > 0$, such that*

$$\sup_{\tau \leqslant t-t^*}\|z_2(t)\|_{\mathcal{H}^{\theta+\sigma}_t} = \sup_{\tau \leqslant t-t^*}\|W(t,\tau)z_2(\tau)\|_{\mathcal{H}^{\theta+\sigma}_t} \leqslant N_2, \tag{55}$$

where $\sigma = \min\{\frac{\theta}{4}, \frac{(n+2)\theta-2n}{n-2}, \frac{n}{2}-\theta\}$ and $t^* = t^*(R_0, \rho, \lambda_1, \|g^\varrho\|)$.

Proof. Note that $f = f_0 + f_1$ implies that

$$f(u) - f_0(v) = f(u) - f(v) + f(v) - f_0(v) = f(u) - f(v) + f_1(v),$$

therefore, taking the inner product of (48) with $2A^\sigma(w_t(t) + \rho w(t))$, we obtain

$$\begin{aligned}
&\frac{\mathrm{d}}{\mathrm{d}t}(\|w\|^2_{\theta+\sigma} + \varepsilon(t)\|w_t\|^2_\sigma + \|\xi^t\|^2_{\mu,\theta+\sigma} + 2\rho\varepsilon(t)\langle w_t, A^\sigma w\rangle + 2\langle f(u) - f_0(v), A^\sigma w\rangle - 2\langle g^\varrho, A^\sigma w\rangle) \\
&\quad + \frac{3\rho}{2}\|w\|^2_{\theta+\sigma} - (\varepsilon'(t) + 2\rho\varepsilon(t))\|w_t\|^2_\sigma + (\delta - 2\rho k_0)\|\xi^t\|^2_{\mu,\theta+\sigma} - 2\rho\varepsilon'(t)\langle w_t, A^\sigma w\rangle \\
&\quad + 2\rho\langle f(u) - f_0(v), A^\sigma w\rangle - 2\rho\langle g^\varrho, A^\sigma w\rangle \\
&\leqslant 2\langle f'(u)u_t - f'(v)v_t, A^\sigma w\rangle + 2\langle f_1'(v)v_t, A^\sigma w\rangle.
\end{aligned} \tag{56}$$

Next, we will deal with each term on the right-hand of (56).

First, by virtue of (18), (10), (49) and (54), we have

$$\begin{aligned}
&2\langle f'(u)u_t - f'(v)v_t, A^\sigma w)\rangle \\
&\leqslant C\Big(\int_\Omega(1+|u|^{\frac{4}{n-2}})^{\frac{n}{\theta-\sigma}}\mathrm{d}x\Big)^{\frac{\theta-\sigma}{n}}\Big(\int_\Omega|u_t|^2\mathrm{d}x\Big)^{\frac{1}{2}}\Big(\int_\Omega|A^\sigma w|^{\frac{2n}{n-2(\theta-\sigma)}}\mathrm{d}x\Big)^{\frac{n-2(\theta-\sigma)}{2n}} \\
&\quad + C\Big(\int_\Omega(1+|v|^{\frac{4}{n-2}})^{\frac{n}{\theta-\sigma}}\mathrm{d}x\Big)^{\frac{\theta-\sigma}{n}}\Big(\int_\Omega|v_t|^2\mathrm{d}x\Big)^{\frac{1}{2}}\Big(\int_\Omega|A^\sigma w|^{\frac{2n}{n-2(\theta-\sigma)}}\mathrm{d}x\Big)^{\frac{n-2(\theta-\sigma)}{2n}} \\
&\leqslant C(1+\|u\|_\theta^{\frac{4}{n-2}})\|u_t\|\|w\|_{\theta+\sigma} + C(1+\|v\|_\theta^{\frac{4}{n-2}})\|v_t\|\|w\|_{\theta+\sigma} \\
&\leqslant \frac{\rho}{8}\|w\|^2_{\theta+\sigma} + C,
\end{aligned} \tag{57}$$

where $\frac{4}{n-2}\cdot\frac{n}{\theta-\sigma}\leqslant\frac{2n}{n-2\theta}$.

Second, by (46), (10) and (49), we get

$$-2\langle f_1'(v)v_t, A^\sigma w)\rangle \leqslant \frac{\rho}{8}\|w\|^2_{\theta+\sigma}+C. \tag{58}$$

Choosing a suitable constant $C>0$, we define the functional:

$$\begin{aligned}\mathcal{M}_2(t)=&\|w\|^2_{\theta+\sigma}+\varepsilon(t)\|w_t\|^2_\sigma+\|\xi^t\|^2_{\mu,\theta+\sigma}+2\rho\varepsilon(t)\langle w_t, A^\sigma w\rangle\\&+2\langle f(u)-f_0(v), A^\sigma w\rangle-2\langle g^\varrho, A^\sigma w\rangle+C,\end{aligned} \tag{59}$$

as ρ is small enough, we know that

$$\frac{1}{2}\|W(t,\tau)z_2(\tau)\|^2_{\mathcal{H}_t^{\theta+\sigma}}\leqslant\mathcal{M}_2(t)\leqslant 2\|W(t,\tau)z_2(\tau)\|^2_{\mathcal{H}_t^{\theta+\sigma}}+C. \tag{60}$$

Indeed, we can easily deduce that

$$\begin{aligned}2\rho\varepsilon(t)|\langle w_t, A^\sigma w\rangle|&\leqslant\rho\|w\|^2_{\theta+\sigma}+\frac{\rho L}{\lambda_1^\theta}\varepsilon(t)\|w_t\|^2_\sigma,\\2|\langle g^\varrho, A^\sigma w\rangle|&\leqslant\rho\|w\|^2_{\theta+\sigma}+\frac{1}{\rho\lambda_1^{\theta-\sigma}}\|g^\varrho\|^2.\end{aligned}$$

Thanks to (10), we have

$$\begin{aligned}&2|\langle f(u)-f_0(v), A^\sigma w\rangle|\\&\leqslant 2|\langle f(u)-f(v), A^\sigma w\rangle|+2|\langle f_1(v), A^\sigma w\rangle|\\&\leqslant C\Big(\int_\Omega(1+|u|^{\frac{4}{n-2}}+|v|^{\frac{4}{n-2}})^{\frac{n}{\theta-\sigma}}dx\Big)^{\frac{\theta-\sigma}{n}}\Big(\int_\Omega|w|^2dx\Big)^{\frac{1}{2}}\Big(\int_\Omega|A^\sigma w|^{\frac{2n}{n-2(\theta-\sigma)}}dx\Big)^{\frac{n-2(\theta-\sigma)}{2n}}\\&\quad+C\Big(\int_\Omega(1+|v|^\gamma)^{\frac{n}{\theta-\sigma}}dx\Big)^{\frac{\theta-\sigma}{n}}\Big(\int_\Omega|v|^2dx\Big)^{\frac{1}{2}}\Big(\int_\Omega|A^\sigma w|^{\frac{2n}{n-2(\theta-\sigma)}}dx\Big)^{\frac{n-2(\theta-\sigma)}{2n}}\\&\leqslant C(1+\|u\|_\theta^{\frac{4}{n-2}}+\|v\|_\theta^{\frac{4}{n-2}})\|w\|_\theta\|w\|_{\theta+\sigma}+C(1+\|v\|_\theta^\gamma)\|v\|_\theta\|w\|_{\theta+\sigma}\\&\leqslant\frac{1}{4}\|w\|^2_{\theta+\sigma}+C,\end{aligned} \tag{61}$$

where $\frac{2n}{n-2\theta}\geqslant\frac{4n}{(n-2)(\theta-\sigma)}>\frac{n\gamma}{\theta-\sigma}$.

It follows from the above estimates that

$$\begin{aligned}&\frac{\mathrm{d}}{\mathrm{d}t}\mathcal{M}_2(t)+\rho\mathcal{M}_2(t)+\frac{\rho}{2}\|w\|^2_{\theta+\sigma}-(\varepsilon'(t)+3\rho\varepsilon(t))\|w_t\|^2_\sigma+(\delta-\rho-2\rho k_0)\|\xi^t\|^2_{\mu,\theta+\sigma}\\&-2\rho(\varepsilon'(t)+\rho\varepsilon(t))\langle w_t, A^\sigma w\rangle\leqslant\frac{\rho}{4}\|w\|^2_{\theta+\sigma}+C.\end{aligned} \tag{62}$$

Obviously, we can gain

$$-2\rho(\varepsilon'(t)+\rho\varepsilon(t))\langle w_t, A^\sigma w\rangle\geqslant-\frac{\rho}{4}\|w\|^2_{\theta+\sigma}-\frac{4\rho L^2}{\lambda_1^\theta}\|w_t\|^2_\sigma. \tag{63}$$

Substituting (63) into (62), we have

$$\frac{\mathrm{d}}{\mathrm{d}t}\mathcal{M}_2(t)+\rho\mathcal{M}_2(t)-(\varepsilon'(t)+3\rho\varepsilon(t)+\frac{4\rho L^2}{\lambda_1^\theta})\|w_t\|^2_\sigma+(\delta-2\rho k_0-\rho)\|\xi^t\|^2_{\mu,\theta+\sigma}\leqslant C. \tag{64}$$

Taking ρ small enough, we know

$$\frac{\mathrm{d}}{\mathrm{d}t}\mathcal{M}_2(t)+\rho\mathcal{M}_2(t)\leqslant C. \tag{65}$$

From (65) and (60), we obtain (55).

We completed the proof. □

To verify the asymptotic compactness of the process $U(t,\tau)$ corresponding to the problem in (15) and (16), we also need the following preliminary results.

For any $\xi_0 \in L^2_\mu(\mathbb{R}^+; V_\theta)$, the Cauchy problem (see [13,29,30])

$$\begin{cases} \xi^t_t = -\xi^t_s + w_t, \quad t \geqslant \tau, \\ \xi^\tau = \xi_0 \end{cases} \tag{66}$$

has a unique solution $\xi^t \in C([\tau,\infty); L^2_\mu(\mathbb{R}^+; V_\theta))$. Then, for (66), we have the explicit expression

$$\xi^t(x,s) = \begin{cases} w(x,t) - w(x,t-s), & \tau \leqslant s < t, \\ w(x,t), & s \geqslant t. \end{cases} \tag{67}$$

Let $\mathfrak{B}_t$ be the time-dependent absorbing set for the process $U(t,\tau)$ corresponding to the problem in (15) and (16) in $\mathcal{H}^\theta_t$ obtained from Theorem 4. Then,

Lemma 7. *For every given $\tau < T$, we set*

$$\mathcal{K}_T := \Pi W(T,\tau)\mathfrak{B}_\tau.$$

Assume that the forcing term $g \in V_{-\theta}$. If the assumptions (12)–(14), (18)–(20), (3) and (40)–(46) hold, then there exists a positive constant $N_4 = N_4(\|\mathcal{B}_0\|_{\mathcal{H}^\theta_\tau})$, such that

(i) *$\mathcal{K}_T$ is bounded in $L^2_\mu(\mathbb{R}^+; V^{\theta+\sigma}) \cap H^1_\mu(\mathbb{R}^+; V^\sigma)$;*
(ii) $\sup_{\xi\in\mathcal{K}_T} \|\xi(s)\|^2_{V^\theta} \leqslant N_4$,

where $\sigma = \min\{\frac{\theta}{4}, \frac{(n+2)\theta-2n}{n-2}, \frac{n}{2}-\theta\}$, $W(T,\tau)$ is a solution operator of (48) and $\Pi : V_{\theta+\sigma} \times V_\sigma \times L^2_\mu(\mathbb{R}^+; V_{\theta+\sigma}) \to L^2_\mu(\mathbb{R}^+; V_{\theta+\sigma})$ is a projection operator.

Proof. From (67), we conclude that

$$\xi^T_s(x,s) = \begin{cases} w_s(x,T-s), & \tau \leqslant s < T, \\ 0, & s \geqslant T. \end{cases} \tag{68}$$

Applying Lemma 6, we know (i) holds.

Using (68) once again, we can easily deduce that

$$\|\xi^T(x,s)\|_{V_\theta} = \begin{cases} \|w(x,T) - w(x,T-s)\|_{V_\theta} \leqslant \|w(x,T)\|_{V_\theta} + \|w(x,T-s)\|_{V_\theta}, & \tau < s < T, \\ \|w(x,T)\|_{V_\theta}, & s \geqslant T. \end{cases} \tag{69}$$

Clearly, it implies (ii) holds. The proof is complete. □

Therefore, applying Lemma 3, we conclude that $\mathcal{K}_T$ is relatively compact in $L^2_\mu(\mathbb{R}^+; V_\theta)$. Moreover, by the compact embedding $V_{\theta+\sigma} \times V_\sigma \hookrightarrow V_\theta \times L^2(\Omega)$, we obtain:

Lemma 8. *Let $\{W(T,\tau)\}_{\tau\leqslant T}$ be the process corresponding to the problem (48). If the assumptions of Lemma 7 hold, then for any $\tau < T$ and given $R > 0$, $W(T,\tau)\mathbb{B}_\tau(R)$ is relatively compact in $\mathcal{H}^\theta_T$.*

Theorem 5. *Let $U(t,\tau) : \mathcal{H}^\theta_\tau \to \mathcal{H}^\theta_t$ be the process generated by the problem in (15) and (16). Assume that $g \in V_{-\theta}$. If (12)–(14), (18)–(20), (3) and (40)–(46) hold, then the process $U(t,\tau)$ possesses an invariant time-dependent global attractor $\mathfrak{A} = \{A_t\}_{t\in\mathbb{R}}$ in $\mathcal{H}^\theta_t$.*

Proof. According to Lemmas 6 and 8, we consider the family $\mathfrak{K} = \{K_t^{\theta+\sigma}\}_{t\in\mathbb{R}}$, where

$$K_t^{\theta+\sigma} = \{z(t) \in \mathcal{H}_t^{\theta+\sigma} : \|z(t)\|_{\mathcal{H}_t^{\theta+\sigma}} \leqslant M\}.$$

By the compact embedding $\mathcal{H}_t^{\theta+\sigma} \hookrightarrow \mathcal{H}_t^{\theta}$ and Lemma 8, $K_t^{\theta+\sigma}$ is compact in $\mathcal{H}_t^{\theta}$. In addition, since the injection constant M is independent of t, the set $\mathfrak{K}$ is uniformly bounded.

It follows from Theorem 4, Lemmas 5 and 6 that $\mathfrak{K}$ is pullback attracting. In fact,

$$\delta_t(U(t,\tau)\mathbb{B}_\tau(R), K_t^{\theta+\sigma}) \leqslant Ce^{-\rho(t-\tau)}, \quad \forall \tau \leqslant t,$$

here, $\delta_t(\cdot,\cdot)$ denotes the Hausdorff semidistance of two subsets of $\mathcal{H}_t^{\theta}$. Hence, the process $U(t,\tau)$ is asymptotically compact, which implies the existence of the unique time-dependent global attractor $\mathfrak{A} = \{A_t\}_{t\in\mathbb{R}}$ of the process $U(t,\tau)$. Finally, the invariance of $\mathfrak{A}$ can be concluded by Lemma 4 (the continuity of the process $U(t,\tau)$ in $\mathcal{H}_t^{\theta}$).

We completed the proof. □

3.4. The Regularity of the Time-Dependent Attractor

Here, we prove that the time-dependent attractor $\mathfrak{A} = \{A_t\}_{t\in\mathbb{R}}$ is bounded in $\mathcal{H}_t^{2\theta}$, where the bound is independent of t.

For any given $\tau \in \mathbb{R}$ and $z(\tau) \in A_\tau$, we give a decomposition of the solution $U(t,\tau)z(\tau)$:

$$U(t,\tau)z(\tau) = z(t) = z_1(t) + z_2(t) = V_1(t,\tau)z_1(\tau) + W_1(t,\tau)z_2(\tau),$$

where

$$V_1(t,\tau)z_1(\tau) = (v(t), v_t(t), \zeta^t(s)), \quad W_1(t,\tau)z_2(\tau) = (w(t), w_t(t), \xi^t(s))$$

solve the equations, respectively,

$$\begin{cases} \varepsilon(t)v_{tt} + A^\theta v + \int_0^\infty \mu(s)A^\theta\zeta^t(s)\mathrm{d}s = 0, \\ \zeta_t^t = -\zeta_s^t + v_t, \\ v(x,t)|_{\partial\Omega} = 0, \;\; \zeta^t(x,t)|_{\partial\Omega} = 0, \\ v(x,\tau) = u_0(x,t), \;\; v_t(x,\tau) = u_1(x,t), \; x \in \Omega, \; t \leqslant \tau, \\ \zeta^\tau(x,s) = u_0(x,\tau) - u_0(x,\tau-s), \; (x,s) \in \Omega \times \mathbb{R}^+, \end{cases} \tag{70}$$

and

$$\begin{cases} \varepsilon(t)w_{tt} + A^\theta w + \int_0^\infty \mu(s)A^\theta\xi^t(s)\mathrm{d}s + f(u) = g(x), \\ \xi_t^t = -\xi_s^t + w_t, \\ w(x,t)|_{\partial\Omega=0}, \;\; \xi^t(x,t)|_{\partial\Omega} = 0, \\ w(x,\tau) = 0, \;\; w_t(x,\tau) = 0, \;\; \xi^\tau(x,s) = 0, \; x \in \Omega, s \in \mathbb{R}^+. \end{cases} \tag{71}$$

As a special case of Lemma 5, we can get

$$\|V_1(t,\tau)z(\tau)\|_{\mathcal{H}_t^\theta} \leqslant Ce^{-\rho_1(t-\tau)}, \quad \forall \tau \leqslant t - \bar{t}_1, \tag{72}$$

where $\bar{t}_1 = \max\{0, \frac{1}{\rho_1}\ln\frac{Q_2(R)}{2\epsilon}\}$.

Lemma 9. *Let $z_2(t)$ be the solution of (71) with initial data $z_2(\tau) \in A_\tau$ satisfying $\|z_2(\tau)\|_{\mathcal{H}_\tau^\theta} = 0$. If the assumptions (12)–(14), (18), (3) and (40)–(46) hold, then $\{A_t\}_{t\in\mathbb{R}}$ is bounded in $\mathcal{H}_t^{2\theta}$ and the bound is independent of t.*

Proof. For $\theta \in (\frac{2n}{n+2}, \frac{n}{2})$, $n \geqslant 3$, we set

$$0 < \sigma_0 = \sigma < \min\{\frac{\theta}{4}, \frac{(n+2)\theta - 2n}{n-2}, \frac{n}{2} - \theta\}, \text{ and } \sigma_0 < \sigma_1 = \min\{\frac{(n+2)(\theta+\sigma_0) - 2n}{n-2}, \theta\}.$$

Taking the inner product of (71) with $2(A^{\sigma_1} w_t + \rho A^{\sigma_1} w)$, we obtain

$$\begin{aligned}
&\frac{\mathrm{d}}{\mathrm{d}t}(\|w\|^2_{\theta+\sigma_1} + \varepsilon(t)\|w_t\|^2_{\sigma_1} + \|\xi^t\|^2_{\mu,\theta+\sigma_1} + 2\rho\varepsilon(t)\langle w_t, A^{\sigma_1} w\rangle + 2\langle f(u) - g, A^{\sigma_1} w\rangle) \\
&\quad + \frac{3}{2}\rho\|w\|^2_{\theta+\sigma_1} - (\varepsilon'(t) + 2\rho\varepsilon(t))\|w_t\|^2_{\sigma_1} + (\delta - 2\rho k_0)\|\xi^t\|^2_{\mu,\theta+\sigma_1} \\
&\quad - 2\rho\varepsilon'(t)\langle w_t, A^{\sigma_1} w\rangle + 2\rho\langle f(u) - g, A^{\sigma_1} w\rangle \\
&\quad \leqslant 2\langle f'(u) u_t, A^{\sigma_1} w\rangle.
\end{aligned} \tag{73}$$

Set

$$\begin{aligned}
\mathcal{M}_3(t) = &\|w\|^2_{\theta+\sigma_1} + \varepsilon(t)\|w_t\|^2_{\sigma_1} + \|\xi^t\|^2_{\mu,\theta+\sigma_1} + 2\rho\varepsilon(t)\langle w_t, A^{\sigma_1} w\rangle \\
&+ 2\langle f(u) - g, A^{\sigma_1} w\rangle + C.
\end{aligned} \tag{74}$$

Similar to (60), for ρ small enough, we have

$$\frac{1}{2}\|W_1(t,\tau)z(\tau)\|^2_{\mathcal{H}_t^{\theta+\sigma_1}} \leqslant \mathcal{M}_3(t) \leqslant 2\|W_1(t,\tau)z(\tau)\|^2_{\mathcal{H}_t^{\theta+\sigma_1}} + C, \tag{75}$$

and

$$\frac{\mathrm{d}}{\mathrm{d}t}\mathcal{M}_3(t) + \rho\mathcal{M}_3(t) + \frac{\rho}{4}\|w\|^2_{\theta+\sigma_1} \leqslant 2\langle f'(u) u_t, A^{\sigma_1} w\rangle + \rho C. \tag{76}$$

Due to the invariance of $\mathfrak{A}$, we have

$$\|U(t,\tau)z(\tau)\|_{\mathcal{H}_t^{\theta+\sigma_0}} \leqslant C,$$

where C is a generic constant depending on the size of A_t in $\mathcal{H}_t^{\theta+\sigma_0}$.

Using the embedding (10), we can deduce that

$$\begin{aligned}
&2\langle f'(u) u_t, A^{\sigma_1} w\rangle \\
&\leqslant C\Big(\int_\Omega (1+|u|^{\frac{4}{n-2}})^{\frac{n}{\theta+\sigma_0-\sigma_1}} \mathrm{d}x\Big)^{\frac{\theta+\sigma_0-\sigma_1}{n}} \cdot \Big(\int_\Omega |u_t|^{\frac{2n}{n-2\sigma_0}} \mathrm{d}x\Big)^{\frac{n-2\sigma_0}{2n}} \\
&\quad \cdot \Big(\int_\Omega |A^{\sigma_1} w|^{\frac{2n}{n-2(\theta-\sigma_1)}} \mathrm{d}x\Big)^{\frac{n-2(\theta-\sigma_1)}{2n}} \\
&\leqslant C(1 + \|u\|^{\frac{4}{n-2}}_{\theta+\sigma_0})\|u_t\|_{\sigma_0}\|w\|_{\theta+\sigma_1} \\
&\leqslant \frac{\rho}{4}\|w\|^2_{\theta+\sigma_1} + C,
\end{aligned} \tag{77}$$

where $\frac{4}{n-2} \cdot \frac{n}{\theta+\sigma_0-\sigma_1} \leqslant \frac{2n}{n-2(\theta+\sigma_0)}$.

Therefore, we conclude that

$$\frac{d}{dt}\mathcal{M}_3(t) + \rho\mathcal{M}_3(t) \leqslant C.$$

Applying the Gronwall lemma and combining with (75), we obtain that there exists $N^{*_1} = N^{*_1}(\mathfrak{A}) > 0$, such that

$$\sup_{\tau \leqslant t-t^*} \|z_2(t)\|_{\mathcal{H}_t^{\theta+\sigma_1}} = \sup_{\tau \leqslant t-t^*} \|W_1(t,\tau)z_2(\tau)\|_{\mathcal{H}_t^{\theta+\sigma_1}} \leqslant N^{*_1}, \tag{78}$$

where $t^* = \max\{0, \frac{1}{\rho_1}\ln\frac{Q_2(R)}{2\epsilon}, \frac{1}{\rho}\ln\frac{2Q_1(R)}{R_0^2}\}$. Namely, $\|W_1(t,\tau)z_2(\tau)\|_{\mathcal{H}_t^{\theta+\sigma_1}}$ is uniformly bounded.

We denote

$$K_t^{\theta+\sigma_1} = \{z(t) \in \mathcal{H}_t^{\theta+\sigma_1} : \|z(t)\|_{\mathcal{H}_t^{\theta+\sigma_1}} \leqslant N_4\}.$$

It follows from (72) and Lemma 9 that

$$\lim_{\tau\to-\infty} \delta_t(U(t,\tau)A_\tau, K_t^{\theta+\sigma_1}) = 0, \forall t \in \mathbb{R}.$$

The invariance of $\mathfrak{A}$ implies that

$$\delta_t(A_t, K_t^{\theta+\sigma_1}) = 0.$$

Therefore, $A_t \subset \overline{K_t^{\theta+\sigma_1}} = K_t^{\theta+\sigma_1}$. We can deduce that A_t is bounded in $\mathcal{H}_t^{\theta+\sigma_1}$ (with a bound independent of $t \in \mathbb{R}$).

For $\theta \in (\frac{2n}{n+2}, \frac{n}{2})$, $n \geqslant 3$, we set

$$\sigma_1 < \sigma_2 = \frac{(n+2)(\theta+\sigma_1) - 2n}{n-2}.$$

Repeating the above process, we obtain that A_t is bounded in $\mathcal{H}_t^{\theta+\sigma_2}$ (with a bound independent of $t \in \mathbb{R}$).

We set $\sigma_{\min} = \min\{\sigma_i\}$, $i = 1, 2, \cdots$. Repeating the above process at most $[\frac{\theta}{\sigma_{\min}} + 1]$ times, we can finally obtain that A_t is bounded in $\mathcal{H}_t^{2\theta}$ (with a bound independent of $t \in \mathbb{R}$). □

3.5. The Asymptotic Regularity of the Solution

By using bootstrap methods, the following results can be obtained.

Lemma 10. *Assume that the forcing term $g \in V_{-\theta}$. Let the assumptions (12)–(14), (18)–(20), (3) and (40)–(46) hold. For any bounded (in $\mathcal{H}_\tau^{\theta+\sigma}$) set $B_\tau^{\theta+\sigma}$, there is a positive constant $N_{\|B_t^{\theta+\sigma}\|_{\mathcal{H}_t^{\theta+\sigma}}}$ that depends on $\|B_t^{\theta+\sigma}\|_{\mathcal{H}_t^{\theta+\sigma}}$, such that for any $\tau \in \mathbb{R}$ and $t_2 \leqslant t^* \leqslant t$,*

$$\|U(t,\tau)z_\tau\|^2_{\mathcal{H}_t^{\theta+\sigma}} \leqslant N_{\|B_{\theta+\sigma}\|_{\mathcal{H}_t^{\theta+\sigma}}}, \text{ as } \tau \leqslant t - t_2 \text{ and } z_\tau \in B_\tau^{\theta+\sigma},$$

where $\sigma = \min\{\frac{\theta}{4}, \frac{(n+2)\theta-2n}{n-2}, \frac{n}{2} - \theta\}$.

Lemma 11. *Let $\sigma < \iota = \min\{\theta, \frac{(n+2)(\theta+\sigma)-2n}{n-2}\}$. Under the assumptions of Lemma 10, for any bounded (in $\mathcal{H}_\tau^{\theta+\iota}$) set $B_\tau^{\theta+\iota}$, there is a positive constant $N_{\|B_t^{\theta+\iota}\|_{\mathcal{H}_t^{\theta+\iota}}}$ that depends on $\|B_t^{\theta+\iota}\|_{\mathcal{H}_t^{\theta+\iota}}$, such that for any $\tau \in \mathbb{R}$ and $t_3 \leqslant t_2 \leqslant t$,*

$$\|U(t,\tau)z_\tau\|^2_{\mathcal{H}_t^{\theta+\iota}} \leqslant N_{\|B_t^{\theta+\iota}\|_{\mathcal{H}_t^{\theta+\iota}}}, \text{ as } \tau \leqslant t - t_3 \text{ and } z_\tau \in B_\tau^{\theta+\iota}.$$

Lemma 12. *Assume that $\sigma < \iota = \min\{\theta, \frac{(n+2)(\theta+\sigma)-2n}{n-2}\}$. Under the assumptions of Lemma 10, for any bounded (in $\mathcal{H}_\tau^{\theta+\iota}$) set $B_\tau^{\theta+\iota}$, there is a positive constant $J_{\|B_t^{\theta+\iota}\|_{\mathcal{H}_t^{\theta+\iota}}}$ that depends on $\|B_t^{\theta+\iota}\|_{\mathcal{H}_t^{\theta+\iota}}$, such that for the solution $z_2(t)$ of Equation (48), for any $\tau \in \mathbb{R}$ and $t_4 \leqslant t_3 \leqslant t$,*

$$\|W(t,\tau)z_\tau\|^2_{\mathcal{H}_t^{\theta+\kappa_0}} \leqslant J_{\|B_t^{\theta+\iota}\|_{\mathcal{H}_t^{\theta+\iota}}}, \text{ as } \tau \leqslant t - t_4 \text{ and } z_\tau \in B_\tau^{\theta+\iota},$$

where $\iota < \kappa_0 = \min\{\theta, \frac{(n+2)(\theta+\iota)-2n}{n-2}\}$.

Theorem 6 (Asymptotic regularity of solution). *Let Ω be a bounded domain in $\mathbb{R}^n (n \geqslant 3)$ with smooth boundary $\partial\Omega$. Under the assumptions of Lemma 11, then there exist a bounded (in $\mathcal{H}_t^{2\theta}$) set $\mathcal{B}_t \subset \mathcal{H}_t^{2\theta}$, a positive constant ν and a monotonically function $Q(\cdot)$, such that for any bounded (in $\mathcal{H}_\tau^\theta$) set $B_\tau \subset \mathcal{H}_\tau^\theta$, any $\tau \in \mathbb{R}$, the following estimate holds:*

$$\delta_t(U(t,\tau)B_\tau, \mathcal{B}_t) \leqslant Q(\|B_\tau\|_{\mathcal{H}_\tau^\theta})e^{-\nu(t-\tau)}, \tag{79}$$

where δ_t is the Hausdorff semidistance in $\mathcal{H}_t^\theta$ and ν is independent of B_τ, g and τ.

Proof. Let $\mathfrak{B}_t = \{\mathbb{B}_t(R_0)\}_{t\in\mathbb{R}}$ be the time-dependent absorbing set in $\mathcal{H}_t^\theta$ obtained from Theorem 4. From Lemmas 5 and 6, we can deduce that there exists a bounded (in $\mathcal{H}_t^{\theta+\sigma}$) subset $A_t^{\theta+\sigma} \subset \mathcal{H}_t^{\theta+\sigma}$, such that

$$\begin{aligned}\delta_t(U(t,\tau)\mathbb{B}_\tau(R_0), A_t^{\theta+\sigma}) &\leqslant \delta_t(V(t,\tau)\mathbb{B}_\tau(R_0), A_t^{\theta+\sigma}) \\ &\leqslant Q_2(R_0)e^{-\rho_1(t-\tau)}.\end{aligned} \tag{80}$$

In regard to $A_\tau^{\theta+\sigma}$, from Lemmas 5 and 12, it is easy to know that there exists a bounded set $A_t^{\theta+\kappa_0}$ in $\mathcal{H}_t^{\theta+\kappa_0}$, such that

$$\begin{aligned}\delta_t(U(t,\tau)A_\tau^{\theta+\sigma}, A_t^{\theta+\kappa_0}) &\leqslant \delta_t(V(t,\tau)\mathcal{A}_\tau^{\theta+\sigma}, A_t^{\theta+\kappa_0}) \\ &\leqslant Q_2(\|A_\tau^{\theta+\sigma}\|_{\mathcal{H}_\tau^\theta})e^{-\rho_1(t-\tau)},\end{aligned} \tag{81}$$

where ρ_1 is positive and only depends on $\|A_{\theta+\sigma}\|_{\mathcal{H}_\tau^\theta}$, and $\kappa_0 = \min\{\theta, \frac{(n+2)(\theta+\iota)-2n}{n-2}\}$.

From (38), (80), (81) and Lemma 2, we obtain

$$\delta_t(U(t,\tau)\mathbb{B}_\tau(R_0), A_t^{\theta+\kappa_0}) \leqslant CQ_2(R_0)e^{-\rho_2(t-\tau)}, \tag{82}$$

where C and ρ_2 are both positive constants.

Fix $\kappa_0 = \min\{\theta, \frac{(n+2)(\theta+\iota)-2n}{n-2}\}$ and $\sigma = \min\{\frac{\theta}{4}, \frac{(n+2)\theta-2n}{n-2}, \frac{n}{2} - \theta\}$. By a finite number of steps (no more than $[\frac{\theta}{\kappa_0} + 1]$ steps), we can deduce that there exists a bounded (in $\mathcal{H}_t^{2\theta}$) set $\mathcal{B}_t \subset \mathcal{H}_t^{2\theta}$, such that

$$\delta_t(U(t,\tau)\mathbb{B}_\tau(R_0), \mathcal{B}_t) \leqslant Q(R_0)e^{-\nu(t-\tau)}, \tag{83}$$

where ν is dependent of R_0.

For any bounded (in $\mathcal{H}_\tau^\theta$) set B_τ, by Theorem 4, there exists a t_0 such that

$$U(t,\tau)B_\tau \subset \mathbb{B}_t(R_0), \text{ as } \tau \leqslant t - t_0. \tag{84}$$

Therefore,

$$\delta_t(U(t,\tau)B_\tau, \mathbb{B}_t(R_0)) \leqslant N_3 e^{\nu t_0} e^{-\nu(t-\tau)}, \tag{85}$$

where $N_3 = \sup\{\|U(t,\tau)B_\tau\|_{\mathcal{H}_t^\theta}, \tau \leqslant t - t_0\} < \infty$.

By Lemma 2 once more, we can deduce that (79). The proof is complete. □

4. Conclusions

For the undamped second-order abstract evolution equation with fading memory, when the nonlinear term satisfies the critical exponential growth, the existence and asymptotic regularity of the time-dependent attractor, as well as the asymptotic regularity of the solutions, can be obtained by using the process theory, asymptotic prior technique and decomposition technique. This result improves and generalizes some known results (see [9,23,30]).

We will continue to study the asymptotic behavior of the solutions of the equation in the strong topological space $V_{2\theta} \times V_\theta \times L^2_\mu(\mathbb{R}^+; V_{2\theta})$, and it is expected that corresponding research results will be obtained.

Author Contributions: Formal analysis, X.W.; Methodology, D.H.; Writing—original draft, C.G. All authors have read and agreed to the published version of the manuscript.

Funding: This work was partly supported by the NSFC Grant (11961060, 11961059, 12061062).

Institutional Review Board Statement: Not applicable.

Informed Consent Statement: Not applicable.

Data Availability Statement: Not applicable.

Conflicts of Interest: The authors declare no conflict of interest.

References

1. Cavalcanti, M.; Fatori, L.; Ma, T. Attractors for wave equations with degenerate memory. *J. Differ. Equ.* **2016**, *260*, 56–83. [CrossRef]
2. Ghisi, M.; Gobbino, M.; Haraux, A. Finding the exact decay rate of all solutions to some second order evolution equations with dissipation. *J. Funct. Anal.* **2016**, *271*, 2359–2395. [CrossRef]
3. Liu, G. Global nonexistence for abstract evolution equations with dissipation. *Nonlinear Anal. Real World Appl.* **2015**, *22*, 225–235. [CrossRef]
4. Liang, F.; Guo, Z. Asymptotic behavior for second order stochastic evolution equations with memory. *J. Math. Anal. Appl.* **2014**, *419*, 1333–1350. [CrossRef]
5. Coleman, B.D.; Noll, W. Foundations of Linear Viscoelasticity. *Rev. Mod. Phys.* **1961**, *33*, 239–249. [CrossRef]
6. Conti, M.; Pata, V. Weakly dissipative semilinear equations of viscoelasticity. *Commun. Pure Appl. Anal.* **2005**, *4*, 705–720. [CrossRef]
7. Dafermos, C.M. Asymptotic stability in viscoelasticity. *Arch. Ration. Mech. Anal.* **1970**, *37*, 297–308. [CrossRef]
8. Fabrizio, M.; Morro, A. *Mathematical Problems in Linear Viscoelasticity*; Society for Industrial and Applied Mathematics: Philadelphia, PA, USA, 1992.
9. Giorgi, C.; Rivera, J.E.M.; Pata, V. Global Attractors for a Semilinear Hyperbolic Equation in Viscoelasticity. *J. Math. Anal. Appl.* **2001**, *260*, 83–99. [CrossRef]
10. Wang, X.; Duan, F.X.; Hu, D.D. Attractors for a class of abstract evolution equations with fading memory. *Math. Probl. Eng.* **2017**, *6*, 1–16. [CrossRef]
11. Arrieta, J.; Carvalho, A.N.; Hale, J.K. A damping hyperbolic equation with critical exponent. *Comm. Partial. Differ. Equ.* **1992**, *17*, 841–866. [CrossRef]
12. Ma, Q.Z.; Zhong, C.K. Existence of strong global attractors for hyperbolic equation with linear memory. *Appl. Math. Comput.* **2004**, *157*, 745–758. [CrossRef]
13. Pata, V.; Zucchi, A. Attractors for a damped hyperbolic equation with linear memory. *Adv. Math. Sci. Appl.* **2001**, *11*, 505–529.
14. Temam, R. *Infinite Dimensional Dynamical System in Mechanics and Physics*, 2nd ed.; Spring: New York, NY, USA, 1997.
15. Xu, L.; Ma, Q.Z. Existence of random attractors for the floating beam equation with strong damping and white noise. *Bound. Value Probl.* **2015**, *1*, 126. [CrossRef]
16. Conti, M.; Pata, V.; Temam, R. Attractors for processes on time-dependent spaces. Applications to wave equations. *J. Differ. Equations* **2013**, *255*, 1254–1277. [CrossRef]
17. Plinio, F.D.; Duane, G.S.; Temam, R. Time-dependent attractor for the oscillon equation, *Discrete Contin Dyn. Syst.-A* **2012**, *29*, 141–167.
18. Conti, M.; Pata, V. Asymptotic structure of the attractor for processes on time-dependent spaces. *Nonlinear Anal. Real World Appl.* **2014**, *19*, 1–10. [CrossRef]
19. Luo, X.D.; Ma, Q.Z. The existence of time-dependent attractor for wave equation with fractional damping and lower regular forcing term. *Discrete Contin. Dyn. Syst.-B* **2021**. [CrossRef]
20. Meng, F.J.; Liu, C.C. Necessary and sufficient conditions for the existence of time-dependent global attractor and application. *J. Math. Phys.* **2017**, *58*, 032702. [CrossRef]
21. Meng, F.J.; Yang, M.H.; Zhong, C.K. Attractors for wave equations with nonlinear damping on time-dependent space. *Discrete Contin. Dyn. Syst.-B* **2016**, *21*, 205–225. [CrossRef]
22. Meng, F.J.; Wu, J.; Zhao, C.X. Time-dependent global attractor for extensible Berger equation, *J. Math. Anal. Appl.* **2019**, *469*, 1045–1069. [CrossRef]
23. Liu, T.T.; Ma, Q.Z. Time-dependent asymptotic behavior of the solution for plate equations with linear memory. *Discrete Contin. Dyn. Syst.-B* **2017**, *22*, 1–22. [CrossRef]
24. Liu, T.T.; Ma, Q.Z. Time-dependent attractor for plate equations on Rn. *J. Math. Anal. Appl.* **2019**, *479*, 315–332. [CrossRef]

25. Ding, T.; Liu, Y.-F. Time-dependent global attractor for the nonclassical diffusion equations. *Appl. Anal.* **2014**, *94*, 1439–1449. [CrossRef]
26. Ma, Q.Z.; Wang, X.P.; Xu, L. Existence and regularity of time-dependent global attractors for the nonclassical reaction-diffusion equations with lower forcing term. *Bound. Value Probl.* **2016**, *1*, 10. [CrossRef]
27. Wang, X.; Yang, L.; Zhong, C.K. Attractors for the nonclassical diffusion equations with fading memory. *J. Math. Anal. Appl.* **2010**, *362*, 327–337. [CrossRef]
28. Zelik, S.V. Asymptotic regularity of solutions of a nonautonomous damped wave equation with a critical growth exponent. *Comm. Pure Appl. Anal.* **2004**, *3*, 921–934. [CrossRef]
29. Borini, S.; Pata, V. Uniform attractors for a strongly damped wave equations with linear memory. *Asymptot. Anal.* **1999**, *20*, 263–277.
30. Gatti, C.; Miranville, A.; Pata, V.; Zelik, S.V. Attractors for semilinear equations of viscoelasticity with very low dissipation. *Rocky Mountain J. Math.* **2008**, *38*, 1117–1138. [CrossRef]

mathematics

MDPI

Article

A Solution Procedure Combining Analytical and Numerical Approaches to Investigate a Two-Degree-of-Freedom Vibro-Impact Oscillator

Nicolae Herisanu [1,2,*] and Vasile Marinca [1,2]

1 Department of Mechanics and Strength of Materials, University Politehnica Timisoara, 300222 Timisoara, Romania; vasile.marinca@upt.ro

2 Center for Fundamental Technical Research, Romanian Academy-Branch of Timisoara, 300223 Timisoara, Romania

* Correspondence: nicolae.herisanu@upt.ro

Abstract: In this paper, a new approach is proposed to analyze the behavior of a nonlinear two-degree-of-freedom vibro-impact oscillator subject to a harmonic perturbing force, based on a combination of analytical and numerical approaches. The nonlinear governing equations are analytically solved by means of a new analytical technique, namely the Optimal Auxiliary Functions Method (OAFM), which provided highly accurate explicit analytical solutions. Benefiting from these results, the application of Schur principle made it possible to analyze the stability conditions for the considered system. Various types of possible motions were emphasized, taking into account possible initial conditions and different parameters, and the explicit analytical solutions were found to be very useful to analyze the kinetic energy loss, the contact force, and the stability of periodic motions.

Keywords: vibro-impact; Optimal Auxiliary Functions Method; stability

Citation: Herisanu, N.; Marinca, V. A Solution Procedure Combining Analytical and Numerical Approaches to Investigate a Two-Degree-of-Freedom Vibro-Impact Oscillator. *Mathematics* **2021**, *9*, 1374. https://doi.org/10.3390/math9121374

Academic Editor: Dan B. Marghitu

Received: 30 April 2021
Accepted: 11 June 2021
Published: 14 June 2021

1. Introduction

Vibro-impact processes are widely used in mechanical-engineering applications and devices such as hammer-like devices, rotor-casing dynamical systems, wheel–rail interaction of high-speed railway couches, heat exchangers, fuel elements of nuclear reactors, gears, piping systems [1], stiction and electric short circuits of MEMS [2], noise and wear-producing processes, grinding, foraging, drilling, punching, tamping, pile cutting a variety of objects, and vibrating machinery or structures with stops or clearance [3].

Dynamics of vibro-impact systems has received great attention in the last decades. Some numerical or experimental results were obtained, but adequate analytical predictions need to be further studied. For the first time, Masri [4] analyzed the stability of the impact damper by using the perturbation method. The finite-difference method was used by Kobrinski [5], Babitsky [6], Brîndeu [7], and Okolewska et al. [8]. The nonlinear equation of a single-degree-of-freedom oscillator with an impact damper was analyzed approximately using the method of Kryloff and Bogoliuboff by Cronin and Van [9]. Shaw [10] studied the symmetric double-impact motion, both harmonic and subharmonic, by means of a mapping that related conditions at subsequent impacts. A mechanical system with one degree of freedom was investigated by Ivanov [11] using standard Poincare-Bendixon theory, Lyapunov's second method, and Zhuravlev transformations. The exact solutions of the vibro-impact oscillator with two degrees of freedom, including localized states and their bifurcation structure, were obtained by Aziz et al. [12]. Stability and bifurcation for the unsymmetrical, periodic motion of a horizontal impact oscillator under a periodic excitation were employed through four mappings based on two switch planes by Luo [13]. De Souza and Caldas [14] implemented the Ott–Grebogi–Yorke (OGY) method to stabilize desired unstable periodic orbits by applying a small perturbation on an available control parameter and introducing an impact map for the dynamical variables. Chaos and periodic motion of a cantilever beam system with impacts was found readily by Emans et al. [15]. In addition, nonlinear behavior such as coexistence of attractors was found even at modest oscillation levels during investigations with realistic parameters.

The dynamics of an impact oscillator with viscoelastic and Hertzian contact was examined by Mann et al. [16] using the presence of multiple periodic attractors, subharmonics, quasi-periodic motions, and chaotic oscillations. Avramov [17] considered the forced vibrations of an impact Duffing oscillator using Zhuravlev's transformation and the multiple-scale method. The stability and bifurcations of periodic vibrations were explored. The control of vibro-impact dynamics of a single-sided Hertzian contact forced oscillator was considered by Bichri et al. [18]. The multiple-scale technique was applied to study the slow dynamic and the frequency-response curves near the primary resonance. Based on Zhuravlev and Ivanov transformations, Grace et al. [19] developed an analytical model of a ship's roll motion interacting with ice. Extensive numerical simulations were carried out for all initial conditions covered by the ship's grazing orbit for different values of excitation amplitude and frequency of the external wave roll moment.

Jovic et al. [20] considered the motion trajectory of a vibro-impact system based on the oscillator moving along a rough parabolic line in the vertical plane, under the action of extreme force with nonlinearity of the bond originates of sliding Coulomb's type friction force. The study of Askari and Tahani [2] was focused on the effect of mechanical shock on dynamic pull-in instability of electrically actuated microbeams through an alternative reduced-order model and using the fourth-order Runge–Kutta method. By setting up a Poincaré map and using a shooting method, Lin et al. [21] obtained a fixed point of periodic motion in the system, period doubling bifurcation, and Hopf bifurcation. The existing and stability conditions of period-1 motion in a single-degree-of-freedom oscillator with double-side constraints were studied by Wang et al. [22]. A smaller shock gap than impact gap could make the periodic motion more stable.

Reboucas et al. [23] examined limitations of the coefficient of restitution model using experimental observations from a simple vibro-impact setup, and the effect of the magnitude response of gap deviation during an experimental sweep. Numerical simulation using Zhuravlev and Ivanov transformations was obtained. A triboelectric energy harvester on a three-degree-of-freedom vibro-impact oscillator was investigated by Fu et al. [24]. The symmetric mass configurations of the oscillator were more beneficial to energy harvesting than the asymmetric cases. Zukovic et al. [25] explored a vibro-impact system consisting of a crank-slider mechanism with one attached oscillator. The cases with an ideal and a non-ideal excitation were analyzed, and analytical and numerical solutions were obtained for the vibrating system with the ideal excitation. For the system with non-ideal excitation, the average value of the frequency was obtained.

The present paper analyzes the motion of a horizontal vibro-impact system with two degrees of freedom in the case when the external coercive force and viscous damping force are known and a periodic vibro-impact model is realized in the system. A variety of possible cases of motion of the vibro-impact system is presented. This process is very complex, and a nonlinear differential equation for one of the masses and a linear differential equation for the other mass is needed in the analysis. It is very difficult to obtain an exact solution for these types of equations, but an analytical approximate solution with a high accuracy will be given by means of OAFM [26–31]. The condition for periodic motion is obtained and the stability of periodic motion is studied.

2. Dynamical Model of the Vibro-Impact System

The considered dynamical model of the vibro-impact system under investigation is depicted in Figure 1, which details the two-degree-of-freedom system with gaps.

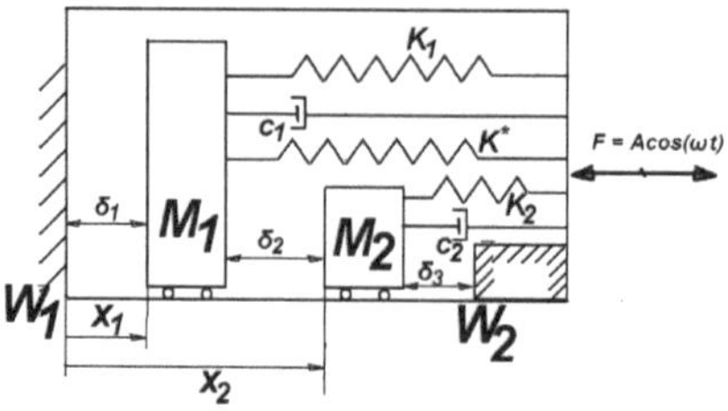

Figure 1. The model of the vibro-impact oscillator.

The system under investigation is composed of a horizontal direction periodical force $F = A\cos\omega t$ and two oscillators of mass M_1 and M_2, which also move in the horizontal direction. These two masses are initially separated by three gaps: δ_1 between the wall W_1 and the mass M_1, δ_2 between the masses M_1 and M_2, and δ_3 between the mass M_2 and the wall W_2. The mass M_1 is connected to a linear spring K_1, nonlinear spring K^*, and the damper c_1, and the mass M_2 is connected to linear spring K_2 and damper c_2. The oscillator M_2 collides with the rigid wall W_2 and with the mass M_1. The coefficient of restitution between M_1 and M_2 is R, and R_2 is the coefficient of restitution between M_1 and W_1 and M_2 and W_2. The displacement between the mass M_1 and vertical wall W_1 is x_1, and the displacement between the mass M_2 and the wall W_1 is x_2.

The equations of motion between any two successive collisions are given as:

$$M_1 x''_1 + c_1(x'_1 - F') + K_1(x_1 - F) + K^*(x_1 - F)^3 = 0, \tag{1}$$

$$M_2 x''_2 + c_2(x'_2 - F') + K_2(x_2 - F) = 0, \tag{2}$$

where the prime denotes the derivative with respect to time t.

By means of the following transformations:

$$\omega t = \Omega\tau \;;\; \Omega_1 = (K_1/M_1)^{1/2} \;;\; \Omega_2 = (K_2/M_2)^{1/2} \;;\; \tau = \Omega_1 t \;;\; x_1 = \delta_1 X \;;\; x_2 = \delta_1 Y \;;\; \alpha_1 = c_1/(2M_1\Omega_1), \\ \alpha_2 = c_2/(2M_2\Omega_1); \beta = K^*\delta_1^2/K_1 \;;\; a = A/\delta_1 \;;\; k = \tfrac{\Omega_2}{\Omega_1}, \tag{3}$$

Equations (1) and (2) can be rewritten in the dimensionless variables, after some simple manipulations:

$$\ddot{X} + 2\alpha_1\dot{X} + (1 + \tfrac{3}{2}a^2)X + \beta X^3 = -2\alpha_1 a\Omega\sin(\Omega\tau) + (a + 3a\beta X^2 + \tfrac{3}{4}\beta a^3)\cos(\Omega\tau) - \tfrac{3}{2}a^2\beta X\cos(2\Omega\tau) + \tfrac{\beta}{4}a^3\cos(3\Omega\tau), \tag{4}$$

$$\ddot{Y} + 2\alpha_2 k\dot{Y} + k^2 Y = ak^2\cos(\Omega\tau) - 2\alpha_2 ka\Omega\sin(\Omega\tau), \tag{5}$$

where the dot denotes the derivative with respect to variable τ.

Between two successive collisions, the initial conditions of Equations (4) and (5) are:

$$X(0) = 1 \;,\; \dot{X}\,(0) = V_1 \;,\; Y\,(0) = 1 + \frac{\delta_2}{\delta_1} \;,\; \dot{Y}\,(0) = V_2, \tag{6}$$

where V_1 and V_2 are known parameters.

Equations (4)–(6) are second-order nonlinear differential equations, and therefore are very difficult to be analytically solved. In the following, we present a relatively new approach, namely the Optimal Auxiliary Functions Method, to obtain an approximate analytical solution in an efficient manner.

3. Basics of the Optimal Auxiliary Functions Method (OAFM)

Following basic concepts of OAFM presented in [26–31], in this work, we consider a general nonlinear equation in the form:

$$L[U(\tau)] + N[U(\tau)] = 0, \tag{7}$$

where L and N are the linear differential and nonlinear differential operators, τ is the independent variable, and $U(\tau)$ is an unknown function. The initial or boundary conditions are:

$$B\left(U, \frac{dU(\tau)}{d\tau}\right) = 0. \tag{8}$$

In order to obtain an approximate solution $\overline{U}(\tau)$, we consider that this approximate solution can be expressed in the form:

$$\overline{U}(\tau) = U_0(\tau) + U_1(\tau, C_1, C_2, \ldots, C_n), \tag{9}$$

in which U_0 (initial approximation) and U_1 (the first approximation) will be determined as described below, and with $C_1, C_2, \ldots, C_n$ being the unknown parameters whose values will be optimally determined.

Substituting Equation (9) into Equation (7), one obtains:

$$L[U_0(\tau)] + L[U_1(\tau)] + N[U_0(\tau) + U_1(\tau, C_1, C_2, \ldots, C_n)] = 0. \tag{10}$$

The initial approximation $U_0(\tau)$ can be obtained from the linear equation:

$$L[U_0(\tau)] = 0 \ , \quad B\left(U_0(\tau), \tfrac{dU_0(\tau)}{d\tau}\right) = 0 \ , \tag{11}$$

and therefore, the first approximation is obtained from Equations (10) and (11):

$$L[U_1(\tau, C_1, C_2, \ldots, C_n)] + N[U_0(\tau) + U_1(\tau, C_1, C_2, \ldots, C_n)] = 0 \ , \quad B\left(U_1(\tau), \tfrac{dU_1(\tau)}{d\tau}\right) = 0 \ . \tag{12}$$

The nonlinear term from Equation (12) can be expanded in the form:

$$N[U_0(\tau) + U_1(\tau, C_i)] = N[U_0(\tau)] + \sum_{k=1}^{p} \frac{U_1^k(\tau, C_i)}{k!} N^{(k)}[U_0(\tau)]. \tag{13}$$

To accelerate the convergence of the first approximation $U_1(\tau, C_i)$ and therefore of the approximate solution as well, and also to avoid the difficulties that can appear in solving the nonlinear differential Equation (12), we propose another expression, such that Equation (12) can be written as follows:

$$U_1(\tau, C_1, C_2, \ldots, C_n) + \sum_{k=1}^{n} C_i f_i(\tau) = 0 \ , \quad B\left(U_1, \tfrac{\partial U_1}{\partial \tau}\right) = 0, \tag{14}$$

where C_i are arbitrary unknown parameters and $f_i(\tau)$ are auxiliary functions depending on the initial approximation $U_0(\tau)$, on the functions which appear in $N[U(\tau)]$ or $N[U_0(\tau)]$ or are combinations of such expressions. These auxiliary functions $f_i(\tau)$ are very important and are not unique. It should be emphasized that we have much freedom to choose such auxiliary functions, because it is known that the nonlinear differential equation does not have a unique solution. The parameters C_i, which appear into Equations (9) and (14), can be optimally identified via rigorous methods such as the Galerkin method, the least square method, the Ritz method, the collocation method, or by minimizing the square residual error. In this last case, if:

$$R(\tau, C_i) = L[\overline{U}(\tau)] + N[\overline{U}(\tau)] \ , \quad \tau \in D \ , \tag{15}$$

where $\overline{U}(\tau)$ is given by Equations (9), (11), and (14), then

$$J(C_1, C_2, \ldots, C_n) = \int_{(D)} R^2(\tau, C_i) d\tau. \quad (16)$$

The values of the parameters C_i are obtained from the following system:

$$\frac{\partial J}{\partial C_1} = \frac{\partial J}{\partial C_2} = \ldots = \frac{\partial J}{\partial C_n} = 0. \quad (17)$$

It is clear that in this way, the approximate solution $\overline{U}(\tau)$ is well determined after the identification of the optimal values of the initially unknown convergence-control parameters C_i.

4. Application of the Optimal Auxiliary Functions Method to Equations (4)–(6)

The approximate solutions and exact solutions of Equations (4)–(6) are of the following forms:

$$\overline{X}(\tau) = X_0(\tau) + X_1(\tau, C_i) \ , \quad i = 1, 2, \ldots, p \ , \quad (18)$$

$$\overline{Y}(\tau) = Y_0(\tau) + Y_1(\tau, C_i + C_j) \ , \quad j = 1, 2, \ldots, q \ . \quad (19)$$

For the nonlinear differential Equation (4) and for the linear Equation (5), the linear operators are, respectively:

$$L(X) = \ddot{X} + 2\alpha_1 \dot{X} + \left(1 + \frac{3}{2}a^2\right) X, \quad (20)$$

$$L(Y) = \ddot{Y} + 2\alpha_2 k \dot{Y} + k^2 Y. \quad (21)$$

The initial approximations $X_0(\tau)$ and the exact solution $Y_0(\tau)$ are obtained from the linear equations:

$$\ddot{X}_0 + 2\alpha_1 \dot{X}_0 + \left(1 + \frac{3}{2}a^2\right) X_0 = 0 \ , \quad X_0(0) = 1 \ , \quad \dot{X}_0(0) = V_1 \ , \quad (22)$$

$$\ddot{Y}_0 + 2\alpha_2 k \dot{Y}_0 + k^2 Y_0 = 0 \ , \quad Y_0(0) = 1 + \tfrac{\delta_2}{\delta_1} \ , \quad \dot{Y}_0(0) = V_2 \ , \quad (23)$$

and therefore:

$$X_0(\tau) = e^{-\alpha_1 \tau}\left(C_1 \cos\sqrt{\overline{\Omega}^2 - \alpha_1^2}\tau + C_2 \sin\sqrt{\overline{\Omega}^2 - \alpha_1^2}\right), \overline{\Omega}^2 = 1 + \frac{3}{2}a^2, \quad (24)$$

$$Y_0(\tau) = e^{-\alpha_2 k \tau}\left[C_3 \cos\sqrt{k^2 - k\alpha_2}\tau + C_4 \sin\sqrt{k^2 - k\alpha_2}\tau\right]. \quad (25)$$

The nonlinear operator for Equation (4) is obtained from Equations (7) and (20):

$$N(X) = \beta X^3 + 2\alpha_1 a \Omega \sin(\Omega\tau) - (a + 3a\beta X^2 + \frac{3}{4}\beta a^3)\cos(\Omega\tau) + \frac{3}{2}a^2 \beta X \cos(2\Omega\tau) - \frac{\beta}{4}a^3 \cos(3\Omega\tau), \quad (26)$$

and the corresponding nonlinear operator for Equation (5) is obtained from Equations (7) and (29) as:

$$N(Y) = ak^2 \cos(\Omega\tau) - 2\alpha_2 k a \Omega \sin(\Omega\tau). \quad (27)$$

Taking into account the initial conditions (6) for the first variable $X(\tau)$ and the initial conditions for (22) for the initial approximation $X_0(\tau)$, one can write the initial conditions for the first approximation $X_1(\tau)$:

$$X_1(0) = \dot{X}_1(0) = 0. \quad (28)$$

The initial conditions for $Y(\tau)$ are:

$$Y_1(0) = 1 + \frac{\delta_2}{\delta_1}, \dot{Y}_1(0) = V_2. \tag{29}$$

From the Equations (26) and (28) and from the Equations (27) and (29), we can choose the first approximation in the forms:

$$X_1(\tau) = Q_1[\cos(3\Omega\tau) - \cos(\Omega\tau)] + Q_2[\sin(3\Omega\tau) - 3\sin(\Omega\tau)], \tag{30}$$

$$Y_1(\tau) = Q_3\cos(\Omega\tau) + Q_4\sin(\Omega\tau). \tag{31}$$

The approximate solution of Equations (4) and (6) and exact solution of Equations (5) and (6) are respectively:

$$\overline{X}(\tau) = e^{-\alpha_1\tau}\left(\cos\sqrt{\overline{\Omega}^2 - \alpha_1^2}\tau + \frac{\alpha_1 + V_1}{\sqrt{\overline{\Omega}^2 - \alpha_1^2}}\sin\sqrt{\overline{\Omega}^2 - \alpha_1^2}\right) + Q_1[\cos(3\Omega\tau) - \cos(\Omega\tau)] + Q_2[\sin(3\Omega\tau) - 3\sin(\Omega\tau)], \tag{32}$$

$$Y(\tau) = e^{-\alpha_2 k\tau}\left[C_3\cos\sqrt{k^2 - k\alpha_2}\tau + C_4\sin\sqrt{k^2 - k\alpha_2}\tau\right] + Q_3\cos(\Omega\tau) + Q_4\sin(\Omega\tau). \tag{33}$$

The parameters Q_1 and Q_2 can be determined by the Galerkin method, Q_3, Q_4 can be determined by identification of coefficients, and C_3, C_4 can be determined from Equation (29):

$$\begin{gathered}
Q_1 = -\frac{9a}{8+12a^2}, \quad Q_2 = \frac{(36a^3-3a)\Omega\alpha_1}{8(2+3a^2)^2}, \quad C_3 = 1 + \frac{\delta_2}{\delta_1} - \frac{ak^2(k^2-\Omega^2)+2a\alpha_2^2k^2\Omega^2}{(\Omega^2-k^2)^2+4\alpha_2^2k^2\Omega^2}, \\
C_4 = \frac{V_2}{\sqrt{k^2-k\alpha_2}} - \frac{a\alpha_2 k\Omega^4}{\sqrt{k^2-k\alpha_2}\left((\Omega^2-k^2)^2+4\alpha_2^2k^2\Omega^2\right)} + \alpha_2 k\left(1 + \frac{\delta_2}{\delta_1} - \frac{ak^2(k^2-\Omega^2)+2a\alpha_2^2k^2\Omega^2}{(\Omega^2-k^2)^2+4\alpha_2^2k^2\Omega^2}\right) \\
Q_3 = \frac{(k^2-\Omega^2)ak^2+4\alpha_2^2ak^2\Omega^2}{(\Omega^2-k^2)+4\alpha_2^2k^2\Omega^2}, \quad Q_4 = \frac{2\alpha_2 ak\Omega^3}{(\Omega^2-k^2)+4\alpha_2^2k^2\Omega^2}
\end{gathered} \tag{34}$$

The approximate solution of Equations (4)–(6) and exact solution are obtained using Equations (32)–(34):

$$\begin{aligned}
\overline{X}(\tau) = e^{-\alpha_1\tau}&\left(\cos\sqrt{\overline{\Omega}^2 - \alpha_1^2\tau} + \frac{\alpha_1 + V_1}{\sqrt{\overline{\Omega}^2 - \alpha_1^2}}\sin\sqrt{\overline{\Omega}^2 - \alpha_1^2\tau}\right) + \frac{9a}{8+12a^2}[\cos(\Omega\tau) - \cos(3\Omega\tau)] + \\
&+ \frac{(36a^3-3a)\Omega\alpha_1}{8(2+3a^2)^2}[\sin(3\Omega\tau) - 3\sin(\Omega\tau)],
\end{aligned} \tag{35}$$

$$\begin{aligned}
Y(\tau) = e^{-\alpha_2 k\tau}&\left[\left(1 + \frac{\delta_2}{\delta_1} - \frac{ak^2(k^2-\Omega^2)+2a\alpha_2^2k^2\Omega^2}{(\Omega^2-k^2)^2+4\alpha_2^2k^2\Omega^2}\right)\cos\sqrt{k^2 - k\alpha_2}\tau + \left(\frac{V_2}{\sqrt{k^2-k\alpha_2}} - \frac{a\alpha_2 k\Omega^4}{\sqrt{k^2-k\alpha_2}\left((\Omega^2-k^2)^2+4\alpha_2^2k^2\Omega^2\right)} + \right.\right. \\
&\left.\left. + \alpha_2 k\left(1 + \frac{\delta_2}{\delta_1} - \frac{ak^2(k^2-\Omega^2)+2a\alpha_2^2k^2\Omega^2}{(\Omega^2-k^2)^2+4\alpha_2^2k^2\Omega^2}\right)\right)\sin\sqrt{k^2 - k\alpha_2}\tau\right] + \\
&+ \frac{ak^2(k^2-\Omega^2)+2a\alpha_2^2k^2\Omega^2}{(k^2-\Omega^2)^2+4\alpha_2^2k^2\Omega^2}\cos\Omega\tau + \frac{a\alpha_2 k\Omega^3}{(k^2-\Omega^2)^2+4\alpha_2^2k^2\Omega^2}\sin\Omega\tau,
\end{aligned} \tag{36}$$

Now, to prove the accuracy and the effectiveness of our method, we consider a set of numerical values for the physical parameters involved in the governing equations; i.e., $M_1 = 0.072$, $M_2 = 0.086$, $K_1 = 110$, $K_2 = 141$, $a = 3$.

Figure 2 presents a comparison between the approximate solution (35) and corresponding numerical integration results obtained by means of a fourth-order Runge–Kutta method.

This comparison emphasizes the high accuracy of the proposed analytical solution obtained through OAFM. Figure 3 presents the corresponding phase portraits.

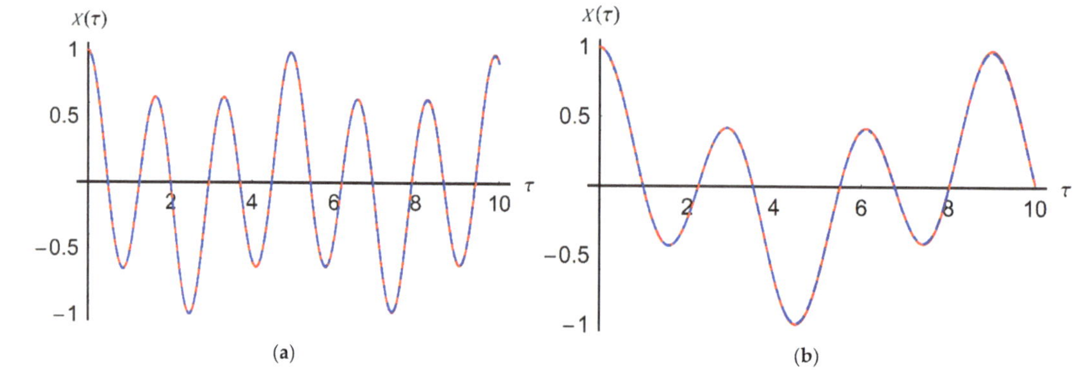

Figure 2. Comparison between the approximate analytical solution (35) and numerical integration results: (**a**) for ω = 49.61 and $\delta_1 = \delta_2$ = 0.02; (**b**) for ω = 27.26 and $\delta_1 = \delta_2$ = 0.04. ______—numerical solution; _ _ _ _—analytical solution.

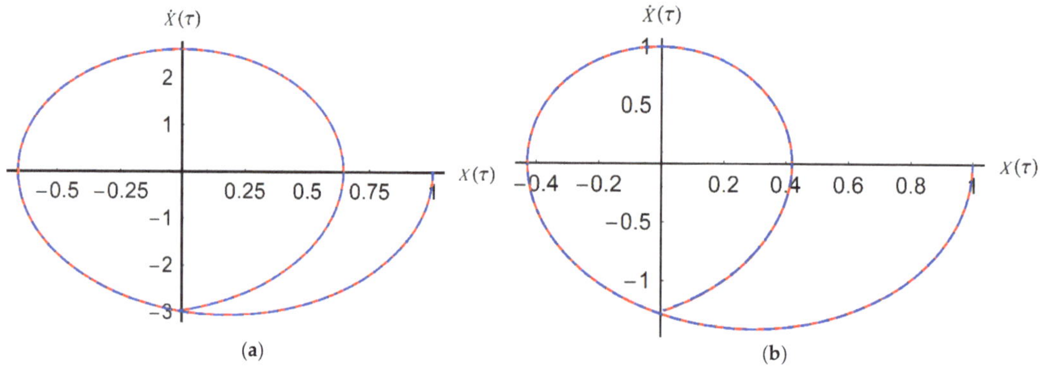

Figure 3. Comparison of the phase portraits for analytical solution (35) and numerical integration results: (**a**) for ω = 49.61 and $\delta_1 = \delta_2$ = 0.02; (**b**) for ω = 27.26 and $\delta_1 = \delta_2$ = 0.04. ______—numerical solution; _ _ _ _—analytical solution.

5. Analysis of Non-Periodic Motion

In this section, we present some possible situations, depending on the motions of the masses M_1 and M_2 of the dynamical system considered in Section 2. Depending on the perturbing force F and initial conditions, we present the following possible cases of non-periodic motions.

Case 5.1. The oscillator of mass M_1 is moving from the static equilibrium position to the position of the impact with the wall W_1. If the moment of impact M_1 with W_{1is} τ_1, then this case takes place if $X(\tau) \leq 1$, $X(\tau_1) = 0$ and $\dot{X}(\tau_1) \leq 0$. In particular, if $\dot{X}(\tau_1) = 0$, then the oscillator M_1 stops at the wall W_1. The velocity of the mass M_1 after the impact will be $\dot{X}(\tau_1^+) = -R_2\dot{X}(\tau_1^-)$, where the symbols "+" and "−" indicate the moments after and before the impact, respectively, and R_2 is the coefficient of restitution. For the oscillator of mass M_2, the following situations are possible (types of movement):

(a) The oscillator M_2 is moving between the oscillator of mass M_1 and the wall W_2 without impact and without sticking motion. The condition is equivalent to the existence of the inequalities $X(\tau) < Y(\tau) < 1 + \delta_2/\delta_1 + \delta_3/\delta_1$ and $\tau < \tau_{1.}$ It follows that there exists a moment of time τ_1^* when M_2 stops, $\dot{Y}(\tau_1^*) = 0$.

(b) The oscillator M_2 impacts the oscillator M_1 before the collision of the mass M_1 with the wall W_1. In this case, there exists a moment $\tau_2 < \tau$ so that $X(\tau_2) = Y(\tau_2)$, $\dot{X}(\tau_2) < 0, \dot{Y}(\tau_2) < 0$, and $\dot{X}(\tau_2) \neq \dot{Y}(\tau_2)$. After this collision, the velocities can be written as:

$$\begin{gathered}\dot{X}(\tau_2^+) = \frac{M_1 - RM_2}{M_1+M_2}\dot{X}(\tau_2^-) + \frac{M_2(1+R)}{M_1+M_2}\dot{Y}(\tau_2^-), \\ \dot{Y}(\tau_2^+) = \frac{M_1(1+R)}{M_1+M_2}\dot{X}(\tau_2^-) + \frac{M_2 - RM_1}{M_1+M_2}\dot{Y}(\tau_2^-)\ ;\ \dot{X}(\tau_2^-) = \dot{X}(\tau_2)\ ;\ \dot{Y}(\tau_2^+) = \dot{Y}(\tau_2)\ ,\end{gathered} \tag{37}$$

(c) The oscillator of mass M_1 is sticking to M_2 (sticking motion). The condition is written as $X(\tau_2) = Y(\tau_2)$ and $\dot{X}(\tau_2) = \dot{Y}(\tau_2)$. The contact force corresponding to masses M_1 and M_2 is obtained from Equations (4) and (5):

$$\begin{gathered}F_{12}(\tau) = 2a\Omega(k\alpha_2 - \Omega\alpha_1)\sin\Omega\tau + (a - ak^2 + 3a\beta X^2 + \tfrac{3}{4}\beta a^3)\cos\Omega\tau - \tfrac{3}{2}a^2\beta X\cos 2\Omega\tau \\ + \tfrac{\beta}{4}a^3\cos 3\Omega\tau + 2\alpha_2\dot{Y} - 2\alpha_1\dot{X} + k^2 Y - (1 + \tfrac{3}{2}a^2)X - \beta X^3.\end{gathered} \tag{38}$$

The sticking takes place only if $F_{12} > 0$.

(d) The oscillator of mass M_2 impacts the wall W_2 at the moment $\tau = \tau_3$. The conditions are $Y(\tau_3) = 1 + \delta_2/\delta_1 + \delta_3/\delta_1$ and $\dot{Y}(\tau_3) \geq 0$. If $\dot{Y}(\tau_3) = 0$, then the oscillator M_2 stops without impacting the wall W_2.

Case 5.2. The oscillator of mass M_1 is moving without impact from the position of static equilibrium between the wall W_1 and the oscillator of mass M_2. The oscillator of mass M_1 stops at the moment $\tau = \tau'_1$ before reaching the wall W_1, then stops at the moment $\tau = \tau''_1$, before reaching the oscillator of mass M_2, so that $\dot{X}(\tau'_1) = \dot{X}(\tau''_1) = 0$. For the oscillator of mass M_2, one can identify two situations:

(a) The oscillator of mass M_2 is moving without impact between the oscillator of mass M_1 and the wall W_2, so that one can write $X(\tau) < Y(\tau) < 1 + \delta_2/\delta_1 + \delta_3/\delta_1$. It results that there exist two moments: τ'_2, when the oscillator of mass M_2 stops before the oscillator M_2; and τ''_2, when the oscillator M_2 stops before the wall W_2, so that $\dot{Y}(\tau'_2) = \dot{Y}(\tau''_2) = 0$.

(b) The oscillator of mass M_2 moving between the oscillator of mass M_1 without impact and the wall W_2 with impact, so that $X(\tau) < Y(\tau) \leq 1 + \delta_2/\delta_1 + \delta_3/\delta_1$. The moment of impact of the oscillator M_2 with the wall W_2 is τ_4. It follows the condition $Y(\tau_4) = 1 + \delta_2/\delta_1 + \delta_3/\delta_1$ and $\dot{Y}(\tau_4^-) > 0$. After the impact, the following expression takes place: $\dot{Y}(\tau_4^+) = -R\dot{Y}(\tau_4^-)$.

Case 5.3. In this last case, the oscillator of mass M_1 is moving from the static equilibrium position toward the oscillator of mas M_2, $X(\tau) \geq 1$. For the oscillator of mass M_2, the following five situations are possible:

(a) The oscillator of mass M_2 is moving from the position of static equilibrium $Y(0) = 1 + \delta_2/\delta_1$ toward the oscillator of mass M_1 without impact. The conditions are $X(\tau) < Y(\tau)$, $Y(\tau) < 1 + \delta_2/\delta_1$, and $\dot{Y}(0) \leq 0$. In this subcase, there exists a moment τ_5 so that $\dot{Y}(\tau_5) = 0$.

(b) The oscillator of mass M_2 is moving from the position of static equilibrium toward the oscillator of mass M_1 with impact. The corresponding conditions for this subcase are $X(\tau) \leq Y(\tau) < 1 + \delta_2/\delta_1$, and there exists a moment τ_6 when $X(\tau_6) = Y(\tau_6)$ and $\dot{Y}(\tau_6) < 0$. After the impact, the velocities of the oscillators will be:

$$\begin{gathered}\dot{X}(\tau_6^+) = \frac{M_1 - RM_2}{M_1+M_2}\dot{X}(\tau_6) + \frac{M_2(1+R)}{M_1+M_2}\dot{Y}(\tau_6), \\ \dot{Y}(\tau_6^+) = \frac{M_1(1+R)}{M_1+M_2}\dot{X}(\tau_6) + \frac{M_2 - RM_1}{M_1+M_2}\dot{Y}(\tau_6).\end{gathered} \tag{39}$$

(c) The oscillator of mass M_2 is moving from the position of static equilibrium toward the wall W_2 without impact with the oscillator of mass M_1 or with the wall W_2. This subcase takes place if $X(\tau) \geq 1$, $Y(\tau) \geq 1 + \delta_2/\delta_1$, and $X(\tau) < Y(\tau)$. It is clear that there exist two moments τ'_7 and τ''_7 when $\dot{X}(\tau'_7) = 0$, which means the oscillator of mass M_1 stops; and $\dot{Y}(\tau''_7) = 0$, which means the oscillator of mass M_2 stops.

(d) The oscillator of mass M_2 is moving from the position of static equilibrium toward the wall W_2, but with impact between the two oscillators. The impact takes place before the oscillator of mass M_2 to reach at the wall W_2 and without sticking. The conditions will be $X(\tau) \geq 1, Y(\tau) \geq 1 + \delta_2/\delta_1$, and there exists a moment τ_7 when $X(\tau_7) = Y(\tau_7)$, $\dot{X}(\tau'_7) > 0$, $\dot{Y}(\tau_7) \geq 0$, and $\dot{X}(\tau_7) > \dot{Y}(\tau_7)$. If the impact of the oscillators takes place with sticking, then the condition is similar to (38), $F_{12}(\tau) > 0, \tau > \tau_7$.

(e) The oscillator of mass M_2 is moving toward the wall W_2, and the impact with W_2 takes place at the moment τ_8, but there is no impact between the two oscillators. Therefore, the following conditions should be satisfied:

$$X(\tau) < Y(\tau) \leq 1 + \frac{\delta_2}{\delta_1} + \frac{\delta_3}{\delta_1} \;;\; Y(\tau_8) = 1 + \frac{\delta_2}{\delta_1} + \frac{\delta_3}{\delta_1} \;;\; \dot{Y}(\tau_8) > 0. \tag{40}$$

After the impact, the velocity of the oscillator of mass M_2 is $\dot{Y}(\tau_8^+) = -R\dot{Y}(\tau_8^-)$.

It should be stated that after the two oscillators undergo one of the movements studied before, the movement of each oscillator should be studied by taking into account the conditions mentioned for the corresponding case. For example, after Case 5.1a takes place, therefore in the condition $\dot{X}(\tau_1) < 0$ (the oscillator of mass M_1 does not stop), $\dot{X}(\tau_1^+) = -R\dot{X}(\tau_1^-)$, and $\dot{Y}(\tau_1^*) = 0$, then the approximate solutions (35) and (36) will be written as:

$$\begin{aligned}\overline{X}(\tau) = {} & e^{-\alpha_1(\tau-\tau_1)}\frac{R_2\dot{X}(\tau_1^-)}{\sqrt{\overline{\Omega}^2-\alpha_1^2}}\sin\sqrt{\overline{\Omega}^2-\alpha_1^2}(\tau-\tau_1) + \frac{9a}{8+12a^2}[\cos\Omega(\tau-\tau_1) - \cos 3\Omega(\tau-\tau_1)] + \\ & + \frac{(36a^3-3a)\Omega\alpha_1}{8(2+3a^2)^2}[\sin 3\Omega(\tau-\tau_1) - 3\sin\Omega(\tau-\tau_1)] \;,\quad \tau \geq \tau_1 \;,\end{aligned} \tag{41}$$

$$\begin{aligned}Y(\tau) = {} & e^{-\alpha_2 k(\tau-\tau_1^*)}\left[(\gamma - Q_3)\cos\sqrt{k^2-k\alpha_2}(\tau-\tau_1^*) + \frac{\alpha_2 k(\gamma-Q_3)-\Omega Q_4}{\sqrt{k^2-k\alpha_2}}\sin\sqrt{k^2-k\alpha_2}(\tau-\tau_1^*)\right] + \\ & + Q_3\cos\Omega(\tau-\tau_1^*) + Q_4\sin\Omega(\tau-\tau_1^*),\end{aligned} \tag{42}$$

where τ_1^* is obtained from the equation $\dot{Y}(\tau_1^*) = 0$; $\gamma = Y(\tau_1^*)$, $Y(\tau_1^*)$ is given by Equation (36); and Q_3 and Q_4 are given by Equation (34).

Figure 4 shows the contact force of (38).

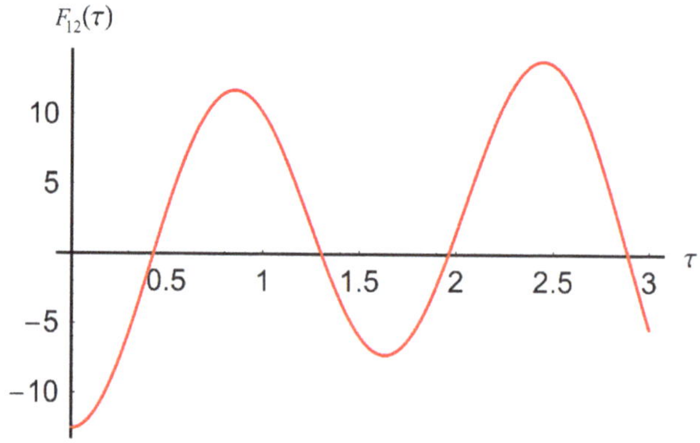

Figure 4. The contact force.

It can be seen that the contact force has two maximum points at $\tau = 0.8$ and $\tau = 2.4$, and is null for four points: $\tau = 0.4$, $\tau = 1.3$, $\tau = 1.9$, and $\tau = 2.9$.

Another specific element for the impact phenomenon is the kinetic energy loss. At the moment of impact, the kinetic energy is transformed into work, caloric energy, etc. The kinetic energy loss is given by the difference between the kinetic energy at the beginning and at the end of impact:

$$\Delta T = \frac{1}{2}M_1\dot{X}^2(\tau^-) + \frac{1}{2}M_2\dot{Y}^2(\tau^-) - \frac{1}{2}M_1\dot{X}^2(\tau^+) - \frac{1}{2}M_2\dot{Y}^2(\tau^+). \tag{43}$$

In Case 5.1b, the kinetic energy loss becomes:

$$\Delta T = \frac{M_1 M_2}{2(M_1 + M_2)}\left[\dot{X}(\tau) - \dot{Y}(\tau)\right]^2(1 - R^2). \tag{44}$$

Figure 5 presents the kinetic energy loss for Case 5.1b.

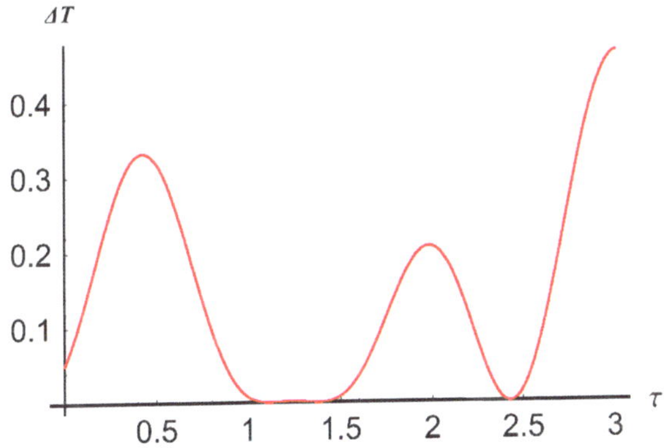

Figure 5. The kinetic energy loss for Case 5.1b.

6. Analysis of Periodic Motion

The solutions (35) and (36) are written for the initial conditions in (6), in the absence of impact. After some time, the free vibration described by (35) and (36) disappear, so that the movement becomes periodic with the same period of the perturbing force F, which is $T = 2\pi/\Omega$.

The problem of the stability of periodic motion is of interest only in the case of the impact between the two oscillators, since the impact with the walls W_1 and W_2 is assumed to be elastic. In the general case, instead of the conditions in (6), we consider that the approximate solution $\overline{X}(\tau)$ and the exact solution $Y(\tau)$ are:

$$\overline{X}(\tau) = e^{-\alpha_1\tau}\left(C_1\cos\sqrt{\overline{\Omega}^2 - \alpha_1^2}\tau + C_2\sin\sqrt{\overline{\Omega}^2 - \alpha_1^2}\tau\right) + \frac{9a^2}{8+12a^2}[\cos(\Omega\tau + \phi) - \cos(3\Omega\tau + 3\phi)] + \\ + \frac{(36a^3 - 3a)\Omega\alpha_1}{8(2+3a^2)^2}[\sin(3\Omega\tau + 3\phi) - \sin(\Omega\tau + \phi)], \tag{45}$$

$$Y(\tau) = e^{-\alpha_2 k\tau}\left[C_3\cos\sqrt{k^2 - k\alpha_2}\tau + C_4\sin\sqrt{k^2 - k\alpha_2}\tau\right] + Q_3\cos(\Omega\tau + \phi) + Q_4\sin(\Omega\tau + \phi). \tag{46}$$

where Q_3 and Q_4 are given by Equation (34), and C_1–C_4 and ϕ are determined from initial conditions.

Solutions (45) and (46) are stable if a negligible change of a small perturbation takes place in the initial conditions and delay at any time interval. If the change is significant, then the solutions become unstable.

In this section, we consider small perturbations for τ, C_1–C_4, and delay ϕ: $\Delta\tau$, ΔC_1, …, ΔC_4, $\Delta\phi$, so that ([4–7,32]):

$$\overline{X}(\tau+\Delta\tau)=e^{-\alpha_1(\tau+\Delta\tau)}\left[(C_1+\Delta C_1)\cos\sqrt{\overline{\Omega}^2-\alpha_1^2}(\tau+\Delta\tau)+(C_2+\Delta C_2)\sin\sqrt{\overline{\Omega}^2-\alpha_1^2}(\tau+\Delta\tau)\right]+ \\ +\frac{9a^2}{8+12a^2}[\cos(\Omega\tau+\Omega\Delta\tau+\phi+\Delta\phi)-\cos(3\Omega\tau+3\Omega\Delta\tau+3\phi+3\Delta\phi)]+ \\ +\frac{(36a^3-3a)\Omega\alpha_1}{8(2+3a^2)^2}[\sin(3\Omega\tau+3\Omega\Delta\tau+3\phi+3\Delta\phi)-\sin(\Omega\tau+\Omega\Delta\tau+\phi+\Delta\phi)], \quad (47)$$

$$Y(\tau+\Delta\tau)=e^{-\alpha_2k(\tau+\Delta\tau)}\left[(C_3+\Delta C_3)\cos\sqrt{k^2-k\alpha_2}(\tau+\Delta\tau)+(C_4+\Delta C_4)\sin\sqrt{k^2-k\alpha_2}(\tau+\Delta\tau)\right]+ \\ +Q_3\cos(\Omega\tau+\Omega\Delta\tau+\phi+\Delta\phi)+Q_4\sin(\Omega\tau+\Omega\Delta\tau+\phi+\Delta\phi), \quad (48)$$

Taking into account the approximations:

$$e^{-\alpha\Delta\tau}\approx 1-\alpha\Delta\tau\ ;\quad \cos(\alpha\tau+\Delta\tau)=\cos\alpha\tau-\Delta\tau\sin\alpha\tau\ ;\quad \sin\ (\alpha\tau+\Delta\tau)=\sin\alpha\tau+\Delta\tau\cos\alpha\tau, \quad (49)$$

and the modifications of the solutions $\overline{X}$ and Y:

$$\Delta\overline{X}(\tau)=\overline{X}(\tau+\Delta\tau)-\overline{X}(\tau)\ ;\quad \Delta Y(\tau)=Y(\tau+\Delta\tau)-Y(\tau)\ , \quad (50)$$

after linearization of Equations (47) and (48) can be written as:

$$\Delta\overline{X}(\tau)=e^{-\alpha_1\tau}\Big[-\Delta\tau(\alpha 1C_1\cos\sqrt{\overline{\Omega}^2-\alpha_1^2}\tau+C_1\sqrt{\overline{\Omega}^2-\alpha_1^2}\sin\sqrt{\overline{\Omega}^2-\alpha_1^2}\tau+\alpha_1C_2\sin\sqrt{\overline{\Omega}^2-\alpha_1^2}\tau- \\ -C_2\sqrt{\overline{\Omega}^2-\alpha_1^2}\cos\sqrt{\overline{\Omega}^2-\alpha_1^2}\tau)+\Delta C_1\cos\sqrt{\overline{\Omega}^2-\alpha_1^2}\tau+\Delta C_2\sin\sqrt{\overline{\Omega}^2-\alpha_1^2}\tau\Big]+\frac{9a^2}{8+12a^2}[(\Omega\Delta\tau+\Delta\phi)(3\sin(3\Omega\tau+3\phi)- \\ -\sin(\Omega\tau+\phi)]+\frac{(108a^3-9a)\Omega\alpha_1}{8(2+3a^2)^2}(\Omega\Delta\tau+\Delta\phi)[3\cos(3\Omega\tau+3\phi)-\cos(\Omega\tau+\phi)], \quad (51)$$

$$\Delta Y(\tau)=e^{-\alpha_2k\tau}\Big[-\Delta\tau(\alpha_2kC_3\cos\sqrt{k^2-k\alpha_2}\tau+C_3\sqrt{k^2-k\alpha_2}\sin\sqrt{k^2-k\alpha_2}\tau-C_4\sqrt{k^2-k\alpha_2}\cos\sqrt{k^2-k\alpha_2}\tau+ \\ +\alpha_2kC_4\sin\sqrt{k^2-k\alpha_2}\tau)+\Delta C_3\cos\sqrt{k^2-k\alpha_2}\tau+\Delta C_4\sin\sqrt{k^2-k\alpha_2}\tau\Big]+ \\ +Q_4(\Omega\Delta\tau+\Delta\phi)\cos(\Omega\tau+\phi)-Q_3(\Omega\Delta\tau+\Delta\phi)\sin(\Omega\tau+\phi), \quad (52)$$

After differentiating Equations (51) and (52), one obtains:

$$\Delta\dot{\overline{X}}(\tau)=e^{-\alpha_1\tau}\Big[\Delta\tau(\overline{\Omega}^2C_1\cos\sqrt{\overline{\Omega}^2-\alpha_1^2}\tau+2\alpha_1C_1\sqrt{\overline{\Omega}^2-\alpha_1^2}\sin\sqrt{\overline{\Omega}^2-\alpha_1^2}\tau+\overline{\Omega}^2C_2\sin\sqrt{\overline{\Omega}^2-\alpha_1^2}\tau- \\ -2\alpha_1C_2\cos\sqrt{\overline{\Omega}^2-\alpha_1^2}\tau)-\Delta C_1(\alpha_1\cos\sqrt{\overline{\Omega}^2-\alpha_1^2}\tau+\sqrt{\overline{\Omega}^2-\alpha_1^2}\sin\sqrt{\overline{\Omega}^2-\alpha_1^2}\tau)+ \\ +\Delta C_2(\sqrt{\overline{\Omega}^2-\alpha_1^2}\cos\sqrt{\overline{\Omega}^2-\alpha_1^2}\tau-\alpha_1\sin\sqrt{\overline{\Omega}^2-\alpha_1^2}\tau\Big]+\frac{9a\Omega}{8+12a^2}(\Omega\Delta\tau+\Delta\phi)[(9\cos(3\Omega\tau+3\phi)-\cos(\Omega\tau+\phi)]+ \\ +\frac{(108a^3-9a)\Omega^2\alpha_1}{8(2+3a^2)^2}(\Omega\Delta\tau+\Delta\phi)[\sin(\Omega\tau+\phi)-9\sin(3\Omega\tau+3\phi)], \quad (53)$$

$$\Delta\dot{Y}(\tau)=e^{-\alpha_2k\tau}\Big[\Delta\tau(k^2C_3\cos\sqrt{k^2-k\alpha_2}\tau+2\alpha_2kC_3\sqrt{k^2-k\alpha_2}\sin\sqrt{k^2-k\alpha_2}\tau+k^2C_4\sin\sqrt{k^2-k\alpha_2}\tau- \\ -2\alpha_2kC_3\cos\sqrt{k^2-k\alpha_2}\tau)-\Delta C_3(\alpha_2k\cos\sqrt{k^2-k\alpha_2}\tau+\sqrt{k^2-k\alpha_2}\sin\sqrt{k^2-k\alpha_2}\tau)+\Delta C_4(\sqrt{k^2-k\alpha_2}\cos\sqrt{k^2-k\alpha_2}\tau- \\ -\alpha_2k\sin\sqrt{k^2-k\alpha_2}\tau\Big]-\Omega Q_3(\Omega\Delta\tau+\Delta\phi)\cos(\Omega\tau+\phi)-\Omega Q_4(\Omega\Delta\tau+\Delta\phi)\sin(\Omega\tau+\phi). \quad (54)$$

For the perturbed motion described by Equations (51) and (52), the initial conditions at two moments—after the n-th impact and before the $(n+1)$-th impact—are written as:

$$n:\quad \tau=0\ ,\quad \Delta\tau_n=0\ ,\quad \Delta\overline{X}_n=\Delta Y_n=0\quad ,\ \Delta\dot{\overline{X}}_n=\Delta\dot{\overline{X}}_n^+\quad ,\ \Delta\dot{Y}_n=\Delta\dot{Y}_n^+\ , \quad (55)$$

$$n+1:\quad \tau=\frac{2n\pi}{\Omega}+\Delta\tau_n\ ,\quad \Delta\tau_n=\Delta T_n=\frac{\Delta\phi_{n+1}-\Delta\phi_n}{\Omega}\ ,\quad \Delta\overline{X}_n=\Delta Y_n=0\quad ,\ \Delta\dot{\overline{X}}_n=\Delta\dot{\overline{X}}_n^-\quad ,\ \Delta\dot{Y}_n=\Delta\dot{Y}_n^-\ , \quad (56)$$

where

$$\Delta\overline{X}_n=\Delta\overline{X}\left(\frac{2n\pi}{\Omega}+\Delta\tau\right)\ ,\quad \Delta Y_n=\Delta Y\left(\frac{2n\pi}{\Omega}+\Delta\tau^*\right)\ . \quad (57)$$

Then, the velocities after the impact become:

$$\begin{aligned} \Delta\dot{\overline{X}}_n^+(0) &= \tfrac{M_1-RM_2}{M_1+M_2}\Delta\dot{\overline{X}}_n^-\left(\tfrac{2n\pi}{\Omega}+\Delta T_n\right)+\tfrac{M_2(1+R)}{M_1+M_2}\Delta\dot{Y}_n^-\left(\tfrac{2n\pi}{\Omega}+\Delta T_n^*\right), \\ \Delta\dot{Y}_n^+(0) &= \tfrac{M_1(1+R)}{M_1+M_2}\Delta\dot{\overline{X}}_n^-\left(\tfrac{2n\pi}{\Omega}+\Delta T_n\right)+\tfrac{M_2-RM_1}{M_1+M_2}\Delta\dot{Y}_n^-\left(\tfrac{2n\pi}{\Omega}+\Delta T_n^*\right). \end{aligned} \tag{58}$$

From Equations (51)–(53), one obtains:

$$\Delta\overline{X}_n(0)=\Delta C_{1n}+\left[\frac{9a(3\sin 3\phi-\sin\phi)}{8+12a^2}+\frac{(108a^3-9a)\Omega\alpha_1(3\cos 3\phi-\cos\phi)}{8(2+3a^2)^2}\right]\Delta\phi_n, \tag{59}$$

$$\Delta Y_n(0)=\Delta C_{3n}+(Q_4\cos\phi-Q_3\sin\phi)\Delta\phi_n^*. \tag{60}$$

From Equations (53)–(55), one obtains:

$$\Delta\dot{\overline{X}}_n^+(0)=-\alpha_1\Delta C_{1n}+\sqrt{\overline{\Omega}^2-\alpha_1^2}\Delta C_{2n}+\left[\frac{9a\Omega(9\cos 3\phi-\cos\phi)}{8+12a^2}+\frac{(108a^3-9a)\Omega^2\alpha_1(\sin\phi-9\sin 3\phi)}{8(2+3a^2)^2}\right]\Delta\phi_n, \tag{61}$$

$$\Delta\dot{Y}_n^+(0)=-\alpha_2 k\Delta C_{3n}+\sqrt{k^2-k\alpha_2}\Delta C_{4n}-(\Omega Q_3\cos\phi+\Omega Q_4\sin\phi)\Delta\phi_n^*. \tag{62}$$

Analogously, from Equations (51), (52), and (56), one obtains:

$$\begin{aligned} \Delta\overline{X}_n\left(\tfrac{2n\pi}{\Omega}+\Delta\tau_n\right) &= e^{-\frac{2n\pi\alpha_1}{\Omega}}\left\{-\left[\left(\alpha_1 C_1-C_2\sqrt{\overline{\Omega}^2-\alpha_1^2}\right)\cos\frac{2n\pi\sqrt{\overline{\Omega}^2-\alpha_1^2}}{\Omega}+\left(C_1\sqrt{\overline{\Omega}^2-\alpha_1^2}+\right.\right.\right. \\ &\left.\left.+\alpha_1 C_2\right)\sin\frac{2n\pi\sqrt{\overline{\Omega}^2-\alpha_1^2}}{\Omega}\right]\frac{\Delta\phi_{n+1}-\Delta\phi_n}{\Omega}+\Delta C_1\cos\frac{2n\pi\sqrt{\overline{\Omega}^2-\alpha_1^2}}{\Omega}+\Delta C_2\sin\frac{2n\pi\sqrt{\overline{\Omega}^2-\alpha_1^2}}{\Omega}\Bigg\}+ \\ &+\left[\tfrac{9a}{8+12a^2}(3\sin 3\phi-\sin\phi)+\tfrac{(108a^3-9a)\Omega\alpha_1}{8(2+3a^2)^2}(3\cos 3\phi-\cos\phi)\right]\Delta\phi_{n+1}, \end{aligned}$$

$$\begin{aligned} \Delta Y_n\left(\tfrac{2n\pi}{\Omega}+\Delta\tau_n\right) &= e^{-\frac{2n\pi\alpha_2 k}{\Omega}}\left\{-\left[\left(\alpha_2 kC_3-C_4\sqrt{k^2-k\alpha_2}\right)\cos\frac{2n\pi\sqrt{k^2-k\alpha_2}}{\Omega}+\left(C_3\sqrt{k^2-k\alpha_2}+\right.\right.\right. \\ &\left.\left.+\alpha_2 kC_4\right)\sin\frac{2n\pi\sqrt{k^2-k\alpha_2}}{\Omega}\right]\frac{\Delta\phi_{n+1}^*-\Delta\phi_n^*}{\Omega}+\Delta C_3\cos\frac{2n\pi\sqrt{k^2-k\alpha_2}}{\Omega}+\Delta C_4\sin\frac{2n\pi\sqrt{k^2-k\alpha_2}}{\Omega}\Bigg\}+ \\ &+[Q_4\cos\phi-Q_3\sin\phi]\Delta\phi_{n+1}^*. \end{aligned}$$

From Equations (53), (54), and (56), the results are:

$$\begin{aligned} \Delta\dot{\overline{X}}_n^-\left(\tfrac{2n\pi}{\Omega}+\Delta T_n\right) &= e^{-\frac{2n\pi\alpha_1}{\Omega}}\left\{\left[\left(\overline{\Omega}^2 C_1-2\alpha_1 C_2\right)\cos\frac{2n\pi\sqrt{\overline{\Omega}^2-\alpha_1^2}}{\Omega}+(2\alpha_1 C_1+\overline{\Omega}^2 C_2)\sin\frac{2n\pi\sqrt{\overline{\Omega}^2-\alpha_1^2}}{\Omega}\right]\frac{\Delta\phi_{n+1}-\Delta\phi_n}{\Omega}-\right. \\ &-\Delta C_1\left(\alpha_1\cos\frac{2n\pi\sqrt{\overline{\Omega}^2-\alpha_1^2}}{\Omega}+\sqrt{\overline{\Omega}^2-\alpha_1^2}\sin\frac{2n\pi\sqrt{\overline{\Omega}^2-\alpha_1^2}}{\Omega}\right)+\Delta C_2\left(\sqrt{\overline{\Omega}^2-\alpha_1^2}\cos\frac{2n\pi\sqrt{\overline{\Omega}^2-\alpha_1^2}}{\Omega}-\right. \\ &\left.\left.-\alpha_1\sin\frac{2n\pi\sqrt{\overline{\Omega}^2-\alpha_1^2}}{\Omega}\right)\right\}+\left[\tfrac{9a\Omega}{8+12a^2}(9\cos 3\phi-\cos\phi)+\tfrac{(108a^3-9a)\Omega^2\alpha_1}{8(2+3a^2)^2}(\sin\phi-9\sin 3\phi)\right]\Delta\phi_{n+1}, \end{aligned} \tag{63}$$

$$\begin{aligned} \Delta\dot{Y}_n^-\left(\tfrac{2n\pi}{\Omega}+\Delta T_n^*\right) &= e^{-\frac{2n\pi k\alpha_2}{\Omega}}\left\{\left[(k^2 C_3-2\alpha_2 kC_4)\cos\frac{2n\pi\sqrt{k^2-k\alpha_2}}{\Omega}+(2\alpha_2 kC_3+k^2 C_4)\sin\frac{2n\pi\sqrt{k^2-k\alpha_2}}{\Omega}\right]\frac{\Delta\phi_{n+1}^*-\Delta\phi_n^*}{\Omega}-\right. \\ &-\Delta C_3\left(\alpha_2 k\cos\frac{2n\pi\sqrt{k^2-k\alpha_2}}{\Omega}+\sqrt{k^2-k\alpha_2}\sin\frac{2n\pi\sqrt{k^2-k\alpha_2}}{\Omega}\right)+\Delta C_4\left(\sqrt{k^2-k\alpha_2}\cos\frac{2n\pi\sqrt{k^2-k\alpha_2}}{\Omega}-\right. \\ &\left.\left.-\alpha_2 k\sin\frac{2n\pi\sqrt{k^2-k\alpha_2}}{\Omega}\right)\right\}-\Omega[Q_3\cos\phi+Q_4\sin\phi]\Delta\phi_{n+1}^*. \end{aligned} \tag{64}$$

After the elimination of ΔC_{1n} from Equations (55) and (59), as well as of ΔC_{3n} from Equations (55) and (60), and using the notations:

$$F(a,\phi)=-\frac{9a}{8+12a^2}(3\sin 3\phi-\sin\phi)-\frac{(108a^3-9a)\Omega\alpha_1}{8(2+3a^2)^2}(\cos 3\phi-\cos\phi),$$

$$G(a,\phi) = \frac{ak^2(k^2-\Omega^2)+2\alpha_2^2k^2\Omega^2 a}{(k^2-\Omega^2)^2+4\alpha_2^2k^2\Omega^2}\sin\phi - \frac{\alpha_2 ak\Omega^3}{(k^2-\Omega^2)^2+4\alpha_2^2k^2\Omega^2}\cos\phi,$$
$$H(a,\phi) = \frac{9a\Omega}{8+12a^2}(9\cos 3\phi - \cos\phi) + \frac{(108a^3-9a)\Omega^2\alpha_1}{8(2+3a^2)^2}(\sin\phi - 9\sin 3\phi),$$
$$K(a,\phi) = \frac{\alpha_2 ak\Omega^3}{(k^2-\Omega^2)^2+4\alpha_2^2k^2\Omega^2}\sin\phi - \frac{ak^2(k^2-\Omega^2)+2\alpha_2^2k^2\Omega^2 a}{(k^2-\Omega^2)^2+4\alpha_2^2k^2\Omega^2}\cos\phi,$$

$$C(n) = \cos\frac{2n\pi\sqrt{\overline{\Omega}^2-\alpha_1^2}}{\Omega}\ ;\ S(n) = \sin\frac{2n\pi\sqrt{\overline{\Omega}^2-\alpha_1^2}}{\Omega}\ ;\ C^*(n) = \cos\frac{2n\pi\sqrt{k^2-k\alpha_2}}{\Omega}\ ;\ S^*(n) = \sin\frac{2n\pi\sqrt{k^2-k\alpha_2}}{\Omega},$$

$$\begin{aligned}
L_1 &= \frac{1}{\Omega}e^{-\frac{2n\pi\alpha_1}{\Omega}}\left[\left(\alpha_1C_1-\sqrt{\overline{\Omega}^2-\alpha_1^2}C_2\right)C(n)+\left(C_1\sqrt{\overline{\Omega}^2-\alpha_1^2}+\alpha_1C_2\right)S(n)\right],\\
L_2 &= \frac{1}{\Omega}e^{-\frac{2n\pi k\alpha_2}{\Omega}}\left[\left(\alpha_2kC_3-\sqrt{k^2-k\alpha_2}C_4\right)C^*(n)+\left(C_3\sqrt{k^2-k\alpha_2}+\alpha_2kC_4\right)S^*(n)\right],\\
L_3 &= \tfrac{1}{\Omega}e^{-\frac{2n\pi\alpha_1}{\Omega}}\left[\left(\overline{\Omega}^2C_1-2\alpha_1C_2\right)C(n)+\left(2\alpha_1C_1+\overline{\Omega}^2C_2\right)S(n)\right],\\
L_4 &= \tfrac{1}{\Omega}e^{-\frac{2n\pi k\alpha_2}{\Omega}}\left[(k^2C_3-2\alpha_2kC_4)C^*(n)+(2\alpha_2kC_3+k^2C_4)S^*(n)\right],\\
A &= \frac{M_1-M_2R}{M_1+M_2}\ ;\ B = \frac{M_1(1+R)}{M_1+M_2},\\
D_1 &= \sqrt{\overline{\Omega}^2-\alpha_1^2}[AC(n)-1]-\alpha_1AS(n)\ ;\ D_2 = B\left[\sqrt{\overline{\Omega}^2-\alpha_1^2}C(n)-\alpha_1S(n)\right],\\
D_3 &= (1+R-B)\left[\sqrt{k^2-k\alpha_2}C^*(n)-\alpha_2kS^*(n)\right]\ ;\ D_4 = (1-R-A)\left(\sqrt{k^2-k\alpha_2}C^*(n)-\alpha_2kS^*(n)\right)-\sqrt{k^2-k\alpha_2},\\
D_5 &= \alpha_1F(a,\phi)-H(a,\phi)-AF(a,\phi)\left[\alpha_1C(n)+\sqrt{\overline{\Omega}^2-\alpha_1^2}S(n)\right]-AL_3,\\
D_6 &= AH(a,\phi)+AL_3\ ;\ D_7 = \alpha_2kG(a,\phi)-K(a,\phi)-BF(a,\phi)\left[\alpha_1C(n)+\sqrt{\overline{\Omega}^2-\alpha_1^2}S(n)\right]-L_3,\\
D_8 &= BH(a,\phi)+L_3\ ;\ D_9 = -(1+R-B)G(a,\phi)\left[\alpha_2kC^*(n)+\sqrt{k^2-k\alpha_2}S^*(n)\right]-(1+R-B)L_4,\\
D_{10} &= (1+R-B)H(a,\phi)+L_4(1+R-B)\ ;\ D_{11} = -(1-R-A)G(a,\phi)-L_4\ ;\ D_{12} = (1-R-A)H(a,\phi)+L_4,\\
E_1 &= L_1+C(n)F(a,\phi)\ ;\ E_2 = L_1-F(a,\phi)\ ;\ E_3 = C^*(n)G(a,\phi)\ ;\ E_4 = L_2+G(a,\phi),
\end{aligned} \tag{65}$$

one obtains a linear system of four equations with the unknowns ΔC_{2n}, ΔC_{4n}, $\Delta\phi_n$, and $\Delta\phi^*_n$. Introducing the notations:

$$C_{2n} = \theta_1\lambda^n\ ;\ \Delta C_{4n} = \theta_2\lambda^n\ ;\ \Delta\phi_n = \theta_3\lambda^n\ ;\ \Delta\phi_n^* = \theta_4\lambda^n, \tag{66}$$

one obtains the characteristic equation:

$$a_0\lambda^2+a_1\lambda+a_2 = 0, \tag{67}$$

where

$$\begin{aligned}
a_0 &= S(n)S^*(n)(D_6D_{12}-D_7D_{10})+S(n)(D_4D_6E_4-D_3D_8E_4)+S^*(n)(D_1D_{12}E_2-D_2D_{10}E_2)+E_2E_4D_1D_4,\\
a_1 &= S(n)S^*(n)(D_6D_{11}+D_5D_{12}-D_7D_{10}-D_8D_9)+S(n)(D_3D_8E_3-D_4D_6E_3-D_3D_7E_4+D_4D_5E_4)+\\
&\quad +S^*(n)(D_1D_{11}E_2-D_2D_5E_2-D_1D_2E_1+D_2D_{10}E_1)+(D_2D_3-D_1D_4)(E_2E_3+E_1E_4),\\
a_2 &= S(n)S^*(n)(D_5D_{11}-D_7D_9)+S(n)(D_3D_7E_3-D_4D_5E_3)+S^*(n)(D_2D_9E_1-D_1D_{11}E_1)+E_1E_3(D_1D_4-D_2D_3).
\end{aligned} \tag{68}$$

The analyzed motion is asymptotically stable if the roots of the characteristic Equation (67) verify the conditions:

$$|\lambda_{1,2}| < 1. \tag{69}$$

According to the Schur criterion, the conditions in (69) are satisfied if the following inequalities take place:

$$-1 < \frac{a_2}{a_0} < 1 \; ; \quad -1 < \frac{a_1}{a_0 + a_2} < 1. \tag{70}$$

In order to study the stability conditions in (7), the parameters C_1–C_4 and the delay ϕ should be determined. In this respect, the periodicity conditions will be used:

$$\begin{gathered} \overline{X}(0) = Y(0) = \overline{X}\left(\tfrac{2n\pi}{\Omega}\right) = Y\left(\tfrac{2n\pi}{\Omega}\right), \quad \dot{\overline{X}}(0) = A\dot{\overline{X}}\left(\tfrac{2n\pi}{\Omega}\right) + (1 + R - B)\dot{Y}\left(\tfrac{2n\pi}{\Omega}\right), \\ \dot{Y}(0)^{+} = B\dot{\overline{X}}\left(\tfrac{2n\pi}{\Omega}\right) + (1 - R - A)\dot{Y}^{-}\left(\tfrac{2n\pi}{\Omega}\right). \end{gathered} \tag{71}$$

where A and B were defined in Equation (65).

From Equations (45), (46), and (71), one obtains the system of equations:

$$\begin{gathered} \dot{\overline{X}}(0) = A\dot{\overline{X}}\left(\tfrac{2n\pi}{\Omega}\right) + (1 + R - B)\dot{Y}\left(\tfrac{2n\pi}{\Omega}\right) \; ; \quad \dot{Y}(0) = B\dot{\overline{X}}\left(\tfrac{2n\pi}{\Omega}\right) + (1 - R - A)\dot{Y}\left(\tfrac{2n\pi}{\Omega}\right), \\ C_1 = e^{-\frac{2n\pi\alpha_1}{\Omega}}[C_1 C(n) + C_2 S(n)] \; ; \quad C_3 = e^{-\frac{2n\pi\alpha_2 k}{\Omega}}[C_3 C^*(n) + C_4 S^*(n)], \\ C_1 + \tfrac{9a^2}{8+12a^2}(\cos\phi - \cos 3\phi) + \tfrac{(36a^3 - 3a)\Omega\alpha_1}{8(2+3a^2)^2}(\sin 3\phi - \sin\phi) = C_3 + Q_3\cos\phi + Q_4\sin\phi, \end{gathered} \tag{72}$$

where

$$\begin{gathered} \dot{\overline{X}}(0) = -\alpha_1 C_1 + \sqrt{\overline{\Omega}^2 - \alpha_1^2}C_2 + \tfrac{9a^2\Omega}{8+12a^2}(3\sin 3\phi - \sin\phi) + \tfrac{(36a^3 - 3a)\Omega\alpha_1}{8(2+3a^2)^2}(\cos\phi - 3\cos 3\phi), \\ \dot{Y}(0) = -\alpha_2 k C_3 + \sqrt{k^2 - k\alpha_2}C_4 - \Omega Q_3\sin\phi + \Omega Q_4\cos\phi, \\ \dot{\overline{X}}\left(\tfrac{2\pi}{\Omega}\right) = e^{-\frac{2n\pi\alpha_1}{\Omega}}\left[\left(\sqrt{\overline{\Omega}^2 - \alpha_1^2}C_2 - \alpha_1 C_1\right)C(n) - \left(\alpha_1 C_2 + \sqrt{\overline{\Omega}^2 - \alpha_1^2}C_1\right)S(n)\right] + \\ + \tfrac{9a^2\Omega}{8+12a^2}(3\sin 3\phi - \sin\phi) + \tfrac{(36a^3 - 3a)\Omega^2\alpha_1}{8(2+3a^2)^2}(\cos\phi - 3\cos 3\phi), \\ \dot{Y}\left(\tfrac{2\pi}{\Omega}\right) = e^{-\frac{2n\pi\alpha_1}{\Omega}}\left[\sqrt{k^2 - k\alpha_2}C_4 - \alpha_2 k C_3\right]C^*(n) - \left(\alpha_2 k C_4 + \sqrt{k^2 - k\alpha_2}C_3\right)S^*(n) - \Omega Q_3\sin\phi + \Omega Q_4\cos\phi. \end{gathered} \tag{73}$$

A graphical analysis of the stability conditions is presented in Figures 6–9, taking into account the influence of different parameters; the yellow color denotes a zone of stability.

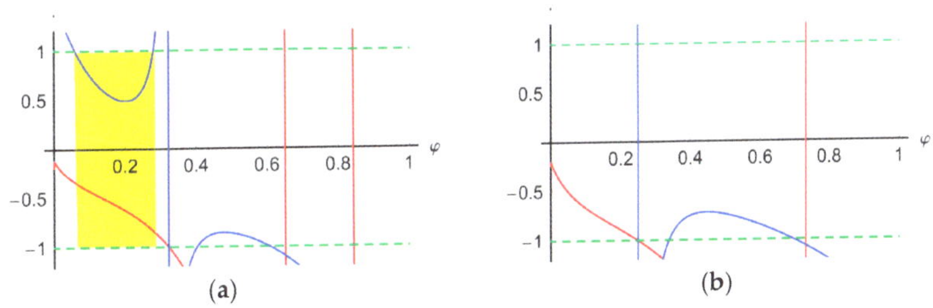

Figure 6. Stability analysis for the variation of ϕ: (**a**) case $n = 1$; (**b**) case $n = 2$.

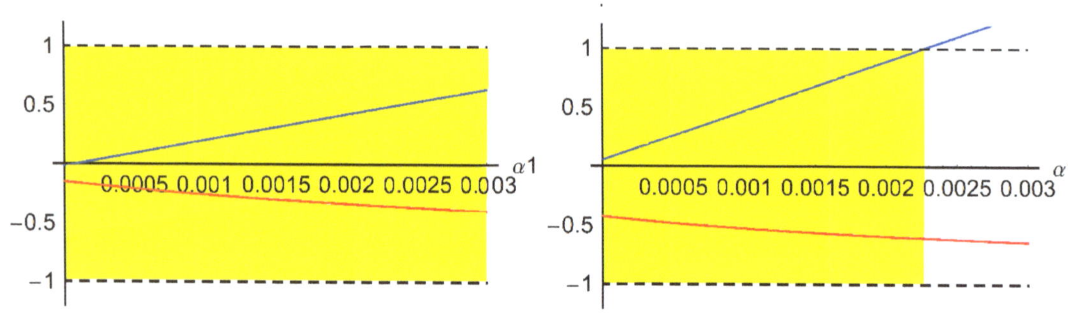

(a) (b)

Figure 7. Stability analysis for the variation of α_1: (**a**) case n = 1; (**b**) case n = 2.

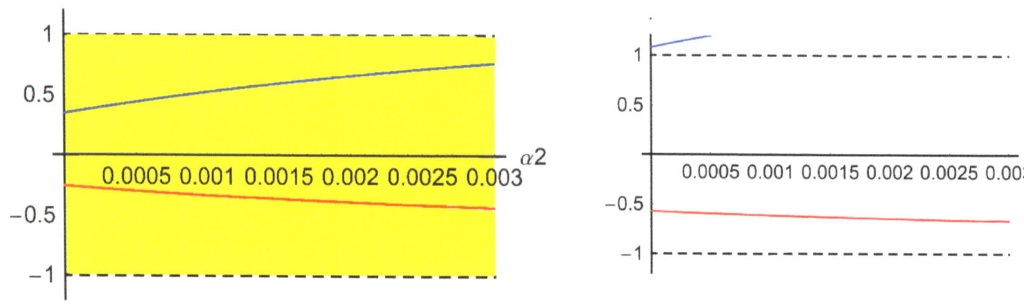

(a) (b)

Figure 8. Stability analysis for the variation of α_2: (**a**) case n = 1; (**b**) case n = 2.

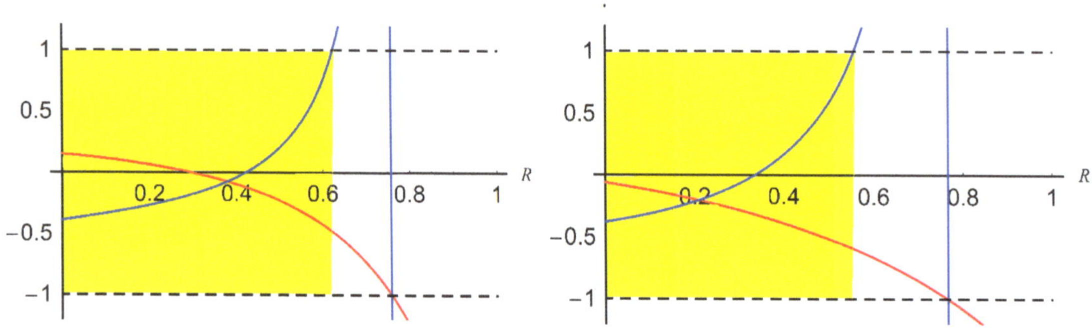

(a) (b)

Figure 9. Stability analysis for the variation of R: (**a**) case n = 1; (**b**) case n = 2.

7. Conclusions

In this paper, we presented a study on the dynamical model of a two-degree-of-freedom vibro-impact oscillator with damping, subject to a perturbing force. Within this system, each of the masses may produce a combination of continuous movements or vibro-impact movements that are quite complicated. The system under study was composed of a linear oscillator and nonlinear one whose movements were determined by means of an approximate analytical solution, very close to the numerical solution. This approximate analytical solution was optimally determined using a new method, namely the Optimal Auxiliary Functions Method (OAFM), which was very efficient in practice. This solution was further used in analyzing the kinetic energy loss, the contact force, and the stability of periodic motions. Several cases were emphasized for possible movements of the system depending on the initial conditions and movement parameters, considering various impact possibilities between the two oscillators or between the oscillators and the walls.

In the case of periodic movements, it was considered that the period of the system was proportional with the period of the perturbing force. The boundary conditions of the system between two periodic subsequent arbitrary impacts were considered. After some analytical and numerical computations and by using the Schur principle, we established the stability conditions of the system. For different values of the movement parameters, a graphical analysis of stability conditions was presented. Different expressions for the velocities of the two oscillators were considered, establishing recurrence relations for periodic solutions.

Limitations of this study include the fact that it did not cover bifurcation- and chaos-related problems, but these aspects will be the subject of future developments. In addition, more research will be necessary to refine and further elaborate the proposed approach to make it applicable to three-degree-of-freedom vibro-impact systems, which would provide a more complete understanding of more complex systems. Moreover, as an alternative to the Schur criterion, use of the Zhukovsky-type stability [33] could be convenient, and the Sommerfeld effect [34,35] could be investigated by means of OAFM.

Author Contributions: Conceptualization, N.H. and V.M.; methodology, V.M.; software, N.H.; validation, N.H., V.M.; formal analysis, V.M.; investigation, N.H., V.M.; resources, N.H.; data curation, N.H., V.M.; writing—original draft preparation, N.H., V.M.; writing—review and editing, N.H.; visualization, N.H.; supervision, N.H.; project administration, N.H. All authors have read and agreed to the published version of the manuscript.

Funding: This research received no external funding.

Institutional Review Board Statement: Not applicable.

Informed Consent Statement: Not applicable.

Data Availability Statement: All the data were presented within the paper.

Conflicts of Interest: The authors declare no conflict of interest.

References

1. Sayed, M.; Hamed, Y.S. Analytical, numerical solutions and study of stability, resonance for nonlinear vibro-impact system under different excitation. *Int. J. Sci. Eng. Res.* **2014**, *5*, 38–45.
2. Askari, A.R.; Tahani, M. An alternative reduced order model for electrically actuated micro-beams under mechanical shock. *Mech. Res. Commun.* **2014**, *57*, 34–39. [CrossRef]
3. Thomsen, J.J.; Fidlin, A. Near-elastic vibro-impact analysis by discontinuous transformations and averaging. *J. Sound Vibr.* **2008**, *12*, 386–407. [CrossRef]
4. Masri, S.F. Analytical and Experimental Studies of Impact Dampers. Ph.D. Thesis, California Institute of Technology, Pasadena, CA, USA, 1965.
5. Kobrinski, A.Y. *Mechanisms with Elastic Couplings. Dynamics and Stability*; Nauka Press: Moscow, Russia, 1964.
6. Babitsky, V.I. *Theory of Vibro-Impact System and Applications*; Springer: Berlin/Heidelberg, Germany, 1988.
7. Brîndeu, L. Stability of the periodic motions of the vibro-impact systems. *Chaos Solitons Fractals* **2000**, *11*, 2493–2503. [CrossRef]
8. Okolewska, B.B.; Czolczynski, K.; Kapitaniak, T. Dynamics of a two-degree-of-freedom cantilever beam with impacts. *Chaos Solitons Fractals* **2009**, *40*, 1991–2006. [CrossRef]
9. Cronin, D.I.; Van, N.K. Substitute for the impact damper. *J. Eng. Ind.* **1975**, *97*, 1295–1300. [CrossRef]
10. Shaw, S.W. The dynamics of a harmonically excited system having rigid amplitude constraints. *J. Appl. Mech.* **1985**, *52*, 453–458. [CrossRef]
11. Ivanov, A.P. Analytical methods in the theory of vibro-impact systems. *J. Appl. Mech.* **1993**, *57*, 221–236. [CrossRef]
12. Aziz, M.A.F.; Vakakis, A.F.; Manevich, L.I. Exact solutions of the problem of the vibro-impact oscillations of a discrete system with two degrees of freedom. *J. Appl. Mech.* **1999**, *63*, 527–530. [CrossRef]
13. Luo, A.C.J. An unsymmetrical motion in a horizontal impact oscillator. *J. Vibr. Acoust.* **2002**, *124*, 453–458. [CrossRef]

14. De Souza, S.L.T.; Caldas, I.L. Controlling chaotic orbits in mechanical systems with impact. *Chaos Solitons Fractals* **2004**, *19*, 171–178. [CrossRef]
15. Emans, J.; Wiercigroch, M.; Krivtsov, A.M. Cumulative effect of structural nonlinearities: Chaotic dynamics of cantilever beam system with impact. *Chaos Solitons Fractals* **2005**, *23*, 1661–1670. [CrossRef]
16. Mann, B.P.; Carter, R.E.; Hazra, S.S. Experimental study of an impact oscillator with viscoelastic and Hertzian contact. *Nonlin. Dyn.* **2007**, *50*, 587–596. [CrossRef]
17. Avramov, K.V. Application of non smooth transformations to analyze a vibroimpact Duffing system. *Int. J. Appl. Mech.* **2008**, *44*, 1173–1179. [CrossRef]
18. Bichri, A.; Belhaq, M.; Liaudet, J.P. Control of vibroimpact dynamics of a single-sided Hertzian contact forced oscillator. *Nonlin. Dyn.* **2010**, *63*, 51–60. [CrossRef]
19. Grace, I.M.; Ibrahim, R.A.; Pilipchuk, V.N.P. Inelastic impact dynamics of ships with one-sided barriers. Part 1: Analytical and numerical investigations. *Nonlin. Dyn.* **2011**, *66*, 589–607. [CrossRef]
20. Jovic, S.; Raicevic, V.; Garic, L. Vibro-impact system based on forced oscillations of heavy mass particle along a rough parabolic line. *Math. Probl. Eng.* **2012**. [CrossRef]
21. Lin, M.; Song, G.; Liu, Y.; Zhu, Y.; Zhang, X. Dynamic behavior analysis of vibro-impact system with two motion limited constraints. In Proceedings of the 4th International Conference on Mechanical and Manufacturing Engineering, Wuhan, China, 15–16 October 2016.
22. Wang, J.; Shen, Y.; Yang, S. Dynamical analysis of a single degree-of-freedom impact oscillator with impulse excitation. *Adv. Mech. Eng.* **2017**, *9*, 1–10. [CrossRef]
23. Reboucas, G.F.S.; Santos, I.E.; Thomsen, J.J. Unilateral vibro-impact systems-experimental observations against theoretical predictions based on the coefficient of restitution. *J. Sound Vibr.* **2019**, *440*, 346–371. [CrossRef]
24. Fu, Y.; Quyang, H.; Davis, R.B. Nonlinear dynamics and triboelectric energy harvesting from a three-degree-of-freedom vibro-impact oscillator. *Nonlin. Dyn.* **2018**, *92*, 1985–2004. [CrossRef]
25. Zukovic, M.; Hajradinovic, D.; Kovacic, I. On the dynamics of vibro-impact systems with ideal and non-ideal excitation. *Meccanica* **2021**, *56*, 439–460. [CrossRef]
26. Marinca, V.; Herisanu, N. The nonlinear thermomechanical vibration of a functionally graded beam on Winkler-Pasternak foundation. *MATEC Web Conf.* **2018**, *148*, 13004. [CrossRef]
27. Herisanu, N.; Marinca, V. An effective analytical approach to nonlinear free vibration of elastically actuated microtubes. *Meccanica* **2021**, *56*, 813–823. [CrossRef]
28. Marinca, V.; Herisanu, N. Optimal Auxiliary Functions Method for a pendulum wrapping on two cylinders. *Mathematics* **2020**, *8*, 1364. [CrossRef]
29. Marinca, V.; Herisanu, N. Construction of analytic solutions to axisymmetric flow and heat transfer on a moving cylinder. *Symmetry* **2020**, *12*, 1335. [CrossRef]
30. Herisanu, N.; Marinca, V. An efficient analytical approach to investigate the dynamics of a misaligned multirotor system. *Mathematics* **2020**, *8*, 1083. [CrossRef]
31. Herisanu, N.; Marinca, V.; Madescu, G. Application of the Optimal Auxiliary Functions Method to a permanent magnet synchronous generator. *Int. J. Nonlinear Sci. Numer. Simul.* **2019**, *20*, 399–406. [CrossRef]
32. Awrejcewicz, J.; Tomczak, K.; Lamarque, C.H. Controlling systems with impacts. *Int. J. Bifurc. Chaos* **1999**, *9*, 547–553. [CrossRef]
33. Shiryaev, A.S.; Khusainov, R.R.; Mamedov, S.N.; Gusev, S.V.; Kuznetsov, N.V. On Leonov's method for computing the linearization of transverse dynamics and analyzing Zhukovsky stability. *Vestnik St. Petersburg Univ. Math.* **2019**, *52*, 334–341.
34. Dudkowski, D.; Jafari, S.; Kapitaniak, T.; Kuznetsov, N.V.; Leonov, G.A.; Prasad, A. Hidden attractors in dynamical systems. *Phys. Rep.* **2016**, *637*, 1–50. [CrossRef]
35. Kiseleva, M.M.; Kondratyeva, N.; Kuznetsov, N.; Leonov, G. *Hidden Oscillations in Electromechanical Systems*; Springer: Cham, Switzerland, 2018.

mathematics

MDPI

Article

Impact Dynamics Analysis of Mobile Mechanical Systems

Sorin Dumitru [1], Andra Constantin [2], Cristian Copilusi [2] and Nicolae Dumitru [2,*]

1 Faculty of Automation, Computers and Electronics, University of Craiova, 200776 Dolj, Romania; sorin.dumitru@edu.ucv.ro
2 Faculty of Mechanics, University of Craiova, 200512 Dolj, Romania; constantin.b6c@student.ucv.ro (A.C.); cristian.copilusi@edu.ucv.ro (C.C.)
* Correspondence: nicolae.dumitru@edu.ucv.ro

Abstract: The current paper focuses on the impact phenomenon analysis, in the case of multi-body mechanical systems undergoing fast motion, due to the presence of some manufacturing and mounting errors or due to some accident during the transport mechanical systems. Thus, the impact phenomenon was analyzed in two cases, the first one consisting of a two bodies, namely, a free-fall body brought in contact with the other considered fixed in space and the second case, which is a complex one, when the analyzed bodies are components of a multi-body mechanical system. The research main objective is to analyze the impact generated between the two bodies through three methods, i.e., the analytical method, a virtual prototyping method accomplished with MSC Adams software and a method based on finite element analysis with Ansys and Abaqus software. A dynamic model of the impact force was developed, which allows to make a comparison of the numerical results obtained through the abovementioned methods. As a multi-body mechanical system, it was considered a mechanism from an internal combustion engine from which the radial clearance between the piston bolt and connecting rod head of the considered mechanism was analyzed.

Keywords: dynamics impact; finite element; radial clearance; modal dynamic analysis

Citation: Dumitru, S.; Constantin, A.; Copilusi, C.; Dumitru, N. Impact Dynamics Analysis of Mobile Mechanical Systems. *Mathematics* **2021**, *9*, 1776. https://doi.org/10.3390/math9151776

Academic Editor: António M. Lopes

Received: 9 June 2021
Accepted: 19 July 2021
Published: 27 July 2021

1. Introduction

The impact phenomenon can be present in the case of multi-body systems and this can be generated by many causes such as mounting and manufacturing errors, discontinuities of kinematic parameters variation or critical situation (for example, vehicle collision, etc.). Vehicle collisions are much more complex than other impact phenomena occurring between the two bodies.

The impact analysis of multi-body systems is of paramount importance since at the impact moment, the system state variables are quickly modified and discontinuities inside of system velocities and accelerations appear. These discontinuities can lead to forces of high values, which vary in a very short period of time. By knowing these forces variation peaks, it is important to take them into account during the mechanical systems design.

Using the research reported in [1], the authors show that there are two methods for solving impact problems of multi-body mechanical systems. The first method is based on a model of a "continuous" contact-force type. Similar models were analyzed in [2] and they underpinned a spring-damper element. Other research like that in [3] uses a Hertzian model for impact analysis. The second method is used more frequently than the first one and this is called the piecewise analysis method, where the motion equations integration stops at the moment of impact.

The research data from [4] present an impact analysis for models with friction from mechanical systems, which contain open kinematic linkages. Two problem types, namely, a bar, which drops free on the ground and a double pendulum system in a free fall, are examined. Thus, by solving the analytical impact problem it can be remarked that this mainly depends on motion equation integration at three impact phases, i.e., before, during and after the impact. This it could be a starting point for the proposed research.

In the research reported in [5] information is provided on whether impact events of mechanical systems can be successfully modeled through simple or low-order finite elements. There are developed simple finite element models which are benchmarked against theoretical impact problems and published experimental impact results. A case study is also presented, a finite element model using simple plastic beam elements. This is further tested to predict stresses and deflections in an experimental structural impact. The established workflow can be transposed to other impact problems because the stress levels predicted by the FEA model were generally within 10% of measured values. This demonstrates that a finite element analysis characterized by simple or low ordered finite elements can be used as a valid engineering tool for a medium-sized manufacturer in developing analyses of mobile mechanical systems or for designing products exhibiting impact requirements.

Impact phenomena can be studied by using simple models, especially experimental ones. This can generate complex problems which are hard to analyze through analytical methods. For example, the methods used in [6] develop analytical and experimental studies on some models based on free-fall drop impact analysis of portable products similar to cell phones. Thus, the research core was represented by a developed simplified analytical procedure for obtaining the dynamic response during impact in case of a free fall dropped object. The developed experimental analysis uses special objects with accelerometers mounted inside. With the use of a special drop tower with a guiding frame for controlled-angle free-fall drop impact, the used objects are dropped at different angles and the acceleration during impact is recorded. The aim of this analysis is to improve the conceptual solutions for impact resistance of portable products. From this research the experimental infrastructure can be noted and for further research the principle can lead to the use of small objects presumed to undergo a free fall drop. Thus, an experimental stand can be elaborated and an accelerometer can be placed inside of the base component. Also for experimental tests video motion analysis with high speed cameras can be used for analyzing impact phenomena.

Impact phenomena occur in various domains and cases, starting from medicine, accidentology, mechanical engineering, civil engineering, military defense, food industry, chemistry or physics. From this different methods and tools can be extracted for impact modeling.

In [7] the civil engineering specialists analyzed the behavior of concrete specimens during impact phenomena. Thus, experimental tests were made by dropping a constant weight hammer from five different heights on concrete specimens placed on special cells equipped with accelerometers. The novelty of this research relies on a finite element model that is made by using the ABAQUS software. Thus, extra attention is paid to the modeling of impact incidence and a non-linear dynamic model was prepared. FEM in the presented research can only be used for giving designers a prior idea about the impact behavior. A similar analysis was conducted in [8] where the authors determined the most suitable drop height for shock testing military equipment. For this, an application of regression analysis and a back-propagation neural network for determining the most suitable drop height for free-fall shock tests was proposed. The mathematical model, elaborated through a regression analysis, was verified through a comparative analysis between experimental tests carried out on several military materials like specific elastomers in order to estimate the optimal height and duration for free-fall shock tests.

Based on free-fall drop objects, several experimental tests were performed in [9] by taking into account the related researches from [10–13]. The authors in [9] performed a quite modern experimental analysis of the impact-induced acceleration by obstacles with the use of flexible ball chains. It can be remarked that impact phenomena have different behavior and depend on various materials. The novel element is represented by the possibility of using modeling of discrete structures as continuous bodies with a complicated constitutive law of impact which includes angle of incidence as a starting point for the elaborated mathematical models.

Bearing in mind the research reported in [14,15] it can be remarked that impact phenomenon modeling can be performed through modern techniques and for these finite element modeling represents an essential tool. For example, in [14] a nonlinear dynamic finite element analysis for a cylinder piston model was performed. The authors developed finite element models analyzed under specific circumstances of collision impact with the ANSYS/LS-DYNA software. The developed mathematical model can be applied to mobile mechanical systems. Several input data for the developed numerical simulations are reported, but in this case the clearance between piston and cylinder was neglected. The obtained results are materialized through impact force under different velocities of the piston impact collar.

An important argument for controlling the input parameters for impact modeling relays on simple numerical simulations like the one reported in [15], where the developed numerical simulations examine the perforation of steel and aluminum plate specimens and then impact load is applied on composite plates with eight layers reinforced by carbon fibers. For this ABAQUS software was used for several numerical models of solid penetration i.e., Mohr-Coulomb model, Johnson-Cook model, Zerilli-Armstrong model, Steinberg-Guinam model and a thermo-mechanical material model. For these it is very important to retain the use of damping coefficient, penetration depth, restitution coefficient, and also the way of controlling the finite elements network.

The impact without friction was analyzed by other studies like those describe in [16–19]. In [20] contact-impact force models, with friction, for cylindrical surfaces and spherical ones are analyzed.

Based on the results reported in [21], a comparison between different contact-impact force models for a crank and connecting rod mechanism were made. In this research the major objective is to analyze the impact phenomena for two major cases:

(1) Impact analysis for two bodies on a free-fall drop, which come in contact together, and
(2) Impact study for a multi-body system from an internal combustion engine mechanism.

For the first case it is important to control the input data for numerical simulations in order to have better accuracy during impact phenomenon modeling. In the second case some input data from the previous case are reconsidered and the obtained results will improve the reliability of the mobile mechanical systems, especially the ones from internal combustion engines, but not limited to these.

The novel element achieved through this research is the numerical result accuracy based on the finite element method for studying a impact phenomena. In order to achieve this several contact analyses it will be proposed, namely with and without friction, with and without damping effect and for different dynamic impact force parameters. Another novel element is represented by the three impact modeling methods used, namely analytical, virtual prototyping and FEM analysis. These methods allow identifying the variation domain for important dynamic factors, namely penetration depth, damping coefficient and force application factor.

In a comparison with [6], the contact between two bodies on a free-fall drop will be better analyzed and controlled due to the mathematical considerations. This can be completed by using modern techniques implemented through finite element software. The desired data can be easily compared through experimental analyses conducted on similar models for analyzing impact phenomena.

2. Impact Analysis of a Link with Planar Surface

In this case, it will analyze a planar mechanical system made up of a bar with a circular cross-section, which will drop freely with one end on a planar surface, being articulated through a ball bearing joint at the other end, as it can be seen in Figure 1.

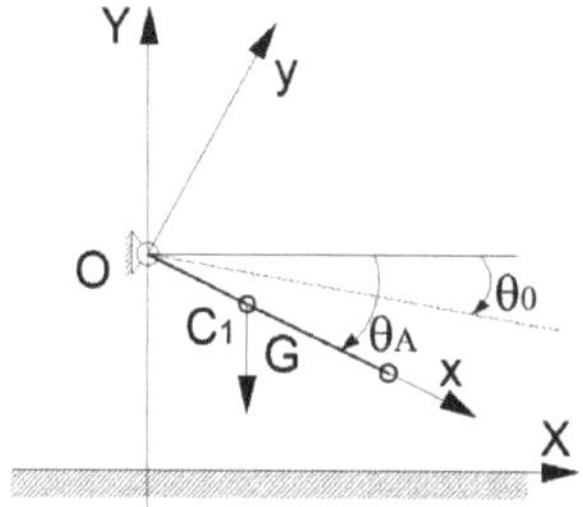

Figure 1. Kinematic model of the planar mechanical system.

This bar has a transversal circular diameter of 4 mm and a length of 240 mm. The planar surface has a rectangular shape with a length of 300 mm, a width of 100 mm and a height of 6 mm. The envisaged bar link will drop free from a height of 194 mm.

For this model it is important to analyze the impact between the bar and the planar surface in three cases, namely, the analytical mode, modeling and numerical simulation with Adams software, the modeling with finite element method through the aid of ANSYS and Abaqus.

2.1. Analytical Method

The analytical method involved an elaborated mathematical model based on the kinematic scheme in Figure 1.

For the schematized model in Figure 1, it will obtain the bar angular speed from dynamics considerations:

$$\frac{l}{2}G\cdot\cos\theta = J_O\cdot\frac{d^2\theta}{dt^2}, \tag{1}$$

where: G represents the bar weight, which actuates in the mass center C_1; J_O is the mass inertia momentum depending on bar rotation center, O and θ is the bar rotation angle:

$$J_O = \frac{m\cdot l^2}{3} = \frac{G\cdot l^2}{3\cdot g}, \tag{2}$$

$$\ddot{\theta} = \frac{d^2\theta}{dt^2} = \frac{G\cdot l}{2}\cdot\frac{\cos\theta}{J_O}, \tag{3}$$

$$\ddot{\theta} = \frac{3g}{2l}\cdot\cos\theta, \tag{4}$$

$$\ddot{\theta} = \omega\cdot\frac{d\omega}{d\theta}. \tag{5}$$

Equation (4) is integrated and it has the following form:

$$\frac{\omega^2}{2} = \frac{3\cdot g}{2\cdot l}\cdot\sin\theta + C. \tag{6}$$

The initial conditions are introduced:

$$t = 0, \theta = \theta_0, \dot{\theta} = \omega = 0. \tag{7}$$

By introducing Equation (7) in (6) we obtain:

$$0 = \frac{3\cdot g}{2\cdot l}\cdot\sin\theta_0 + C. \tag{8}$$

In addition, it will have the following mathematical expressions:

$$C = -\frac{3 \cdot g}{2 \cdot l} \cdot \sin \theta_0. \tag{9}$$

$$\omega^2 = \frac{3 \cdot g}{l} (\sin \theta - \sin \theta_0), \tag{10}$$

where: ω is the bar angular speed, l the bar length, and g the acceleration of gravity.

To determine the impact function, the displacement and speed of the free end of the bar, i.e., point (A), is considered. Therefore, the kinematic constraint equations are:

$$\Phi(q,t) = \begin{cases} X - X_0 - \cos \theta \cdot \frac{l}{2} = 0 \\ Y - Y_0 - \sin \theta \cdot \frac{l}{2} = 0 \\ \theta - \omega \cdot t = 0 \end{cases}. \tag{11}$$

The generalized coordinate is: $q_1 = [x_1, y_1, \varphi_1]^T$.

A proper Jacoby model for Equation (11) is:

$$J_q = \begin{bmatrix} 1 & 0 & \frac{l}{2} \sin \theta \\ 0 & 1 & -\frac{l}{2} \cos \theta \\ 0 & 0 & 1 \end{bmatrix}. \tag{12}$$

Velocities

These will be obtained by differentiating the Equation (11) depending on time, and the obtained results are:

$$\begin{cases} \dot{X}_1 - \dot{X}_0 + \frac{l}{2} \cdot \dot{\theta} \cdot \sin \theta = 0 \\ \dot{Y}_1 - \dot{Y}_0 - \frac{l}{2} \cdot \dot{\theta} \cdot \cos \theta = 0 \\ \dot{\theta} - \omega_1 = 0 \end{cases}. \tag{13}$$

According to Figure 2, the impact function accepted by multibody systems theory [22] will have the following form:

$$F = \begin{cases} 0, if \ldots y > y_1 \\ k(y_1 - y)^e - c_{\max} \cdot \dot{y} \cdot STEP(y, y_1 - d, 1, y_1, 0), if \ldots y \le y_1 \end{cases}. \tag{14}$$

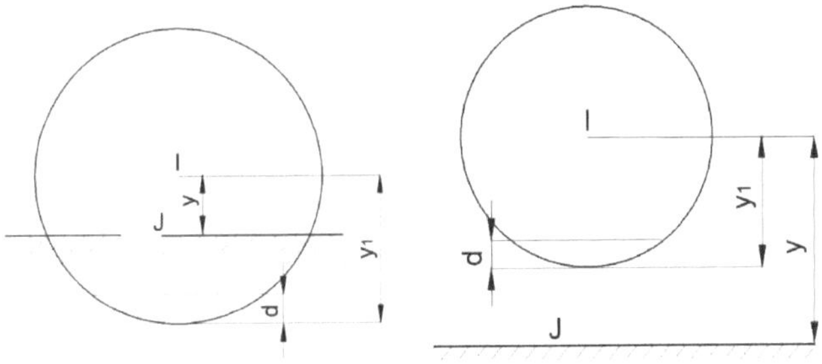

Figure 2. Spherical—planar contact type.

In Equation (14) one can identify the following terms: y—distance depending on time, which is necessary for the impact function calculus, namely: $y_A = y_A(t)$; $\dot{y} = \frac{dy}{dt}$—the point speed at which it will consider that it will actuate the impact force, i.e., speed $\dot{y} = v_A$ (point A speed); y_1—a positive real variable, which can be considered to be the free length of y component displacement; k—stiffness which corresponds to the interaction between the bar and the contact surface; e—is a positive real variable which specifies the deformation characteristics exponent of the applied force; $c_{\max}$—is a non-negative, double-precision variable that specifies the maximum damping coefficient; d—is a positive double-precision variable that specifies the secondary penetration at which ADAMS/Solver applies full damping.

The damping coefficient versus the penetration variation is plotted in Figure 3.

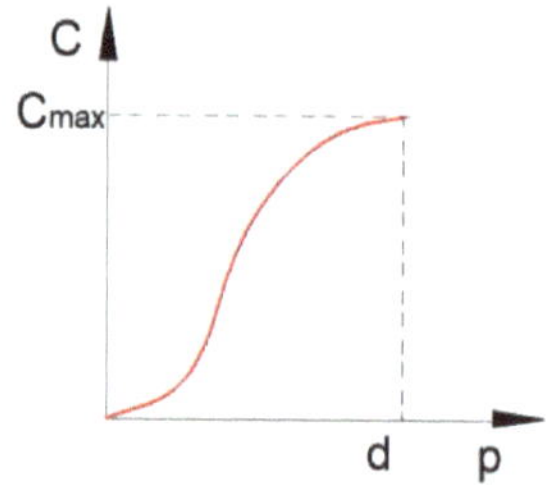

Figure 3. Damping coefficient versus penetration.

The stiffness term k, which can be seen in Equation (14), depends on material properties and the solid bodies' radius when these get in contact together. This is calculated by using the following equation [22]:

$$K = \frac{4}{3\pi(h_i+h_j)}R^{1/2} \quad ,R = \frac{R_i R_j}{R_i+R_j} \quad ,h_k = \frac{1-\nu_k^2}{\pi E_k} \quad ,k = i,j \; , \tag{15}$$

where: R_i, R_j—represent the radii of each body in contact; ν_i, ν_j—represent the transversal contraction coefficients for bodies i and j brought in contact together; E_i, E_j—represent the longitudinal elasticity modulus for body i and body j, respectively.

It is important to know the following conditions:

- if $y \geq y_1$, there will be no penetration at the level of the contact surface, and in this case the impact force will be equal to zero (penetration $p = 0$);
- if $y < y_1$, penetration occurs at the end closer to the i marker, and the impact force is >0 (penetration $p = y_1 - y$);
- when $p < d$, the damping instantaneous coefficient is a cubic STEP function of penetration, labeled p;
- when $p > d$, the instantaneous damping coefficient will be $c_{\max}$.

Thus, the impact function, regarding [22], has the following form:

$$F_{impact} = F(y, \dot{y}, y_1, k, e, c_{\max}, d), \tag{16}$$

The impact function is active when the distance between the markers i and j will be smaller than y_1. Practically, this function enters when the considered bodies collide, otherwise, it will be equal to zero. Also, it will consider that the force in the mathematical expression of (14) has two components:

- a spring or stiffness component and a damping or viscous component:

$$F_1 = k(y_1 - y)^e, \text{ spring (stiffness component)}. \tag{17}$$

Furthermore, the stiffness component will be an opponent term of penetration:

$$F_2 = c_{\max} \cdot \dot{y} \cdot STEP(y, y_1 - d, 1, y_1, 0), \tag{18}$$

- damping (viscous component), which is a function of the speed of penetration.

Damping opposes the direction of relative motion. Alternatively, the mathematical model of the impact force, when the considered bodies collide, will be:

$$F = \max(0, k(y_1 - y)^e - c\dot{y} \cdot STEP(y, y_1 - d, 1, y_1, 0)), \tag{19}$$

Both components of the considered force can be analyzed based on Figures 4 and 5, namely:

$$F_1 = k(y_1 - y)^e;\ F_2 = -c\dot{y}\cdot STEP(y, y_1 - d, 1, y_1, 0). \quad (20)$$

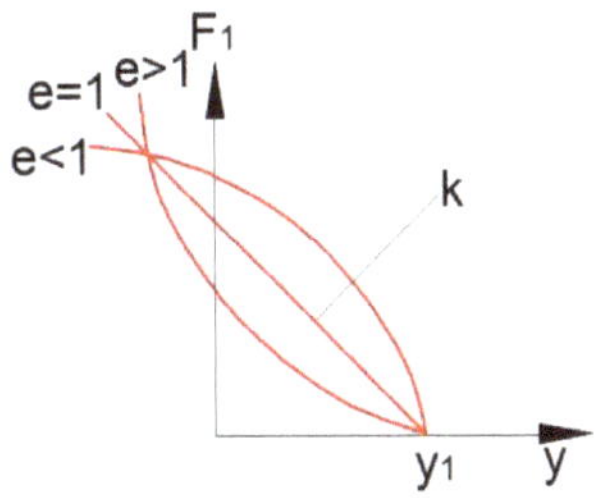

Figure 4. Compression-only spring force from impact function.

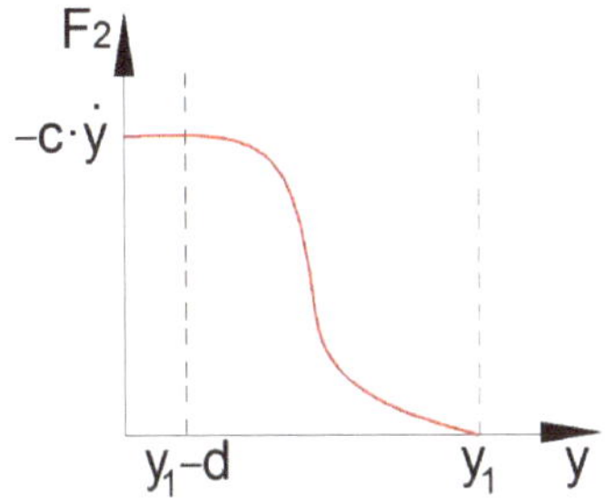

Figure 5. Impact force depending on damping component.

The STEP function approximates the Heaviside step function with a cubic polynomial one, as well as a STEP function, which looks like the one in Figure 6.

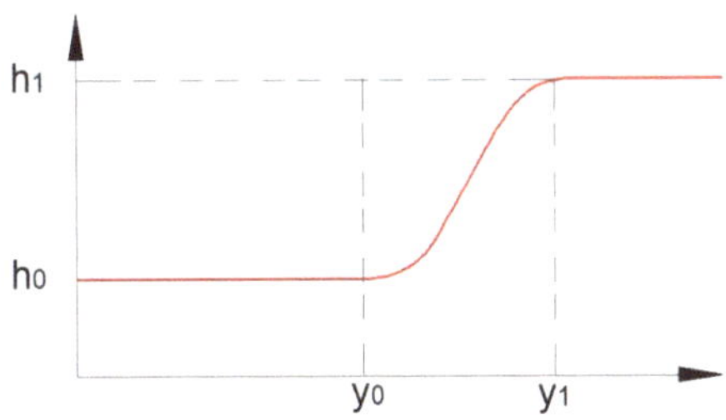

Figure 6. The proposed STEP function.

The equation by which a STEP function will be defined in the following form:

$$STEP = \begin{cases} h_0 : y \leq y_0 \\ h_0 + a\cdot\Delta^2(3 - 2\Delta) : y_0 < y < y_1 \\ h_1 : y \geq y_1 \end{cases}, \quad (21)$$

where:

$$a = h_1 - h_0,\ \Delta = (y - y_0)\cdot(y_1 - y_0). \quad (22)$$

Thus, the impact force mathematical model will be elaborated based on the mathematical expressions in Equation (14). For the proposed system presented in Figure 2, it will obtain: h_0 is equal to zero for this aplication. By comparing Figures 5 and 6, the mathematical expression (21) becomes:

$$STEP = \begin{cases} c_{\max} : y \leq y_1 - d \\ c_{\max} + a \cdot \Delta^2 (3 - 2\Delta) : y_1 - d < y < y_1 \\ 0, y \geq y_1 \end{cases} . \tag{23}$$

where:

$$a = c_{\max},\ \Delta = (y_1 - y_1 + d)/d. \tag{24}$$

Thus, the k stiffness value can be calculated with Equation (15) and the obtained value will be k = 48,049. The presented mathematical models, which correspond to the impact force, can be numerically processed under MAPLE programming environment. Accordingly, the time intervals when the stiffness component and viscous component of impact force are actuated will be identified. These time intervals correspond to a programming sequence, namely Algorithm 1, with the MAPLE software as follows:

Algorithm 1 Programming Sequence for Time Intervals

for t **from** 0 **by** 0.0001 **to** 0.6 **do**
if $(y) \leq (y_1 - d) then$
$F := k \cdot (y_1 - y)^e - vy \cdot c_{\max}; print\ (t, y, F);$
elif $(y_1 - d) < y$ and$y < y_1$ then
$F := k \cdot (y_1 - y)^e - vy \cdot (c_{\max} + a \cdot P^2 \cdot (3 - 2 \cdot P));$
$print\ (t, y, F)$ **else**
$F := 0;$
$print\ (t, y, F)$ **end if**
end do;

Using the initial data and the programming sequence, the processing of the presented mathematical models will lead to the time variation diagrams of both impact force components. Two major cases defined through different damping coefficients will be considered:

(a) Bar length: l = 240 mm; y_1 = 0.8; damping coefficient: $c_{\max}$ = 1; penetration depth: d = 0.01; stiffness: k = 48,049; force exponent: e = 2.2. By considering the diagram in Figure 7, it can be noted that the impact force will have a maximum value of 250 N, and in Figure 8 the viscous component has a maximum value around 3000 N, and these values correspond to c = 1. If we modify the damping coefficient at a value of c = 6, the force stiffness component will reach a value of 450 N, according to Figure 9, and the viscous component reported in Figure 10 will reach a value of 5×10^4 N.

(b) Bar length: l = 240 mm; y_1 = 0.897; damping coefficient: $c_{\max}$ = 6; penetration depth: d = 0.01; stiffness: k = 48,049; force exponent: e = 2.2.

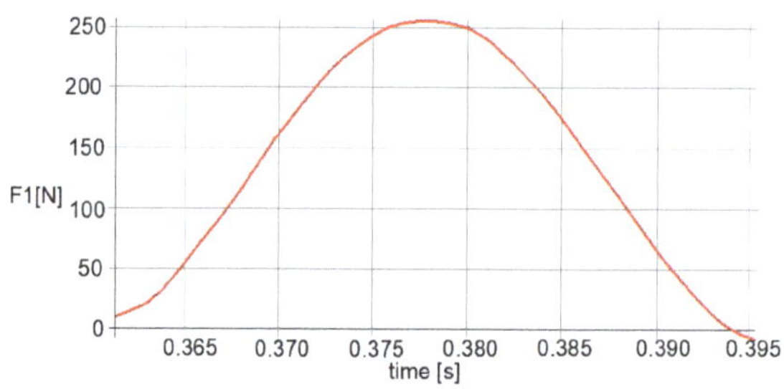

Figure 7. Stiffness component variation during time.

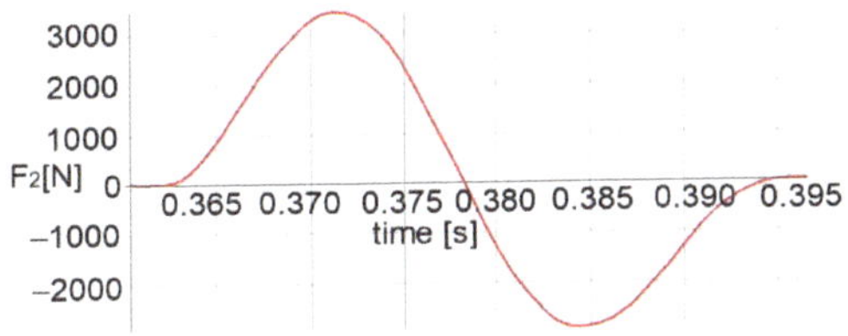

Figure 8. Viscous component variation during time.

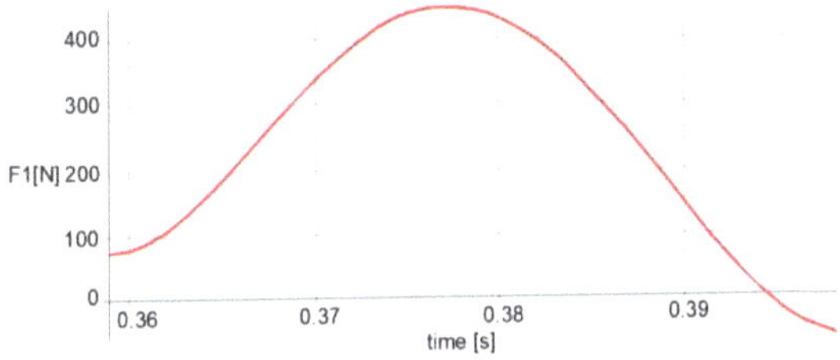

Figure 9. Stiffness component variation with time.

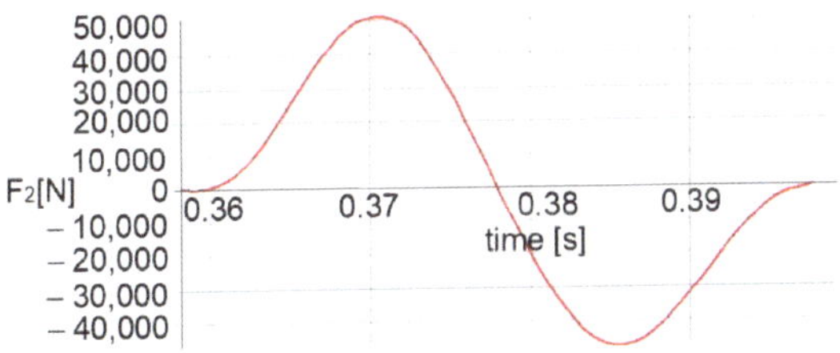

Figure 10. Viscous component variation with time.

2.2. Impact Modeling with MSC Adams

For the impact analysis, a virtual model was designed, equivalent to an experimental testbed, as can be observed in Figure 11. This contains in its structure a bar (1), a prismatic body (2) and a column composed of adjustable length devices (3). The considered bar has free motion on one end under its own weight. With the help of this virtual model the impact between the bar and the prismatic body will be analyzed. This will take place in three virtual environments, namely, MSC Adams, ANSYS and Abaqus software.

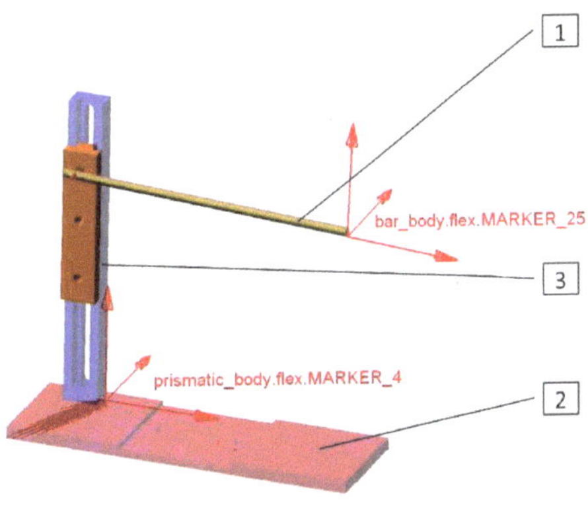

Figure 11. A 3D model of the experimental test bed.

The virtual prototyping accomplished via the Adams program has on its background a kinematic model made up of five parts connected together through kinematic joints. Thus, the adjustable column (3) has devices which allow the user to adjust the drop height of the considered bar (1). By having a closer look at the impact between the abovementioned components from a dynamic viewpoint, these were defined as deformable bodies and were meshed with finite elements such as tetrahedral ones taking into account the connection between interest nodes.

To obtain the impact function corresponding to the elaborated dynamic model, the following steps are to be followed:

The marker which materializes the origin of the impact force (marker_25, in Figure 11) will be identified;

The kinematic parameters (displacements and velocities) for the application point of the impact force (displacement and velocity of marker 25 according to Figure 12) will be obtained through numerical simulations. These parameters are defined as functions depending on time;

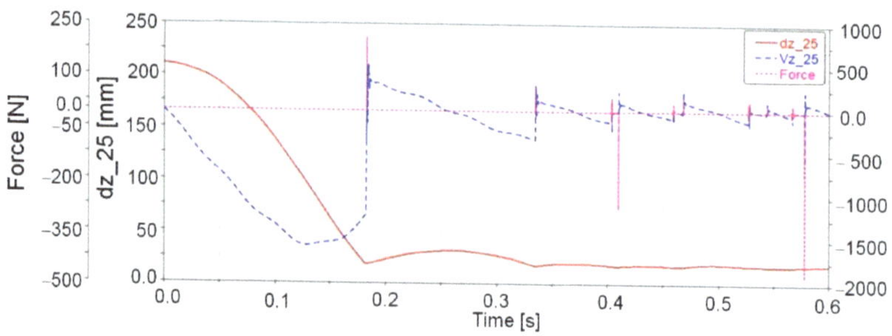

Figure 12. Impact force, displacement and speed time variation (detailed view).

The dynamic model parameters of the impact function, i.e., stiffness—k; damping—c; depth penetration—d; force exponent coefficient—e, will be evaluated.

The stiffness k will be calculated by Equation (15) by taking into account the geometry of the considered components which will be in contact together. The kinematic parameters are monitored in line with a local coordinate system, which is fixed with the prismatic body, defined through marker no. 4 (according to Figure 11).

The impact analysis of both component bodies will be made for two distinct cases, a contact made from rigid bodies and a contact made from deformable ones, respectively.

Thus, in Figure 12 the time variation laws for displacement in the case of marker no. 25 (dz_25), speed (vz_25) and impact force (force) can be seen. The dynamic parameters, which contribute to the impact function definition, have the following values: $k = 48{,}049$, $e = 2.2$, $c = 1$, $d = 0.01$. For differential equations integration the Adams solver GSTIFF I3 was used, with an integration step equal to 1×10^{-3}. In addition, the recorded harmonics as amplitude and number obtained in the case of impact forces variation are generated mainly by the damping phenomenon.

In order to examine the prismatic body behavior at the contact moment with the bar, it is necessary to perform a transversal elastic displacement analysis of this, as shown in Figure 13. Thus, in Figure 13 the prismatic body elastic displacement and the impact force when the damping coefficient is equal to 1 are represented. It can be remarked that the elastic displacement is relatively small, whereas the impact force reaches maximum values.

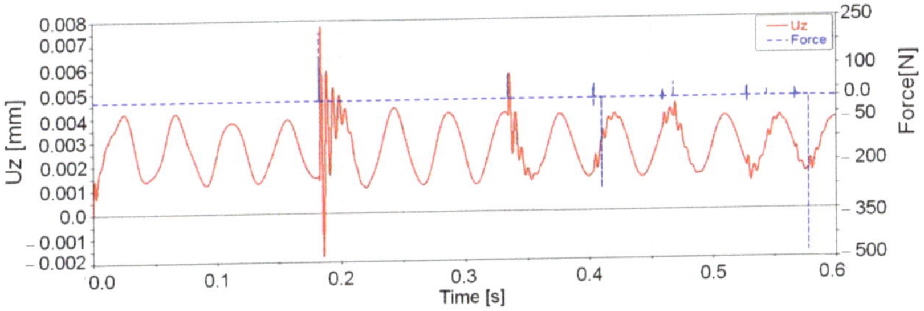

Figure 13. Variation diagrams for impact force and elastic displacement (c = 1) vs. time.

By considering Figures 14 and 15, it can be observed that the impact force and displacement variation occur at the same time intervals. In this case, the contact number between the bar and the prismatic body is smaller than in the previous case.

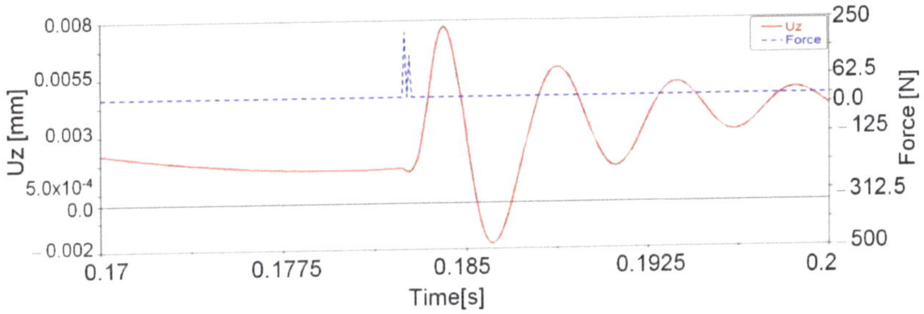

Figure 14. Variation diagrams for impact force and elastic displacement (c = 1) vs. time (detailed view).

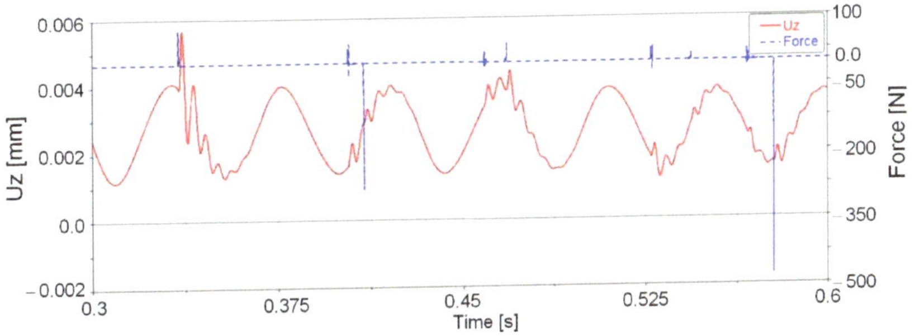

Figure 15. Variation diagrams for impact force and elastic displacement (c = 1) vs. time.

In Figures 16 and 17, the impact force, displacement and speed variation diagrams of the analyzed marker, namely, marker no. 25 (dz_25, vz_25) are shown, when the damping coefficient is equal to c = 6. These diagrams were obtained by taking into account a local coordinate system placed on the prismatic body in marker no. 4, according to Figure 11.

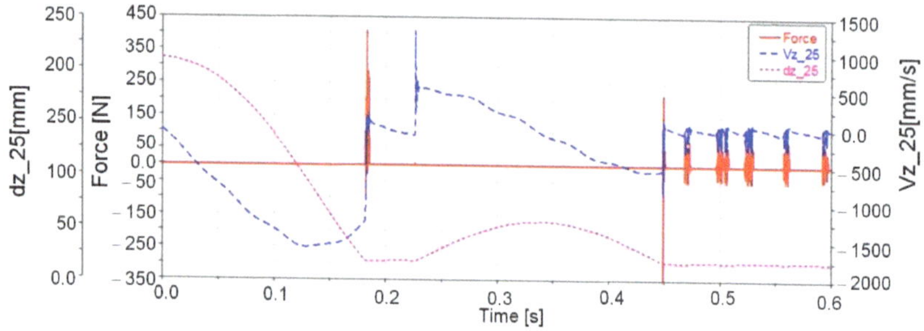

Figure 16. Force, displacement and speed variation diagrams of marker no: 25 vs. time.

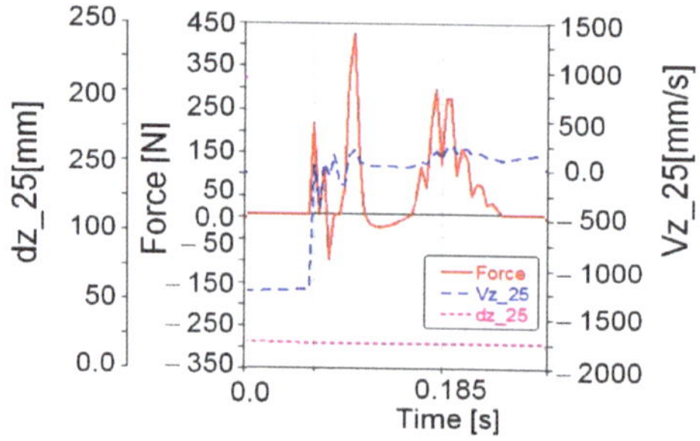

Figure 17. Force, displacement and speed variation diagrams of marker no: 25 vs. time (detailed view at first contact—t = 0.1824 s).

The impact between these components occurs at the time t = 0.1824 s. Thus, it can be observed that the displacement after impact shows a quasilinear variation until the time t = 0.2265 s, when the bar loses contact with the prismatic body component. At the impact moment the bar speed suddenly increases and it records a maximum value of 1325 mm/s. Accordingly, the impact force will occur at the contact moment and it records values ranging from zero to 205.5 N. Between the time interval of 0.1824—0.1862 s, the considered force will record many harmonics and one of them reaches a high value of 406.66 N as it can be seen in the detailed view in Figure 17. This variation has the following dynamic parameters values: $k = 48{,}049$, $e = 2.2$, $c = 6$, $d = 0.01$. In addition, this dynamic analysis was set up for an integration step equal to 1×10^{-4}.

By comparing the variation diagrams in Figures 12 and 16, it can be noted that for a high value of the damping coefficient, the number of harmonics will increase, whereas the transversal elastic displacement will decrease.

Figure 18 shows the force impact and transversal elastic displacement during time of the prismatic body component considered as a solid deformable one. These variations correspond to an impact force dynamic model where the damping coefficient is equal to $c = 6$. A detailed view of these variations is shown in Figure 19, which corresponds to the impact force and elastic displacement during the first impact. Another aspect is represented through the diagram in Figure 20, describing the entire impact force and elastic displacement variations on each impact between the bar and prismatic body components.

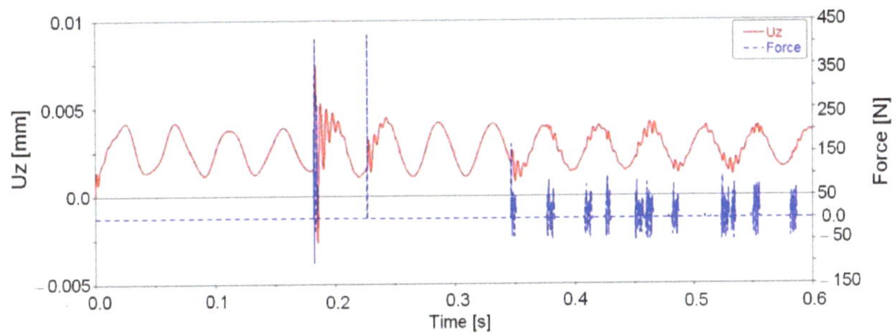

Figure 18. Variation diagrams for impact force and elastic displacement (c = 6) during time.

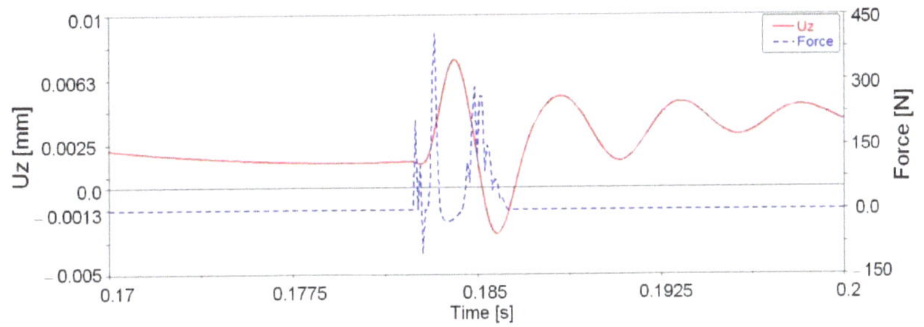

Figure 19. Detailed view of the impact force and elastic displacement c = 6 during first impact vs. time.

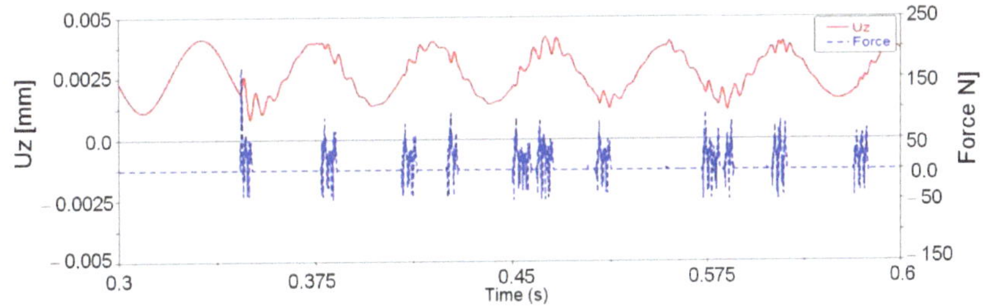

Figure 20. Impact force and elastic displacement c = 6 for entire simulation vs. time.

2.3. Impact Analysis with Finite Element Method

To analyze the impact phenomenon, based on the finite element method, three software programs were used, namely: Abaqus, ANSYS Workbench and ANSYS Mechanical APDL. For the impact phenomenon analysis with ANSYS the following steps were followed: the mechanical system CAD model was imported; the kinematic joints were defined; the entire model was meshed with proper finite elements and material and inertial properties were also defined; there were introduced the contour conditions and the contact between the bar and the prismatic body components was established when that had a free fall at the one end and touched the prismatic body.

The aim of this analysis regards the dynamic parameters, which define the contact between the bar and the prismatic body. Thus, many contact models, such as the ones below, were created:

(a) Frictionless contact, characterized by behavior-symmetric, stabilization damping factor-10, pinball radius—5 mm. In this case, the integration step was set up at 1×10^{-4} which corresponds to the hypothesis of large deflections. Time variation diagrams were identified for the following parameters: prismatic body stress, computed by considering a local coordinate system solitary with this component and dynamic parameters, which define the impact phenomenon between the bar and the prismatic body, namely, penetration depth, friction stress, contact pressure and contact force. For this model, the contact force is represented in Figures 21 and 22.

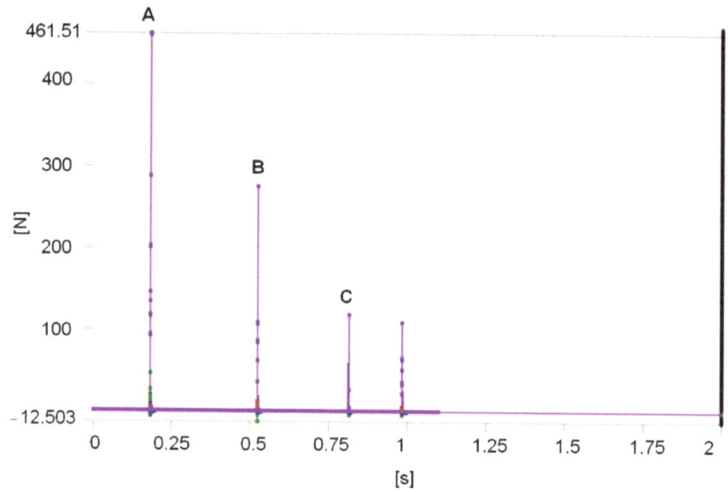

Figure 21. Impact force variation during time.

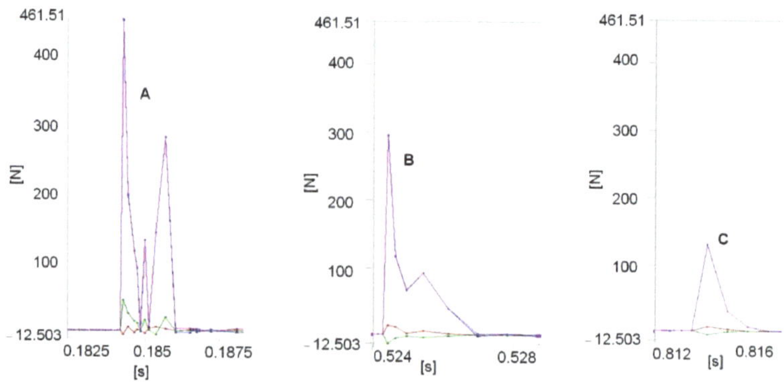

Figure 22. A detailed view of impact force variation during time (Zone A–C).

(b) Frictionless contact when the stabilization damping factor is equal to 0.1 and the integration step is equal to 1×10^{-4}. With reference to the previous case, one it can note a decrease of amplitude for the resultant elastic displacement and an increased number of harmonics for all the monitored kinematic and dynamic parameters after the first impact. Thus, the contact force records a maximum value during the impact of 433. N and afterwards, the values will be small due to the damping phenomenon (Figure 23). Considering the diagrams reported in the first case (case a), it can be observed that after the impact, all the monitored parameters reach absolute values at the same time instants, whereas different variation forms are reported. By having in sight the fact that in all the modeling cases the bar motion in a free fall drop records high amplitudes, the finite element analysis was setup under the large deflections hypothesis. Thus the used finite elements support the contact modeling under this hypothesis. This setup was done in order to evaluate in a correct manner the bar component motion before and during the impact phenomenon.

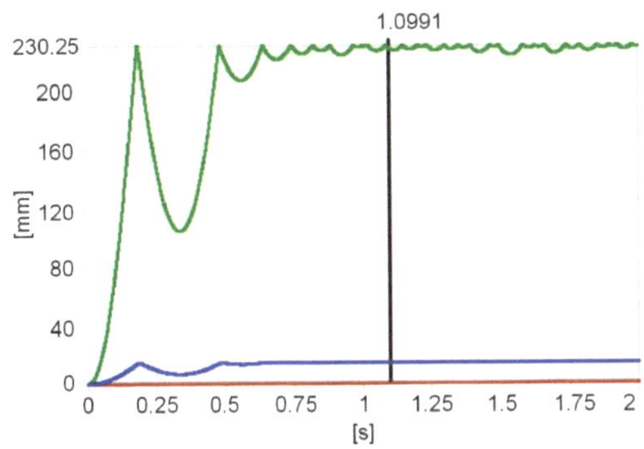

Figure 23. Displacement variation of the considered bar in the global coordinate system vs. time.

It can be observed that during the time interval 0 to 0.18517 s, the bar component will have a free motion under the gravity force until the impact momentum with the prismatic body component (this occurs at time = 0.18426 s). From the reported diagram it can be seen that the impact length period is around 0.002 s (between time = 0.18426–0.18626 s). After the impact, the displacement amplitude of the bar component will decrease due to the damping phenomenon, as it can be seen in Figure 24. The contact problem processing between the bar and prismatic body components take the time variations of the following parameters: contact pressure, sliding distance, friction stress, penetration and contact force.

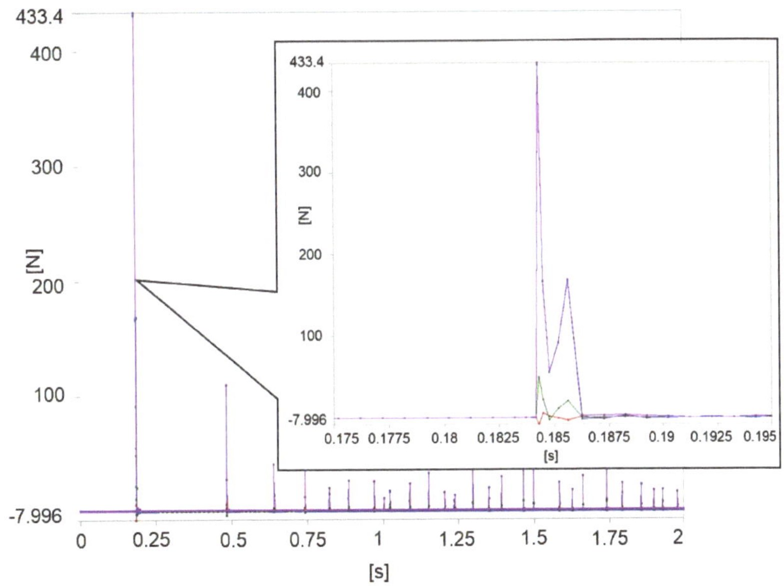

Figure 24. Force impact variation depending on time for the b-case.

(c) Contact type rough, where the stabilization damping factor is equal to 6. For this contact type it was remarked that the impact force has a higher value, i.e., equal to 915 N. In addition, the bar motion amplitude after the first impact is higher than in the other cases.

Thus, the impact was influenced by the quality of the contact surfaces. The obtained results are reported in Figures 25–27.

In Figure 25 can be observed that the bar component displacement amplitudes will not decrease immediately after the impact time period. They reach high values around 230.16 mm. The reason for this behavior is represented by the base component surface, which is rough and the analyzed contact problem was included in this hypothesis of contact definition.

Thus, the impact force records high values at points A, B and C, as shown in Figure 26. A correspondence time interval between the bar component displacement variation and the force at the impact moment and after impact can be spotted.

Figure 27 provides details of the impact force variation with time for the most important contact zones, namely, zone A and zone B. It can be remarked that the time period of the impact is higher than the other time periods. In addition, after the first contact, the time periods were smaller for zones B and C compared to Zone A. High values were recorded for the first contact, namely, 915.73 N (corresponding to zone A), then they decreased to around 515 N (corresponding to Zone B) and to around 735 N (corresponding to Zone C).

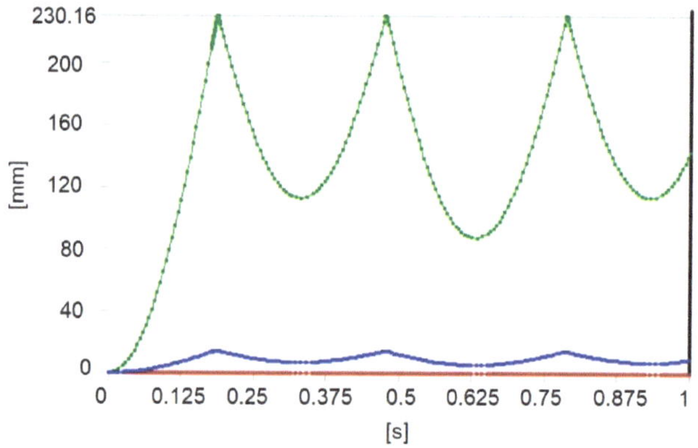

Figure 25. Resultant displacement depending on time for the c-case.

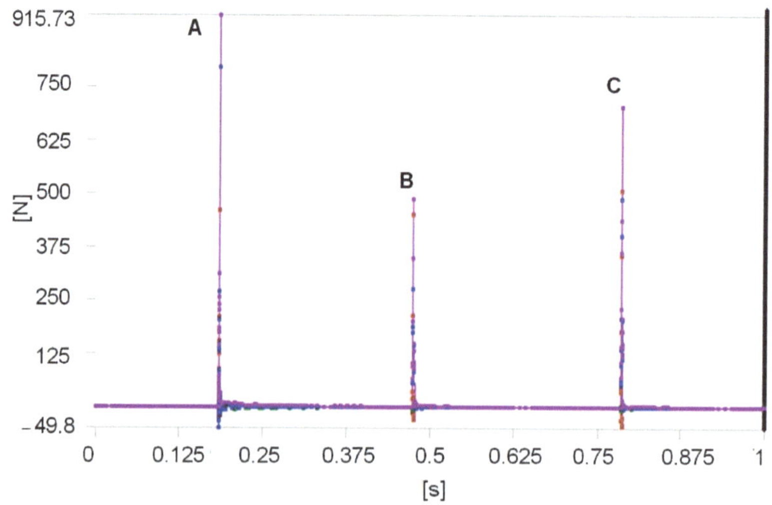

Figure 26. Impact force variation vs. time for the c-case.

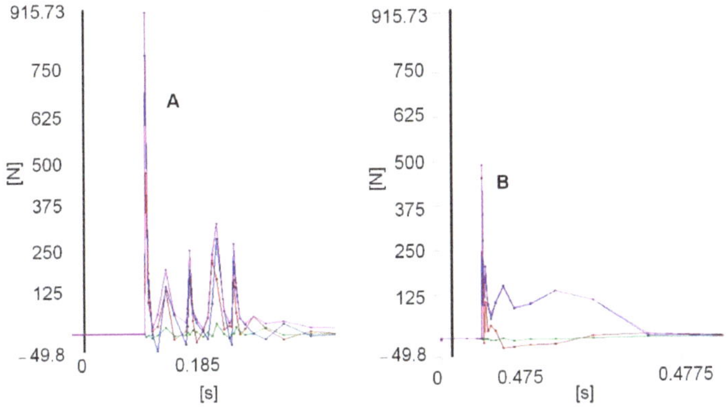

Figure 27. A detailed view of the impact force variation vs. time for the c-case (Zone A and B).

2.4. Impact Modeling with Abaqus

In order to study the impact phenomenon under the Abaqus software environment, the same mechanical system model that was presented in Figure 11 was considered. The virtual prototyping accomplished with the Abaqus software underpins a kinematic model made up of five parts connected together through kinematic joints. There was made a similar setup as in the previous case studies and the main objective was to determine the impact force variation and the elastic displacement during the impact phenomenon. Thus, the obtained results are reported in Figures 28 and 29.

Figure 28 presents the contact force variation during time at the impact moment and after, between the bar and the prismatic body component. This reaches a maximum value of 450 N. In Figure 29, the bar component displacement in a free fall, before and at the impact moment, records values within an amplitude of 200 mm, which will decrease after the first impact.

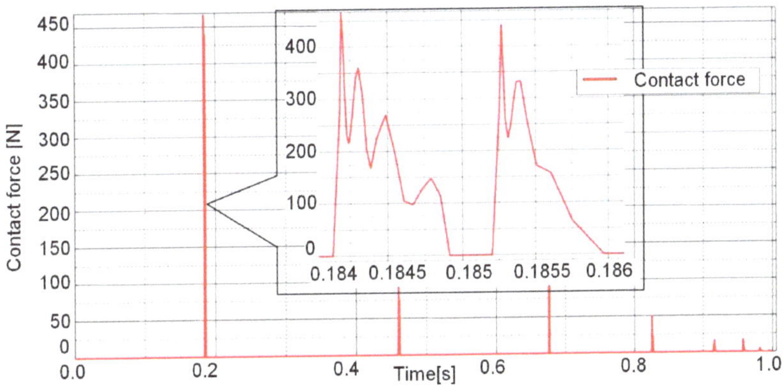

Figure 28. Impact force variation vs. time and a detailed view for the first impact.

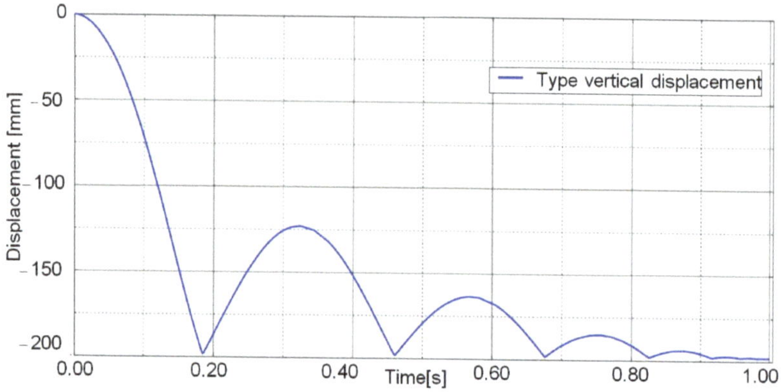

Figure 29. Resultant elastic displacement of the bar component during time.

3. Dynamic Analysis of the Impact in the Case of a Crank-Connecting Rod Mechanism

With the Adams program, a connecting rod-crank mechanism is examined. Due to the modeling of a radial clearance in the rotational joint between the connecting rod and the slide, an impact occurs, which generates a force impacting the dynamics of the mechanism. The radial clearance resultant value in the rotational joint is equal to 1.1 mm. A procedure was designed for generating the torque, under the Adams programming environment, to actuate the entire mechanism in the impact force conditions between the connecting rod and the slide. The following steps are required to create the impact function:

- time variation functions are defined for the interest marker displacement and velocity (the marker that materializes the force point application, i.e., marker 13, placed in the joint rotation center);
- the impact function will be created based on the following kinematic and dynamic parameters: displacement, dy_ MARKER_i, velocity, vy_ MARKER_i, for marker i, stiffness k, exponential coefficient e, damping c, penetration d, as follows:

  IMPACT (.MODEL_1.dy_ MARKER_i, .MODEL_1.vy_ MARKER_i, k, e, c, d)

One will examine the impact force effect developed at the connecting rod-slide contact on the evolution of the mechanism kinematic parameters, in two cases, a mechanism with rigid elements and a mechanism with deformable elements, respectively.

3.1. *Impact Dynamic Analysis with Rigid Elements*

A function is defined in order to identify the motor torque, based on the following expression:

$$M_m = M_0 * (1 - WZ(\text{MARKER\_1, MARKER\_9, MARKER\_10})/\omega_0), \tag{25}$$

where: M_m represents the engine torque (T_m—reported in diagrams); M_0 is the initial torque; ω_0 is an initial angular velocity; WZ represents the resulting angular velocity during the engine torque.

There are three markers placed in the center of rotation of the crank, namely, MARKER_1, MARKER _9, MARKER 10. These will materialize the origins of three reference systems (two reference systems that define the rotation torque and one mobile integrated with the crank). The mechanism dynamics was studied for the dynamic parameters with several sets of values in the impact force structure. These are represented through diagrams as seen in Figures 30–33.

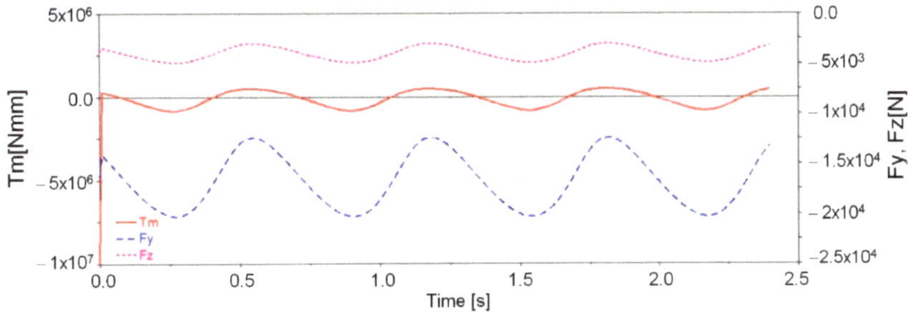

Figure 30. Variation diagrams for the engine torque and impact forces during time for $e = 0.1$.

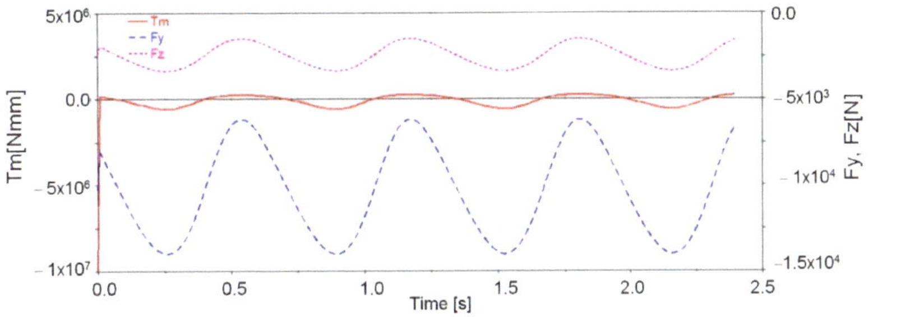

Figure 31. Variation diagrams for the engine torque and impact forces during time for $e = 0.01$.

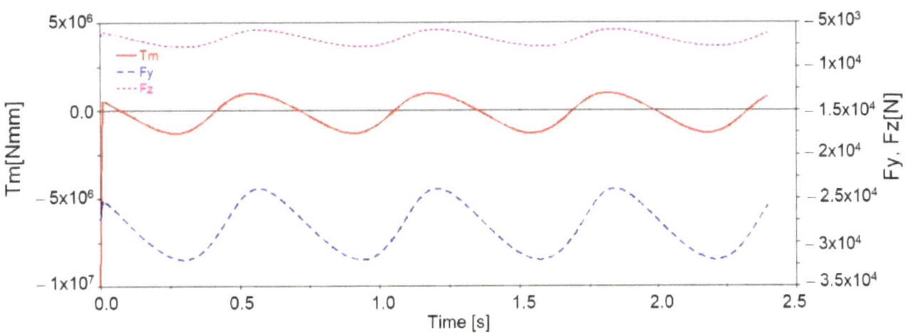

Figure 32. Variation diagrams for the engine torque and impact forces during time for $e = 0.2$.

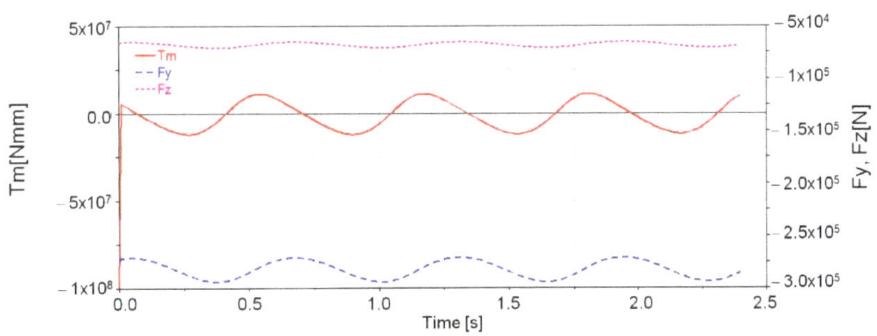

Figure 33. Variation diagrams for the engine torque and impact forces during time for $e = 0.2$.

The functions used for processing the diagrams in Figure 30 are:

Torque:($-$10,000,000 * (1 $-$ WZ(MARKER_1,MARKER_9,MARKER_10)/10);
IMPACT(.MODEL_1.dy_13, .MODEL_1.vy_13, 0.0, 10,000, 0.1, 10, 1×10^{-3})

The functions that were the basis for processing the diagrams in Figures 31–33, are:

Torque: ($-$10,000,000 * (1 $-$ WZ(MARKER_1, MARKER_9, MARKER_10)/10)
IMPACT(.MODEL_1.dy_13, .MODEL_1.vy_13, 0.0, 10,000, 0.01, 10, 1×10^{-3})

The reported results in Figure 33 generate the following functions:

Torque: ($-$10,000,000 * (1 $-$ WZ(MARKER_1, MARKER_9, MARKER_10)/10)
IMPACT(.MODEL_1.dy_13, .MODEL_1.vy_13, 0.0, 10,000, 0.2, 10, 1×10^{-3})
Torque: ($-$100,000,000 * (1 $-$ WZ(MARKER_1, MARKER_9, MARKER_10)/10)
IMPACT(.MODEL_1.dy_13, .MODEL_1.vy_13, 0.0, 100,000, 0.2, 10, 1×10^{-3})

Figures 30–33 show the time variation diagrams for the engine torque and the two components of the impact force along the y and z axes.

Figure 30 indicates that the three components have a cyclic variation, and that each cycle was repeated at an interval of 0.63 s.

The engine torque in each case reaches a maximum value at the start-up, due to the mechanism inertia in the first milliseconds. In addition, as seen in Figure 30, the engine torque maximum value on each cycle is around 4.09×10^5 Nmm and the minimum value around -8.25×10^5 Nmm.

It was found that the component along the y-axis of the force reaches higher values than the one along the z-axis. Thus, the component along the y-axis has a maximum value of 12.375 N, and the one along the z-axis has a maximum value of 3.500 N.

In Figure 31 the impact force components have the same variation shape as the ones in Figure 30, with the specification that the extreme values are lower given the dynamic parameters, except for the force exponent which was smaller in this case. The component along the y axis reaches a maximum value of 6181.81 N and around 1462.5 N along the z axis.

In Figure 32, the engine torque and the impact force dynamic parameters are kept, except for the force exponent, which increased. This led to a significant increase of the two force components extreme values along the y and z axes, i.e., 23,875 N for the y-axis and 5.750 N for the z-axis.

In Figure 33 the structure stiffness in the case of the impact force changes, consequently, a higher value of the initial moment M_0 is required. In this case, the maximum and minimum values of the engine torque and the impact forces components change, resulting in 8.63×10^6 Nmm for the engine torque, the value of the component along the y-axis being around 2.7×10^5 N and the value of the component along the z axis being of 64,925 N.

As a remark, due to the eccentricity of the spindle-bearing coupling type, the two dynamic components (engine torque and impact force), record jumps in an infinitesimal time at the mechanism start-up. We will consider that these jumps can be neglected and find that the mechanism has a correct cyclic operation, even if in some cases the extreme values of the dynamic components are high. For this reason, a dynamic study of the mechanism is required, regarding the elements in contact as deformable.

3.2. *Impact Dynamic Analysis with Deformable Elements*

3.2.1. Mathematical Considerations

A geometrical modeling of the engine mechanism was performed using the SolidWorks software, according to its dimensions and construction geometry.

The mechanism modal dynamic analysis was performed via the ADAMS program and it involved two important steps:

- Evaluating the natural frequencies and vibration modes;
- Elastic dynamic analysis.

The modal analysis mathematical models of a mobile mechanical system with deformable elements were proposed and developed by Craig and Bampton [23].

Thus, the equations defining the flexible body motion are differentiated from Lagrange's equations [23]:

$$\frac{d}{dt}\left(\frac{\partial L}{\partial \dot{\xi}}\right) - \frac{\partial L}{\partial \xi} + \frac{\partial F}{\partial \dot{\xi}} + \left[\frac{\partial \Psi}{\partial \xi}\right]^{-1} \lambda - Q = 0, \quad (26)$$

where: L is the Lagrange defined as: $L = T - V$; T and V are the kinetic energy and potential energy, respectively, F is the energy dissipation function; Ψ are the constraint equations; λ are the constraints Lagrange multipliers; ξ are the generalized coordinates; Q are the generalized applied forces (the applied forces projected on ξ).

The flexible body generalized coordinates are:

$$\xi = \left\{x, y, z, \psi, \theta, \varphi, q_{i\ (i=1\ldots M)}\right\}^T = \{x, \psi, q\}^T. \quad (27)$$

The instantaneous position of a point attached to a node on a flexible body, according to Figure 34, is given by the following mathematical expression:

$$r_{p'} = \overline{r} + \overline{r_u} + \overline{u}, \quad (28)$$

where: $\vec{r}$ is the position vector from the fixed (ground) reference system origin to point O where the local reference system origin is located of the flexible body; $\vec{r_u}$ is the position vector of the undeformed position of point P, according to the local reference system attached to body B; $\vec{u}$—the translational deformation vector of point P, the position vector of the body un-deformed position to its deformed position.

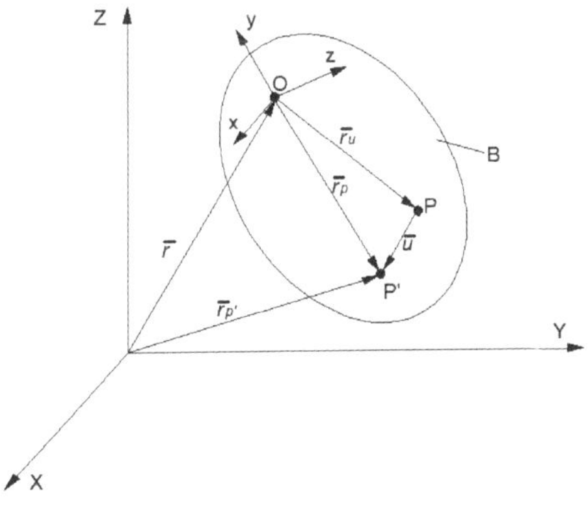

Figure 34. The position vector to a deformed point P′ on a flexible body relative to a local body reference system and to the fixed (ground) reference system.

The elements of the vector $\vec{r}$, x, y and z are the flexible body generalized coordinates. The orientation is evaluated and characterized by the Euler angles set, Ψ, θ and Φ, according with Figure 34.

The translational deformation vector is a modal superposition as follows:

$$u_p = \Phi_p q, \quad (29)$$

where: Φ_p is the part from the modal matrix corresponding to the node P translational degrees of freedom.

The Φ_p matrix dimension is $3 \times M$, where M represents the number of vibration modes. The modal coordinates q_i, ($i = 1, \ldots, M$) are the flexible body generalized coordinates. The angular deformations vector was obtained in a similar mode as the translation deformation vector, using a modal superposition, similar to the following expression:

$$\theta_p = \Phi_p^* q, \tag{30}$$

where Φ^*_p is the modal matrix part corresponding to the rotational degrees of freedom of node P. The Φ^* matrix dimension is $3XM$, where M is the number of modes of vibration [18].

3.2.2. Numerical Simulation

The same planar mechanism as the one with the rigid elements will be analyzed, but in this case, the connecting rod is considered as a deformable element. The connecting rod transformation from a rigid body to a deformable body was based on the finite element method, adopting the following procedure: The connecting rod body is meshed in tetrahedral finite elements, with a finer network in the joints contact areas.

The flexible body, obtained through the mesh network, must have its attachment points corresponding to the kinematic joints. The connection is made through nodes, with a spider type network. The analytical model of the engine torque is the same as Equation (25) namely $\omega_0 = 10$ rad/s, $M_0 = -100{,}000$ Nmm.

Thus, the time variation laws for the mechanism kinematic and dynamic parameters for two operating cycles are identified. Some snapshots during this simulation are shown in Figure 35.

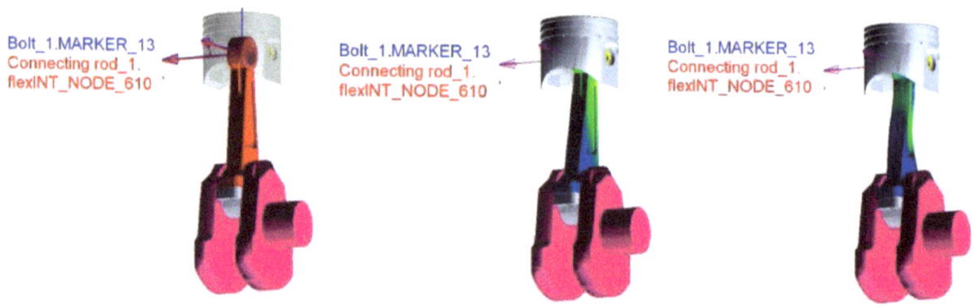

Figure 35. Snapshots during virtual simulations of the considered mechanism with a deformable connecting rod (MARKER_13 Node 610 location).

Figure 35 indicates the location of one node in the center of the bolt as a revolute joint between bolt and connecting rod head. This node is part of the network of hexahedral finite elements in which the mechanism connecting rod is meshed (node 610).

The time variation laws are identified for the kinematic parameters that define the motion of the node 610, i.e., the translational displacement along the x and y axes. These are reported in Figures 36 and 37 in case of the deformable body movement.

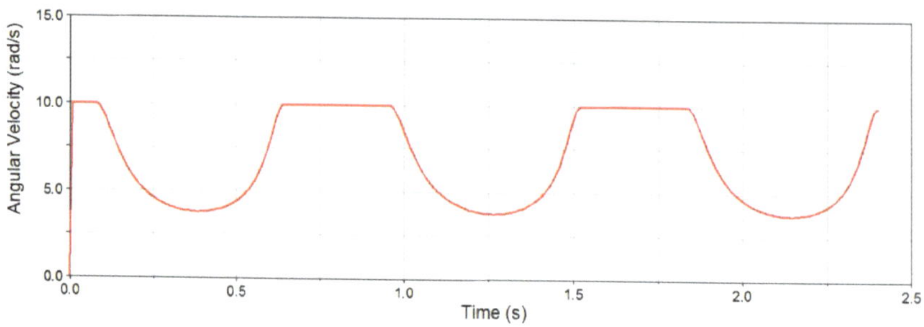

Figure 36. Angular velocity variation of the drive joint during time.

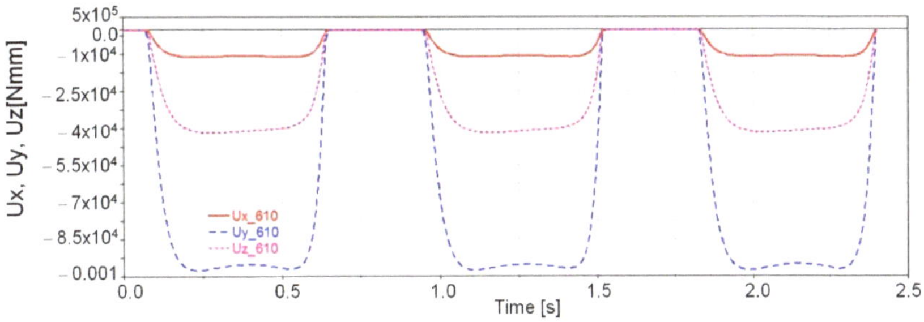

Figure 37. The node 610 elastic displacement vector component variation during time.

The impact structure function is defined as: IMPACT(.MODEL_1.dy_flex13, .MODEL_1.vy_flex13, 0.0, 20,000, 0.1, 10, 1×10^{-3}), with the following values of the dynamic parameters: k = 20,000, $e = 0.1$, $c = 10$, $d = 10^{-3}$.

In addition, as seen in Figure 35, the connecting rod deformed shape for vibration mode no. 3 with a natural frequency records a value of 6.515×10^4 Hz and for vibration mode no.10 with a natural frequency of 1.061×10^4 Hz.

The performed dynamic analysis consists of two major stages:

- modal-dynamic analysis with calculating the engine torque for the mechanism actuation;
- kinematic and dynamic parameters analysis of the generated contact during the impact between the bolt and the connecting rod head.

The modal-dynamic analysis presupposes the natural frequencies evaluation and the component proper vibration modes when this is considered deformable one.

The kinematic parameters involved in the impact force dynamic model definition are the displacement and velocity of the marker 13. This marker is identified as the position with the node 610, so it is located in the bolt center of the revolute joint (between the bolt and the connecting rod). Figure 38 shows the displacement and velocity time variation for marker 13 along the y-axis. Based on Equation (25) and on the numerical simulation performed with the Adams program, the engine torque variation diagram reported in Figure 39 was obtained. In Figure 39 is observed that the engine torque has a periodic variation, with the maximum amplitude around 50,000 Nmm. The jump recorded at start-up results from existence of the impact generated by the shock in the joint.

Figure 39 presents the behavior mode of the impact force on two elements brought in contact, i.e., the connecting rod and the bolt. It is important to note that the effect of this force is visible mainly in the area of maximum eccentricity between the connecting rod and the slide. Figure 39 it shows the time variation diagrams for the considered components along the y and z axes of the impact force. Thus, the engine torque variation mode is close to the impact force variation mode.

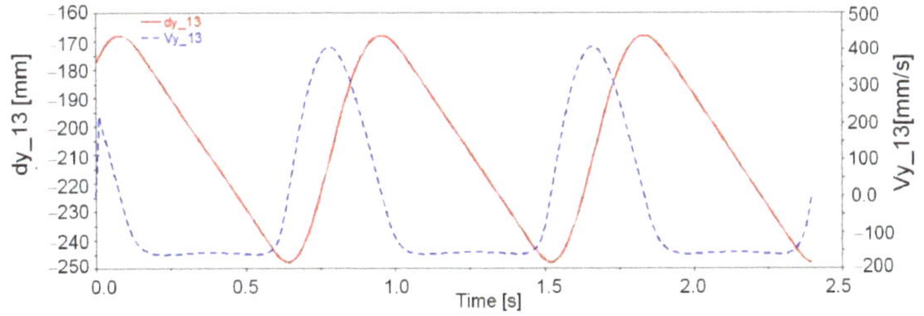

Figure 38. Displacement and velocity variation for the marker no.: 13 depending on time.

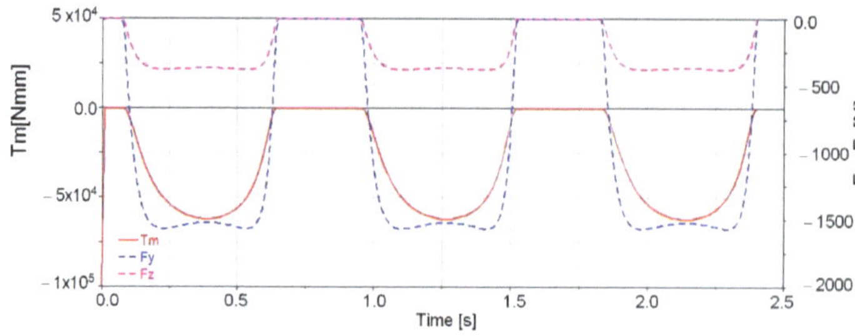

Figure 39. Variation diagrams for the engine torque and impact force components depending on time.

4. Results and Conclusions

The impact phenomenon was examined in two major cases, namely the impact resulting from the contact between the two bodies, when one of them was fixed and the other one was mobile, actuating freely under gravity and the impact generated by two mobile bodies from a multi-body mechanical system structure.

In the first case the impact analysis was done through an analytical method and virtual prototyping with the Adams, ANSYS and Abaqus software. Thus, the following conclusions were drawn:

- The mathematical model corresponding to Equation (16) is compatible with the Adams software theory, in relation to the impact phenomenon analysis. A Maple program sequence was derived, which can be numerically processed based on the created mathematical models by considering two major impact force components, namely, the stiffness component and the viscous component. These component variations with time were presented in Figures 7 and 8.
- The dynamic parameters were processed via the Maple program by considering Equation (16) with the numerical values indicated in Figure 7 and these corresponds to the virtual models obtained with the Adams software (Figures 16 and 17), with Ansys software (Figure 24) and Abaqus software (Figure 28), respectively.
- For the analytical models and numerical processing of the imported models with the Adams software, the bar component motion was monitored in a time interval equal to 0.6 s.
- By analyzing Figure 7 (mathematical model processing) and Figure 12 (Adams software processing) it can be observed that the impact force reaches a maximum value of 250 N in similar time intervals that correspond to the damping coefficient ($c = 1$).
- By modifying the damping coefficient value to $c = 6$, it can be remarked that the impact force maximum value will be modified, as well, along the number of contacts between the bar component and the prismatic body. At the same damping coefficient, by keeping the contact conditions and the integration step of the equations system, which dictate the bar component motion, one can see that the impact force values are not compatible with these cases, namely, analytical case (Figure 9), numerical simulation with the Adams software (Figure 16) and the finite element analysis (Figures 24 and 28).

Thus, the impact analysis based on the finite element method for the proposed model was performed for several cases: contact without friction, contact with friction and a rough type contact. The following kinematic and dynamic parameters were considered: the bar component displacement on three time intervals, i.e., before the impact, during the impact and after the impact, depth penetration; sliding distance, contact pressure and contact force.

Based on these parameters, the analysis was performed for different damping coefficient values and the same materials for contact analysis. In addition, the same integration step for differential equations was preserved. Thus, it was concluded that the variation forms of the impact force are the same for each analyzed case (Figures 22, 24, 27 and 28).

The displacement bar component amplitude is different in terms of value and frequency, depending on the contact and damping coefficient values during motion (Figures 23, 25 and 30).

A parameter which one can consider very important is represented by the prismatic body's elastic displacement variation due to the impact between the bar component and the prismatic body. These variation diagrams are represented and analyzed in Figures 13–15 (via the Adams software). These are very important because they can be compared relatively easy to the ones which were further obtained through experimental analysis.

A crank connecting the rod mechanism was examined, in which a radial clearance was considered in the joint that connects the connecting rod and the piston. Under these conditions, during the engine mechanism operation, a developed impact has major influence on the kinematic parameters that define the movement of the considered bodies brought in contact (the connecting rod and the piston). The dynamics of the mechanism involved two cases, namely, when the kinematic elements are rigid bodies and the case when the connecting rod and the piston are deformable ones.

The impact force dynamic model in the two cases was built on the following parameters: stiffness, force exponent, damping and penetration depth. These parameters were determined by several sets of numerical simulations that were based on the interface between the variation of the impact force and the variation of the engine torque, in the conditions of a cyclic operation of the mechanism. It was found that the results of the numerical simulation, mainly the impact force and the engine torque, are compatible with the mechanism cyclic operation if the elements deformability is taken into account, as well.

Variations of the kinematic parameters were reported, especially in the contact area of the connecting rod-piston, by monitoring the nodes movement on the joint finite elements.

As it was mentioned before, the impact dynamics was analyzed for two cases. Thus, in the first case the impact dynamics between two distinctive bodies were analyzed with the use of the finite element method. This allowed us to determine the kinematic and dynamic factors' influence mode on the impact force size and variation mode. Moreover the friction coefficient influence through contact surface quality and damping factor can be remarked. Thus, multiple tests were performed with different material types, different friction coefficients which were associated with damping factor values sets. By considering these tests, an influence rapport of the mentioned factors which depends with the contact body stiffness respectively impact force amplitude and frequency during and after impact phenomenon was established.

The obtained diagrams through impact phenomenon analysis in the second case show that impact force characteristics such as the amplitude and variation mode were influenced by assembly errors and kinematic parameters variation during motion. By having in sight other results from specific literature like [14], the impact force influence on mechanism torque was analyzed. Thus, the mechanism torque was established through a function definition under the MSC Adams environment. This function allows the user to perform a complete dynamic analysis of the combustion engine mechanism. In addition, the impact function presence which affects the engine mechanism functionality especially when this mechanism starts and the mechanism torque records high values can be observed.

Author Contributions: Conceptualization, S.D.; methodology, N.D.; software, S.D.; validation, N.D., S.D. and C.C.; formal analysis, A.C.; investigation, A.C.; resources, S.D.; data curation, A.C.; writing—original draft preparation, S.D.; writing—review and editing, C.C.; visualization, S.D., C.C. and N.D; supervision, N.D.; project administration, S.D., C.C. and N.D.; funding acquisition, S.D., C.C. and N.D. All authors have read and agreed to the published version of the manuscript.

Funding: This research received no external funding.

Institutional Review Board Statement: Not applicable.

Informed Consent Statement: Not applicable.

Data Availability Statement: Not applicable.

Conflicts of Interest: The authors declare no conflict of interest.

References

1. Ambrósio, J.A.C. Impact of rigid and flexible multibody systems deformation description and contact models. In *Virtual Nonlinear Multibody Systems*; Springer: Dordrecht, The Netherlands, 2003; pp. 57–81.
2. Khulief, Y.A.; Haug, E.J.; Shabana, A.A. *Dynamic Analysis of Large Scale Mechanical Systems with Intermittent Motion*; Technnical Report No. 83-10; University of Iowa, College of Engineering: Iowa City, IA, USA, 1983.
3. Hunt, K.H.; Grossley, F.R.E. Coefficient of Restitution Interpreted as Damping in Vibroimpact. *ASME J. Appl. Mech.* **1975**, *42*, 440–445. [CrossRef]
4. Ahmed, S.; Lankarani, H.M.; Pereira, M.F.O.S. Frictional impact analysis in open-loop multibody mechanical systems. *J. Mech. Des.* **1999**, *121*, 119–127. [CrossRef]
5. McCoy, M.l.; Lankarani, H.; Moradi, R. Use of simple finite elements for mechanical systems impact analysis based on stereomechanics, stress wave propagation, and energy method approaches. *J. Mech. Sci. Technol.* **2011**, *25*, 783–795. [CrossRef]
6. Tempelman, E.; Dwaikat, M.M.S.; Spitás, C. Experimental and Analytical Study of Free-Fall Drop Impact. Testing of Portable Products. *Exp. Mech.* **2012**, *52*, 1385–1395. [CrossRef]
7. Kantar, E.; Erdem, R.; Ozgur, A. Nonlinear Finite Element Analysis of Impact Behavior of Concrete Beam. *Math. Comput. Appl.* **2011**, *16*, 183–193. [CrossRef]
8. Chen, C.-R.; Wu, C.-H.; Lee, H.-T. Determination of Optimal Drop Height in Free-Fall Shock Test Using Regression Analysis and Back-Propagation Neural Network. *Shock Vib. J.* **2014**. [CrossRef]
9. Corbin, N.A.; Hanna, J.A.; Royston, W.R.; Warner, R.B. Impact-induced acceleration by obstacles. *New J. Phys.* **2018**, *20*, 1–8. [CrossRef]
10. Brun, P.-T.; Audoly, B.; Goriely, A.; Vella, D. The surprising dynamics of a chain on a pulley: Lift off and snapping. *Proc. R. Soc. A Math. Phys. Eng. Sci.* **2016**, *472*, 20160187. [CrossRef] [PubMed]
11. Dankowicz, H.; Fotsch, E. On the analysis of chatter in mechanical systems with impacts. *Procedia IUTAM* **2017**, *20*, 18–25. [CrossRef]
12. King, H.; White, R.; Maxwell, I.; Menon, N. Inelastic impact of a sphere on a massive plane: Nonmonotonic velocity dependence of the restitution coefficient. *Europhys. Lett.* **2011**, *93*, 14002. [CrossRef]
13. Muller, P.; Heckel, M.; Sack, A.; Poschel, T. Complex velocity dependence of the coefficient of restitution of a bouncing ball. *Phys. Rev. Lett.* **2013**, *110*, 254301. [CrossRef] [PubMed]
14. Yan, M.; Wang, T. Finite Element Analysis of Cylinder Piston Impact Based on ANSYS/LS-DYNA. In Proceedings of the 2012 International Conference on Mechanical Engineering and Material Science (MEMS 2012), Hong Kong, China, 27–28 July 2012.
15. Žmindák, M.; Pelagić, Z.; Pastorek, P.; Močilan, M.; Vybošt'ok, M. Finite Element Modelling of High Velocity Impact on Plate Structures. *Procedia Eng.* **2016**, *136*, 162–168. [CrossRef]
16. Wittenburg, J. *Dynamics of Systems of Rigid Bodies. Leitfäden der Angewandten Mathematik und Mechanik*; Springer: Berlin/Heidelberg, Germany, 1977.
17. Wehage, R.A. Generalized Coordinate Partitioning in Dynamic Analysis of Mechanical Systems. Ph.D. Thesis, University of Iowa, Iowa, IA, USA, 1980.
18. Haug, E.J.; Wu, S.C.; Yang, S.M. Dynamics of Mechanical Systems with Coulomb Friction, Stiction, Impact and Constraint Addition and Deletion, I. *Mech. Mach. Theory* **1986**, *21*, 401–406. [CrossRef]
19. Lankaram, M.H.N.; Kravesh, P.E. Application of the Canonical Equations of Motion in Problems of Constrained Multibody Systems with Intermittent Motion. In Proceedings of the 14-th ASME Design Automation Conference, Advances in Design Automation, Kissimmee, FL, USA, 25–28 September 1988; Volume 14, pp. 417–423.
20. Flores, P.; Ambrósio, J.; Claro, J.P.; Lankarani, H.M. Contact-impact force models for mechanical systems. In *Kinematics and Dynamics of Multibody Systems with Imperfect Joints*; Springer: Berlin/Heidelberg, Germany, 2008; pp. 47–66.
21. Flores, P.; Ambrosio, J.; Claro, J.C.P.; Lankarani, H.M. Influence of the contact—impact force model on the dynamic response of multi-body systems. *Proc. Inst. Mech. Eng. Part K J. Multi-Body Dyn.* **2006**, *220*, 21–34. [CrossRef]
22. Khemili, I.; Romdhane, L. Dynamic analysis of a flexible slider–crank mechanism with clearance. *Eur. J. Mech. A Solids* **2008**, *27*, 882–898. [CrossRef]
23. Craig, R.R.; Bampton, M.C.C. Coupling of substructures for dynamics analyses. *AIAA J.* **1968**, *7*, 1313–1319. [CrossRef]

mathematics

Article

Approximation Solution for the Zener Impact Theory

Ping-Kun Tsai [1], Cheng-Han Li [1,*], Chia-Chun Lai [1], Ko-Jung Huang [1] and Ching-Wei Cheng [2,*]

[1] Department of Bio-Industrial Mechatronics Engineering, National Chung Hsing University, Taichung 02, Taiwan; tpk8513@gmail.com (P.-K.T.); s39911050@gmail.com (C.-C.L.); g8925008@gmail.com (K.-J.H.)
[2] College of Intelligence, National Taichung University of Science and Technology, Taichung 404, Taiwan
* Correspondence: l634259@gmail.com (C.-H.L.); cwcheng@nutc.edu.tw (C.-W.C.); Tel.: +886-4-2219-5795 (C.-W.C.)

Abstract: Collisions can be classified as completely elastic or inelastic. Collision mechanics theory has gradually developed from elastic to inelastic collision theories. Based on the Hertz elastic collision contact theory and Zener inelastic collision theory model, we derive and explain the Hertz and Zener collision theory model equations in detail in this study and establish the Zener inelastic collision theory, which is a simple and fast calculation of the approximate solution to the nonlinear differential equations of motion. We propose an approximate formula to obtain the Zener nonlinear differential equation of motion in a simple manner. The approximate solution determines the relevant values of the collision force, material displacement, velocity, and contact time.

Keywords: elastic collision; inelastic collision; approximate solution

Citation: Tsai, P.-K.; Li, C.-H.; Lai, C.-C.; Huang, K.-J.; Cheng, C.-W. Approximation Solution for the Zener Impact Theory. *Mathematics* **2021**, *9*, 2222. https://doi.org/10.3390/math9182222

Academic Editor: Dan B. Marghitu

Received: 25 August 2021
Accepted: 8 September 2021
Published: 10 September 2021

1. Introduction

Theoretical research on collisions possesses significant value in engineering application. It is widely used in various industrial engineering fields—including aerospace, machinery, transportation, civil engineering, agriculture, and military applications. Collision is a mechanical phenomenon in which two or more objects in relative motion come into contact with each other for an instant, with changes in speed. Collisions are considered either elastic or inelastic based on the material properties and deformation types of the colliding objects. An elastic collision means that only elastic deformations occur in the colliding objects during collision, regardless of local contact deformation or if the entire deformed structure can be restored. Inelastic collisions mainly occur during crashing of metal objects in flexible systems. During these collisions, the structures of the colliding objects undergo elastic and plastic deformations at the local contact surfaces or overall systems. Therefore, even if the collision speed is low, the local contact stress exceeds the yield limit, thereby causing irreversible plastic deformation.

1.1. Hertz and Zener Impact Theory Related Studies

Hertz proposed a static contact problem for elastic bodies [1], where the spherical contact surface is approximated as a parabola. The two elastic balls contact problem was solved analytically. The colliding objects underwent elastic deformations during the elastic phase, which forms the basis of the Hertz elastic contact theory. The relationships between the contact force, deformation, and compression of the object were established to obtain the elastic collision contact time. Zener [2] extended the Hertz contact theory [1] and the effects of thin plate bending. To describe this model, let us consider the situation of a stationary plate or board impacted by a moving ball. In the Zener model, energy dissipation is considered only by bending waves propagating radially from the contact area [3]. The Zener model is not based on waves propagating through the thickness of the plate [4]; instead, it assumes that—during the collision of a large plate—the bending wave will not return to the contact area after being reflected from the lateral boundary of the plate.

Additionally, it assumes that the radius of curvature of the large flat plate is greater and the contact radius is smaller than the thickness of the plate; that is, the sphere collides with the thin plate. While the sphere is in contact with the surface, quasi-longitudinal waves may propagate through the thickness of the board several times [2,5]. To calculate the rebound velocity of the ball, Zener [2] combined the equations of motion of the ball and the board—as described by Newton's second law—to obtain the relationship between impulse and board displacements. Many scholars have subsequently continued their research based on the theoretical models of Hertz [1] and Zener [2,6]. Hunter [7] applied the kinetic energy transferred from the elastic longitudinal waves, etc., to the Hertz model [5] to consider the impact of small objects on infinite objects. Therefore, Hunter's work was based on Hertz's perfect elastic impact theory, ignoring all the attenuation effects caused by damping [7,8]. Boettcher et al. [8] modified the Hunter model [7] by considering the loss of kinetic energy during impact (Hunter loss) [7,8], they further modified the collision model proposed by Reed [9] to obtain more accurate elastic collision force–time parameters [5,8]. Mueller et al. [4] measured the coefficient of restitution and contact time for comparison with the theoretical predictions of the Hertz and Zener models. The contact time of the Hertz model does not consider the plate thickness as a parameter of influence. Hence, the model predicts the same result for a glass plate of any thickness [4], which is why the model is insufficient for predicting the contact time. The Hertz model can only accurately predict the contact time when the impact of a large plate is being considered [4,10]. Thus, the Hertz model represents a limiting case of the Zener model. In contrast, the contact time predicted by the Zener theoretical model approaches the actual measured value. The aforementioned research on collision is mostly based on the theories of the Hertz and Zener models.

1.2. Main Work and Purpose

In this study, we combined the Hertz contact theory and Zener inelastic collision models to propose a fast and straightforward method of deriving and establishing the Zener inelastic collision model. It is unnecessary to derive Newton's second law from the beginning to obtain the Zener collision equation and hence the collision displacement, contact force, contact time, and speed. Another focus of this article is to provide a detailed review of the equations of the Hertz and Zener's contact theory models for better understanding, along with their possible future applications for theoretical reference.

2. Review of Impact Mechanics

2.1. Hertz Model

This section refers to Hertz (1882) [1] and Timoshenko and Goodier (1970) [11]. Hertz's elastic collision theory [1] assumes that two spheres move linearly along their centers to cause a collision. The distance between the centers of the two spheres reduces gradually, and the speed of movement reduces to a static state. The collision that occurs at this moment causes the two spheres to produce the maximum amount of compression deformations. Finally, the two spheres gradually separate and return to their original conditions. This process is known as an elastic collision.

We assume that the masses of the two spheres are m_1 and m_2. The contact compression force P is generated during collision, and the speed reduces and changes after collision. The velocities of the two spheres during collision are $\vec{v_1}$ and $\vec{v_2}$, as shown in Figure 1, and are expressed by Equation (1) according to Newton's second law of motion.

$$m_1 \frac{d\vec{v_1}}{dt} = -P, \; m_2 \frac{d\vec{v_2}}{dt} = -P. \tag{1}$$

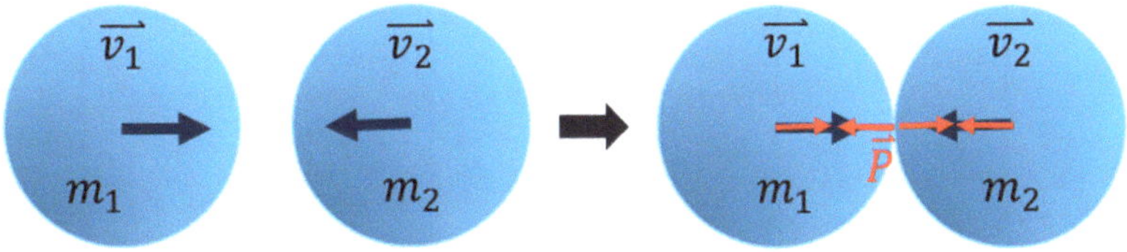

Figure 1. Schematic illustration of the elastic collision of two spheres.

Assuming that the two spheres collide with the compression displacement $\vec{\alpha}$, the relative velocity of the spheres, $\vec{v_r}$, can be expressed by Equation (2) below.

$$\vec{v_r} = \frac{d\vec{\alpha}}{dt} = \vec{v_1} - \vec{v_2} = v_1\hat{\imath} + v_2\hat{\imath} = (v_1 + v_2)\hat{\imath}. \tag{2}$$

From Equations (1) and (2), the relationship after collision of the spheres is

$$\frac{d\vec{v_r}}{dt} = \frac{d^2\vec{\alpha}}{dt^2} = \frac{d\vec{v_1}}{dt} - \frac{d\vec{v_2}}{dt} = -P\left[\frac{m_1 + m_2}{m_1 \cdot m_2}\right] = -\frac{P}{M}, \tag{3}$$

where $M = \frac{1}{\left(\frac{1}{m_1} + \frac{1}{m_2}\right)} = \frac{m_1 \cdot m_2}{m_1 + m_2}$ is the effective (average) mass after collision of the spheres.

When the two spheres come into contact, as shown in Figure 2, the vertical distance between points M and O is z_1, and the vertical distance between points N and O is z_2; hence, $\overline{MN}$ distance can be expressed as Equation (4)

$$\overline{MN} = z_1 + z_2 = r^2\left(\frac{1}{2R_1} + \frac{1}{2R_2}\right) = \frac{r^2(R_1 + R_2)}{2R_1 R_2}. \tag{4}$$

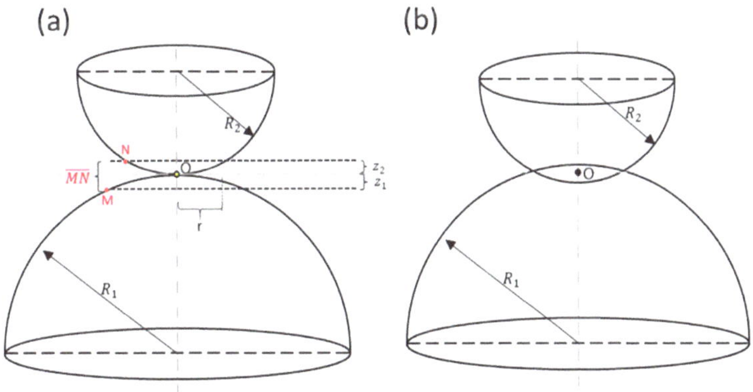

Figure 2. Schematic of compression changes after collision of two balls (**a**) coming into contact with each other and (**b**) colliding with each other.

When the two spheres collide, the elastic deformations w_1 and w_2 of the collision point change as shown in Figure 2a,b, and the contact surfaces of the two spheres are deformed. Therefore, the collision compression distance between the two spheres can be expressed by Equations (5) and (6) and are obtained using Equation (4).

$$\alpha = w_1 + w_2 + z_1 + z_2, \tag{5}$$

$$w_1 + w_2 = \alpha - \beta r^2, \tag{6}$$

where α is the collision compression displacement between the centers of the two spheres after collision and $\beta = \frac{R_1+R_2}{2R_1R_2}$. Assuming that the maximum pressure intensity of a hemispherical collision is q_0 and that it is distributed on the surface of the hemisphere with radius b at the point of contact, the total maximum collision compressive force P can be expressed as in Equation (7).

$$P = \frac{q_0}{b}\frac{2}{3}\pi b^3 = q_0\frac{2}{3}\pi b^2. \tag{7}$$

Here, $\alpha = (k_1+k_2)\frac{a\pi^2 q_0}{2}$, $b = (k_1+k_2)\frac{\pi^2 q_0}{4\beta}$, and $\beta = \frac{R_1+R_2}{2R_1R_2}$ can be obtained from Equations (7) and (6). Then, α, b, and β are applied in Equation (7) to obtain Equation (8).

$$\alpha = \left[\frac{9\pi^2}{16}\cdot\frac{P^2(k_1+k_2)^2(R_1+R_2)}{R_1R_2}\right]^{\frac{1}{3}}. \tag{8}$$

Equation (9) can be obtained from Equation (8) as

$$P = n\alpha^{\frac{3}{2}}, \tag{9}$$

where α is the contact displacement, n is the Hertzian stiffness constant where $n = \frac{4}{3\pi(k_1+k_2)}\sqrt{\frac{R_1R_2}{R_1+R_2}}$.

Equation (9) is substituted with Equation (3) and updated as $d\left(\frac{d\alpha}{dt}\right)^2 = 2\frac{d^2\alpha}{dt^2}d\alpha$ to obtain Equation (10).

$$\frac{1}{2}\left[d\left(\frac{d\alpha}{dt}\right)^2\right] = -\frac{n}{M}\alpha^{\frac{3}{2}}d\alpha. \tag{10}$$

At the beginning of the collision, the compression is 0 and relative velocity is v_0. During the compression process, the relative speed is $v_r = \frac{d\alpha}{dt}$, and Equation (11) can be obtained by integrating Equation (10). At the end of the collision compression, the final relative speed is v_r and maximum compression is α_{max}.

$$\frac{1}{2}\left[\left(\frac{d\alpha}{dt}\right)^2 - v_0{}^2\right] = -\frac{2}{5}\frac{n}{M}\alpha^{\frac{5}{2}} \Rightarrow v_0{}^2 = \frac{4}{5}\frac{n}{M}\alpha_{max}{}^{\frac{5}{2}} \Rightarrow \alpha_{max} = \left[\frac{5}{4}\frac{M}{n}v_0{}^2\right]^{\frac{2}{5}}. \tag{11}$$

Equation (11) gives time t for the distance compression of the centroids of the two spheres and the maximum compression time t_f of the two colliding bodies as

$$\frac{d\alpha}{dt} = \sqrt{v_0{}^2 - \frac{4}{5}\frac{n}{M}\alpha^{\frac{5}{2}}} \Rightarrow dt = \frac{d\alpha}{v_0\sqrt{1-\frac{4}{5}\frac{n}{Mv_0{}^2}\alpha^{\frac{5}{2}}}} \tag{12}$$

Suppose that at the initiation of the collision, the centers of the masses of the two spheres collide with a compression distance $\alpha = 0$, $t = 0$. For $x = 0$ the maximum compression distance of the two collision bodies after collision is $= \alpha_{max}$, $t = t_f$, where $x = 1$, and t_f is the end time of the collision. Let $x = \alpha/\alpha_{max}$, $\alpha_{max}dx = d\alpha$, after integrating Equation (12), the time t required for compression of the distance between the centroids of the two spheres can be obtained through Equation (13).

$$dt = \frac{\alpha_{max}dx}{v_0\sqrt{1-\frac{4}{5}\frac{n}{Mv_0{}^2}(\alpha_{max}x)^{\frac{5}{2}}}} \Rightarrow t = \frac{\alpha_{max}}{v_0}\int_0^1 \frac{dx}{\sqrt{1-x^{\frac{5}{2}}}} \tag{13}$$

Using the initial conditions of Equation (13), we obtain Equation (14).

$$\begin{cases} y = x^{\frac{5}{2}} \Rightarrow dx = \frac{2}{5}y^{-\frac{3}{5}}dy \\ x = 0;\ y = 0 \\ x = 1;\ y = 1 \end{cases} \quad t_f = \frac{\alpha_{max}}{v_0}\frac{2}{5}\int_0^1 y^{-\frac{3}{5}}(1-y)^{-\frac{1}{2}}dy. \tag{14}$$

Using the beta function, $B(m,n) = \int_0^1 x^{m-1}(1-x)^{n-1}dx = \frac{\Gamma(m)\Gamma(n)}{\Gamma(m+n)}$, $m,n > 0$ and gamma function, $\Gamma(n) \int_0^\infty x^{n-1}e^{1x}dx$, $\Gamma(n-1) = n\Gamma(n) = n!$, $n = 1,2,3$. The maximum compression time of the colliding bodies of the two spheres (t_f) is given by Equation (15).

$$t_f = \frac{\alpha_{max}}{v_0}\frac{2}{5}\frac{\Gamma\left(\frac{2}{5}\right)\Gamma\left(\frac{1}{2}\right)}{\Gamma\left(\frac{2}{5}+\frac{1}{2}\right)} = 1.47\frac{\alpha_{max}}{v_0}. \tag{15}$$

The Hertz perfect elastic collision time is $2t_f$ and is given as in Equation (16)

$$2t_f = 2.94\frac{\alpha_{max}}{v_0} = 2.94\left[\frac{5}{4}\frac{M}{n}v_0{}^2\right]^{\frac{2}{5}}\frac{1}{v_0} = 3.215\left(\frac{M}{nv_0{}^{\frac{1}{2}}}\right)^{\frac{2}{5}} = 3.215(T)^{\frac{2}{5}}, \tag{16}$$

where $T = \frac{M}{nv_0{}^{\frac{1}{2}}}$.

2.2. Zener Model

This section refers to Zener (1941) [2]. When a collision occurs between two spheres, it is generally not completely elastic, as assumed in Hertz theory [1]. Because of the energy consumed during a collision, inelastic collision occurs instead. During the impact, pressure is generated in the contact area. Local deformation causes a stress wave, which propagates inward through the object from the excitation point [4]. In addition, the generation of collision waves from the surfaces of the object and body can cause energy loss at the impact area. Elastic waves propagate through solids and exhibit characteristic velocities. Therefore, multiple refractions or reflections of waves at the interface cause energy dissipation. The Hertz model [1] is a conservative force field; hence, it does not consider the energy consumption during actual collision. As a result, there is a gap in the analysis results of actual collisions.

Zener [2] extended the Hertz theory through the bending effect of the board; thus, it is assumed that the board extends infinitely in the horizontal direction. The Zener model describes a sphere hitting a thin plate, and as per the model, bending waves do not return to the contact area after reflecting from the lateral boundary of the plate. Zener [2] combined the sphere and plate motion equations explained by Newton's second law to describe the interaction between the bending effect of the ball colliding with the plate and its process.

First, we assume that the radius of curvature of the plate and the angle between the plate and arrival plate overall are small. However, it is large compared to its thickness. For the theory of thin plates with the reaction to localized impulsivity, the transverse displacement $U(x,y,t)$ of the plate can be represented by the differential equation

$$\left(D\nabla^4 + 2\rho h\frac{d^2}{dt^2}\right)U = f, \tag{17}$$

where f is the surface density of the normal force, ρ is the density, $2h$ is the thickness of the plate, $D = \frac{2}{3}h^3\frac{E}{1-\nu^2}$ the rigidity modulus, and ν is Poisson's ratio of the impacting bodies.

$$\nabla^2 : \frac{\partial^2}{\partial x^2} + \frac{\partial^2}{\partial y^2},\ \nabla^4 : \frac{\partial^4}{\partial x^4} + 2\frac{\partial^4}{\partial x^2\partial y^2} + \frac{\partial^4}{\partial y^4}. \text{Let } U(x,y,t) = \sum_n C_n(t)U_n(x,y), \tag{18}$$

where $C_n(t)$ is an eigenvalue, and $U_n(x,y)$ are normalized eigenfunctions. Substituting Equation (18) into Equation (17) and multiplying by U_k, with dA being integrated over the surface of plate, we get Equation (19).

$$2\rho h{\omega_k}^2 C_k = \int U_k f dA = U_k(0)F(t), \tag{19}$$

where ω_k and U_k are the eigenvalues and normalized eigenfunctions, respectively. $U_k(0)$ is the value of U_k at the point of application of force $F(t)$, which gives $C_k = (2\rho h\omega_k)^{-1}U_k(0)\int_0^t F(t')\sin\omega_k(t-t')dt'$. For the integral of t', as long as t is sufficiently small, the boundaries do not reflect the disturbance. Hence, this integration process does not depend on the shape of the plate and boundary conditions.

$$U(0,t) = \sum_n C_n(t)U_n(0), \tag{20}$$

$$= (2\rho h)^{-1}\sum_n {\omega_n}^{-1}{U_n}^2(0)\int_0^t F(t')\sin\omega_n(t-t')dt', \tag{21}$$

$$= \xi\int_0^t F(t')dt', \tag{22}$$

where the ξ is the bending coefficient given as

$$\xi = \frac{\left(\frac{3\rho}{E'}\right)^{\frac{1}{2}}}{16\rho h^2}, \tag{23}$$

Here, h is the plate thickness, ρ is the density of the plate, and E' is the modulus of elasticity of the plate.

Suppose that when a sphere hits the plate, as it falls freely from a height, the ball and plate collide through an ideal elastic collision, and the total displacement (s) is as shown in Figure 3a. On the other hand, in the case of an inelastic collision, the total displacement caused by the actual collision is (Z), and the irreversible deformation of the plate after collision is (U), as shown in Figure 3b. The total displacements of the elastic and inelastic collisions are as follows.

$$Z = s + U. \tag{24}$$

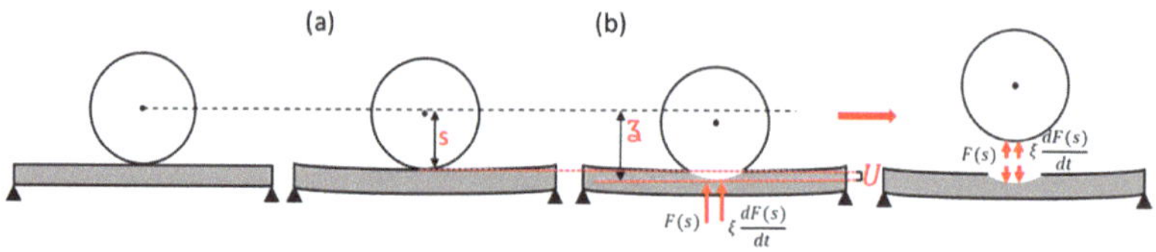

Figure 3. Ball collides with a plate via (**a**) elastic and (**b**) inelastic collisions.

The acceleration obtained using the second differential of the displacement in Equation (24) as well as Equations (1) and (23) gives Zener's nonlinear motion equation as

$$\frac{ds^2}{dt^2} + \frac{1}{M}F(s) + \xi\frac{dF(s)}{dt} = 0. \tag{25}$$

The nonlinear differential equation proposed by Zener [2] is transformed to provide an analytical solution by simplifying the Hertzian contact force as a function of the displacement. Thus, Equation (9) is rewritten as

$$\frac{ds^2}{dt^2} + \frac{1}{M}F(s) + \xi\frac{dF(s)}{dt} = 0. \tag{26}$$

where s is the contact displacement, k is the Hertzian stiffness constant, and Equation (26) can be written in dimensionless form using transformations $\sigma(\tau) = \frac{s(t)}{Tv_0}$ and $\tau = \frac{t}{T}$; σ and τ are dimensionless displacement and time, respectively; t is the time and T is a characteristic time constant that reflects the mechanical material properties of the sphere and plate as

$$T = \left(\frac{M}{nv_0^{\frac{1}{2}}}\right)^{\frac{2}{5}}. \tag{27}$$

Equation (25) can be transformed into a dimensionless differential equation of Equation (28), which is given as [2]

$$\frac{d^2\sigma}{d\tau^2} + \left(1 + \lambda\frac{d}{d\tau}\right)\sigma^{\frac{3}{2}} = 0. \tag{28}$$

According to Zener [2], all parameters that influence the impact are summarized using a dimensionless inelasticity parameter, λ, as per Equation (29)

$$\lambda = \xi k v_0^{\frac{1}{2}} T^{\frac{3}{2}} = \xi M\left(\frac{k}{M} v_0^{\frac{1}{2}} T^{\frac{3}{2}}\right) = \xi M\left(T^{-\frac{5}{2}} \cdot T^{\frac{3}{2}}\right) = \frac{\xi M}{T}. \tag{29}$$

3. Results and Discussion

When the ball collides with the plate material, the inelasticity parameter (λ) between the two and force time during the impact change owing to different material properties. The Zener collision theory equation was numerically analyzed and complex calculations were performed. Furthermore, the force–time collision curve diagram was drawn with λ at 0, 0.5, 1.0, and 1.5, as shown in Figure 4a. Figure 4a shows the process of change in the impact contact time and the magnitude of the contact force. Through the Hertzian contact force as a function of the displacement (Equation (26)), the impact force was converted to material displacement and other parameters. Next, the material inelasticity parameter (λ) was calculated using Equation (29).

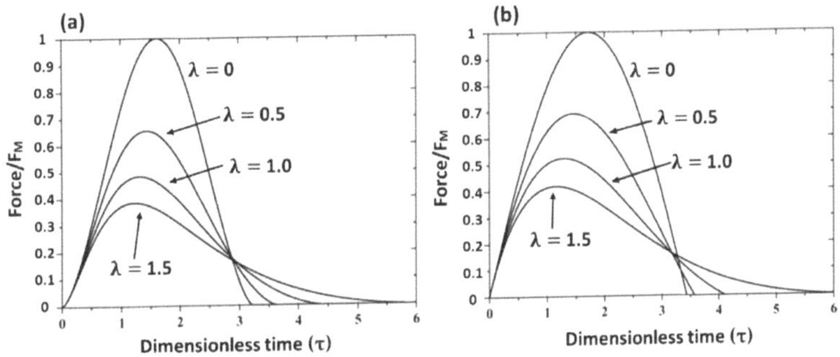

Figure 4. (**a**) Zener standard force–time collision curve [2,12]. (**b**) Boettcher et al. simplified the normalized force–time collision curve graph obtained from the analysis of the Zener equation [12].

Boettcher et al. [12] revised the nonlinear motion equation (Equation (25)) [2] proposed by Zener with a dimensionless parameter, ω_0, to be adjusted (determined by adapting the ω_0 dimensionless analytical displacement–time function), and Hertz's equation (Equation (26)) [1], and established a method that simply analyzes the nonlinear motion equation of the Zener collision [3,12]. Boettcher et al. also simplified the Zener nonlinear collision equation (Equation (28)) to a linear differential equation, which can be regarded as a damped oscillation equation. When $\lambda = 0$, an ideal elastic collision without damping is indicated, whereas $\lambda > 0$ is an inelastic collision and has a damping state [12]. The process obtains the force–time impact curve of $\lambda = 0$, which is a perfect elastic collision equation, and then adjusts ω_0 to obtain $\lambda > 0$, that is, the inelastic collision [3,12]. Figure 4b shows the normalized impact force curve obtained by Boettcher et al. after a simplified analysis of Zener's nonlinear motion equation [12]. This work is based on the nonlinear motion equation provided by Zener [2], which establishes the force–time collision curve equation in different ways.

As observed in Figure 4, the sphere collides with the plate, and the contact force changes with time. Additionally, the force ascends from zero to the maximum value and then decreases gradually. It is considered to be an ideal elastic collision when the inelasticity parameter is zero. The impact process is not affected by the friction force and causes energy attenuation. Consequently, the force–time variation curve is a symmetrical graph. When the inelasticity parameter is larger ($\lambda \neq 0$), it is an inelastic collision. Energy loss was generated during the collision between the sphere and the plate, and the contact force was smaller, while the contact time was longer. As seen in Figure 4, the force–time collision curve based on Equation (28) correlates well with the dimensionless displacement (σ) and time (τ). As shown in Figure 5, the force–time curve is a graph of the quadratic function. First, we used the approximation method, and the curve was drawn based on Equation (30).

$$\frac{Force}{F_m} = \sigma^{\frac{3}{2}} = \left[\tau\left(1 - \frac{\tau}{\tau_{el}}\right)\right]^{\frac{3}{2}}. \tag{30}$$

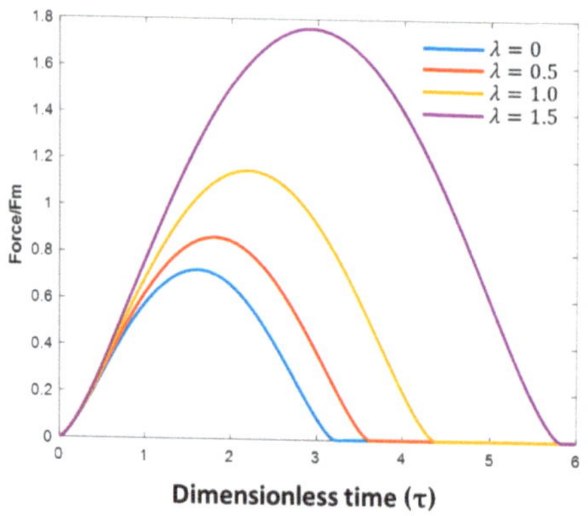

Figure 5. Force–time change quadratic function curve.

The relationship between the elastic collision time (τ) and the inelasticity parameter (λ) of the plate is based on the equation $\tau_{el} = 2.762 + 0.4568e^{1.27\lambda}$ provided by Müller et al. [4]. Additionally, the displacement (σ) can be converted into force using Equation (26). Equation (30) provides the left–right symmetrical quadratic function graph. When the elasticity parameter is 0, it is an elastic collision curve. Consequently, the inelasticity parameter value ($\lambda \neq 0$) and impact force are greater and the contact time is longer, as shown in Figure 5.

However, in the actual collision process, $\lambda \neq 0$. After the sphere collides with the plate, the force on the plate shows an exponential attenuation as the sphere leaves the plate. Therefore, we multiply the σ in Equation (30) by $e^{-\lambda\tau}$ to obtain Equation (31), as

$$\frac{Force}{F_m} = \sigma^{\frac{3}{2}} = \left[\tau\left(1 - \frac{\tau}{\tau_{el}}\right)e^{-\lambda\tau}\right]^{\frac{3}{2}}. \tag{31}$$

Equation (31), as seen from Figure 6a, shows an exponential decay change with time in the force process of different inelasticity parameter materials. The selection range of $e^{-\lambda\tau}$ coefficient is from $e^{-0.3\lambda\tau}$ to $e^{-0.5\lambda\tau}$. When multiplied by $e^{-0.4\lambda\tau}$, the force is an attenuated process of the contact time. As per the comparison curve between Figure 6b and the σ of

Equation (30) multiplied by $e^{-0.4\lambda\tau}$ and Figure 4a, Zener's standard force–time collision curve is the closest.

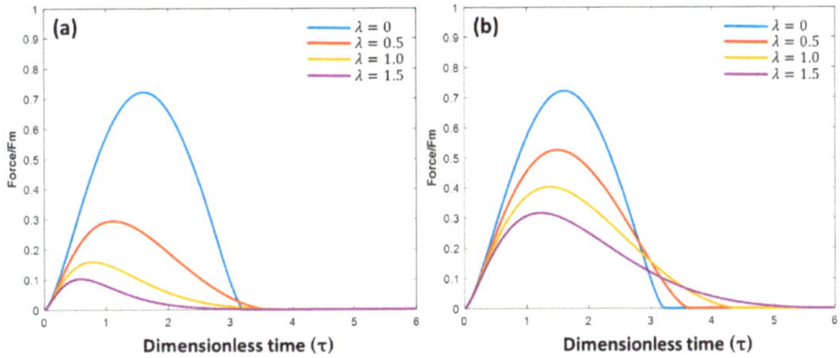

Figure 6. The σ of Equation (30) is multiplied by (**a**) $e^{-\lambda\tau}$ and (**b**) $e^{-0.4\lambda\tau}$.

Figure 6b shows that $\lambda = 0$, and the maximum collision force ($Force/F_m$) is not equal to 1. Therefore, the correction Equation (31), multiplied by $\frac{4}{\tau_{el}}$, aims to correct the change in contact time and force of different inelasticity parameters, and the force ($Force/F_m$) of $\lambda = 0$ can be corrected to 1, as shown in Figure 7a.

$$\frac{Force}{F_m} = \sigma^{\frac{3}{2}} = \left[\frac{4\tau}{\tau_{el}}\left(1 - \frac{\tau}{\tau_{el}}\right)e^{-0.4\lambda\tau}\right]^{\frac{3}{2}}. \tag{32}$$

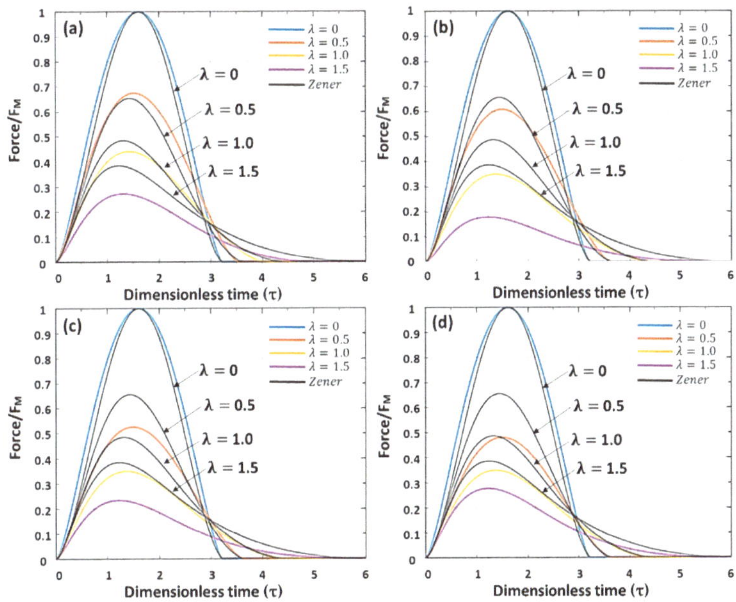

Figure 7. (**a**) The σ of Equation (31) is multiplied by $\frac{4}{\tau_{el}}$. Multiplying the σ in Equation (32) by (**b**) $(\lambda^1 + (1-\lambda))$, (**c**) $(\lambda^{1.3} + (1-\lambda))$, (**d**) $(\lambda^{1.5} + (1-\lambda))$.

The force–time collision curve of Equation (32) is quite close to the original Zener force–time diagram (Figure 4), as shown in Figure 7a. However, when the inelasticity parameters are 1 and 1.5, the maximum force curve is very different from the Zener curve;

hence, we multiply the σ in Equation (32) by $(\lambda^n + (1-\lambda))$ to modify the same, as shown in Equation (33).

$$\frac{Force}{F_m} = \sigma^{\frac{3}{2}} = \left[(\lambda^n + (1-\lambda))\frac{4\tau}{\tau_{el}}\left(1 - \frac{\tau}{\tau_{el}}\right)e^{-0.4\lambda\tau}\right]^{\frac{3}{2}}. \quad (33)$$

When the σ of Equation (32) is multiplied by $(\lambda^n + (1-\lambda))$, the force–time curve with n > 0 and $\lambda = 1$ does not change, as shown in Figure 7b–d. However, when the value of n in Equation (33) becomes larger, and $\lambda = 0.5$, the impact force curve gradually becomes smaller, whereas for $\lambda = 1.5$, it gradually becomes larger, as shown in Figure 7b–d. Additionally, the value of n gradually increases from 1.0, 1.3, and 1.5 for comparison. Finally, the value of n = 1.3 is the maximum force peak value of the force–time curve drawn using Equation (33) established in this study is between the peak force value of the inelasticity parameter 0.5 to 1.5 and the Zener value, and the peaks of the force–time curve are equidistant, as shown in Figure 7c.

As shown in Figure 7c, the force–time curve peak values of the inelasticity parameters 0.5–1.5 are approximately equidistant from the Zener force–time curve peak values. Therefore, the correction Equation (33) adjusts the peak force of the force–time curve and multiplies it by $(1 + \mathrm{n} \times \lambda)^2$ to provide Equation (34).

$$\frac{Force}{F_m} = \sigma^{\frac{3}{2}} = (1 + \mathrm{n} \times \lambda)^2\left[(\lambda^{1.3} + (1-\lambda))\frac{4\tau}{\tau_{el}}\left(1 - \frac{\tau}{\tau_{el}}\right)e^{-0.4\lambda\tau}\right]^{\frac{3}{2}}. \quad (34)$$

When $\lambda = 0$, $(1 + \mathrm{n} \times \lambda)^2$, n is any value, the force is 1, and the force–time curve remains constant. Thus, the inelasticity parameter is between 0.5 and 1.5. When the value of n in Equation (34) is larger, the force–time curve of the inelasticity parameter gradually increases the force value, as shown in Figure 8. When n = 0.2, it is nearly identical to the time–force curve of the original Zener, and the curve drawn with inelasticity parameters of 1–1.5 uses Equation (34). Therefore, n = 0.2 is the most suitable. As shown in Figure 9a,b, the force–time collision curve given using Equation (34) and that given by Boettcher et al. are different from the analytical methods that provide graphs and equations that are considerably similar to the diagrams of the force–time collisions proposed by Zener.

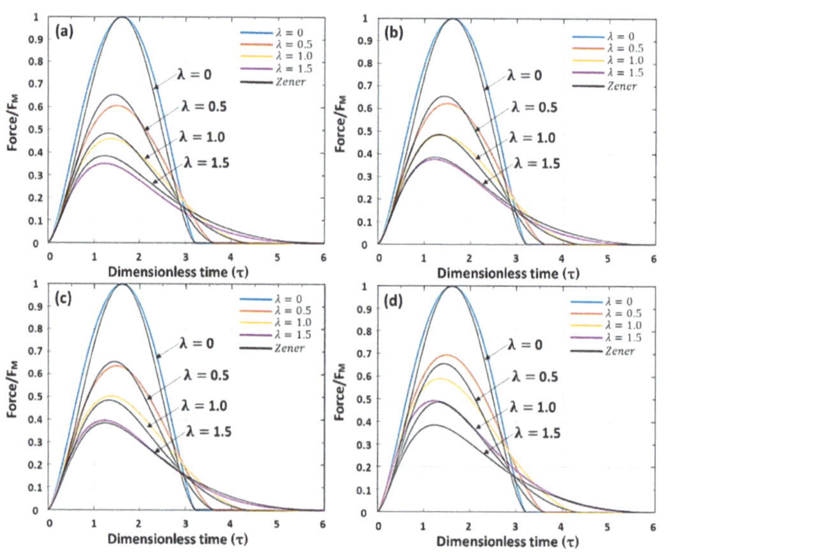

Figure 8. Equation (33) multiplied by $(1 + \mathrm{n} \times \lambda)^2$, when (**a**) n = 0.15, (**b**) n = 0.18, (**c**) n = 0.2, and (**d**) n = 0.3.

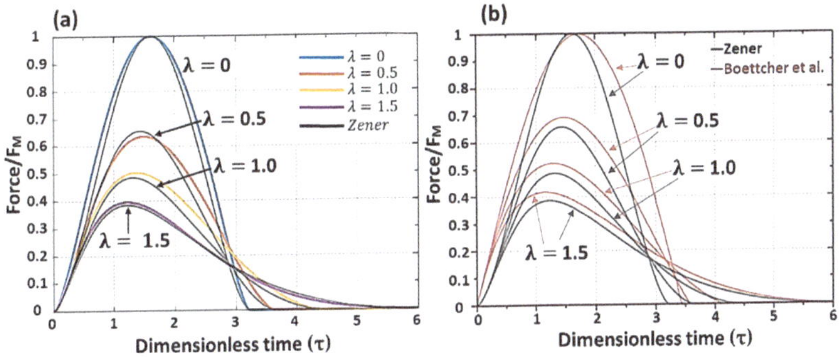

Figure 9. (**a**) Force–time collision curves of the equation established in this study and that of the Zener standard; (**b**) Force–time collision curves of the equations established by Boettcher et al. and Zener standard [2,12].

4. Conclusions

The nonlinear collision motion equation proposed by Zener [2] is the force–time collision curve of the inelasticity parameter ($0 \leq \lambda \leq 1.5$) considered in the collision process through numerical analysis. In this study, we have proposed an approximate solution to Zener's inelastic collision theory, which is expressed as

$$\frac{Force}{F_m} = \sigma^{\frac{3}{2}} = (1+0.2\lambda)^2\left[\left(\lambda^{1.3}+(1-\lambda)\right)\frac{4\tau}{\tau_{el}}\left(1-\frac{\tau}{\tau_{el}}\right)e^{-0.4\lambda\tau}\right]^{\frac{3}{2}}.$$

The contribution of the present study to literature includes an approximate solution that is simple and can be used directly without complex calculations. The contact time, magnitude of the material impact force and displacement, and inelasticity parameter can be obtained simply by substituting the parameters into the equation. Using the approximate solution, the inelasticity parameter can be obtained from the thickness, density, and modulus of elasticity of the material, and the relationship between thickness and contact time can be obtained by simple and fast calculation. Furthermore, the method can be applied to nondestructive impact quality inspections of commercial materials, such as internal defects and thickness differences of liquid crystal glass panels. It can also be applied to the quality inspection of golf club heads in the future. In summary, the use of the proposed formula does not require complex numerical calculations and can be verified easily to obtain the collision-related parameters. Furthermore, the proposed approximate solution can be adjusted for application based on user requirements.

Author Contributions: P.-K.T., C.-H.L., C.-C.L., K.-J.H. and C.-W.C. analyzed the data; P.-K.T., C.-H.L. and C.-W.C. wrote the paper; P.-K.T., C.-H.L. and C.-W.C. reviewed and edited the manuscript. All authors have read and agreed to the published version of the manuscript.

Funding: This research was funded by the Ministry of Science and Technology, Taiwan under grant number MOST 110-2321-B-025-001.

Acknowledgments: The authors thank the Ministry of Science and Technology of the Republic of China for their financial support and also thank the John Wiley and Sons publishing house "*Energiedissipation aufgrund von Biegewellen bei Stoßvorgängen gegen dünne Platten*" images copyright approvement.

Conflicts of Interest: The authors declare no conflict of interest.

References

1. Hertz, H. Ueber die Berührung fester elastischer Körper. *J. Für Die Reine Und Angew. Math.* **1882**, *1882*, 156–171. [CrossRef]
2. Zener, C. The Intrinsic Inelasticity of Large Plates. *Phys. Rev.* **1941**, *59*, 669–673. [CrossRef]
3. Boettcher, R.; Russell, A.; Mueller, P. Energy dissipation during impacts of spheres on plates: Investigation of developing elastic flexural waves. *Int. J. Solids Struct.* **2017**, *106–107*, 229–239. [CrossRef]

4. Müller, P.; Böttcher, R.; Russell, A.; Trüe, M.; Aman, S.; Tomas, J. Contact time at impact of spheres on large thin plates. *Adv. Powder Technol.* **2016**, *27*, 1233–1243. [CrossRef]
5. Jebelisinaki, F.; Boettcher, R.; van Wachem, B.; Mueller, P. Impact of dominant elastic to elastic-plastic millimeter-sized metal spheres with glass plates. *Powder Technol.* **2019**, *356*, 208–221. [CrossRef]
6. Koller, M.G.; Kolsky, H. Waves produced by the elastic impact of spheres on thick plates. *Int. J. Solids Struct.* **1987**, *23*, 1387–1400. [CrossRef]
7. Hunter, S.C. Energy absorbed by elastic waves during impact. *J. Mech. Phys. Solids* **1957**, *5*, 162–171. [CrossRef]
8. Boettcher, R.; Kunik, M.; Eichmann, S.; Russell, A.; Mueller, P. Revisiting energy dissipation due to elastic waves at impact of spheres on large thick plates. *Int. J. Impact Eng.* **2017**, *104*, 45–54. [CrossRef]
9. Reed, J. Energy losses due to elastic wave propagation during an elastic impact. *J. Phys. D Appl. Phys.* **1985**, *18*, 2329–2337. [CrossRef]
10. Stevens, A.B.; Hrenya, C.M. Comparison of soft-sphere models to measurements of collision properties during normal impacts. *Powder Technol.* **2005**, *154*, 99–109. [CrossRef]
11. Timoshenko, S.P.; Goodier, J.N. *Theory of Elasticity*, 3rd ed.; McGraw-Hill: New York, NY, USA, 1970.
12. Böttcher, R.; Müller, P.; Trüe, M.; Russell, A.; Tomas, J. Energiedissipation aufgrund von Biegewellen bei Stoßvorgängen gegen dünne Platten. *Chem. Ing. Tech.* **2016**, *88*, 1002–1011. [CrossRef]

mathematics

MDPI

Article

Fractional-Order Discrete-Time SIR Epidemic Model with Vaccination: Chaos and Complexity

Zai-Yin He [1], Abderrahmane Abbes [2], Hadi Jahanshahi [3], Naif D. Alotaibi [4] and Ye Wang [5,6,*]

1 School of Mathematics, Hunan University, Changsha 410082, China; hzy@zjhu.edu.cn
2 Department of Mathematics, University of Jordan, Amman 11942, Jordan; abder.abbes@gmail.com
3 Department of Mechanical Engineering, University of Manitoba, Winnipeg, MB R3T 5V6, Canada; jahanshahi.hadi90@gmail.com
4 Department of Electrical and Computer Engineering, Faculty of Engineering, King Abdulaziz University, Jeddah 21589, Saudi Arabia; ndalotabi@kau.edu.sa
5 Department of Mathematics, Huzhou University, Huzhou 313000, China
6 Institute for Advanced Study Honoring Chen Jian Gong, Hangzhou Normal University, Hangzhou 311131, China
* Correspondence: 03019@zjhu.edu.cn

Abstract: This research presents a new fractional-order discrete-time susceptible-infected-recovered (SIR) epidemic model with vaccination. The dynamical behavior of the suggested model is examined analytically and numerically. Through using phase attractors, bifurcation diagrams, maximum Lyapunov exponent and the $0-1$ test, it is verified that the newly introduced fractional discrete SIR epidemic model vaccination with both commensurate and incommensurate fractional orders has chaotic behavior. The discrete fractional model gives more complex dynamics for incommensurate fractional orders compared to commensurate fractional orders. The reasonable range of commensurate fractional orders is between $\gamma = 0.8712$ and $\gamma = 1$, while the reasonable range of incommensurate fractional orders is between $\gamma_2 = 0.77$ and $\gamma_2 = 1$. Furthermore, the complexity analysis is performed using approximate entropy $(ApEn)$ and C_0 complexity to confirm the existence of chaos. Finally, simulations were carried out on MATLAB to verify the efficacy of the given findings.

Keywords: discrete SIR epidemic model; commensurate order; incommensurate order; chaos; complexity

check for updates

Citation: He, Z.Y.; Abbes, A.; Jahanshahi, H.; Alotaibi, N.D.; Wang, Y. Fractional-Order Discrete-Time SIR Epidemic Model with Vaccination: Chaos and Complexity. *Mathematics* **2022**, *10*, 165. https://doi.org/10.3390/math10020165

Academic Editors: Alexandra M.S.F. Galhano, António M. Lopes and Carlo Cattani

Received: 17 December 2021
Accepted: 2 January 2022
Published: 6 January 2022

1. Introduction

Epidemiology is a topic of research in the biological sciences, which explores all the elements that influence whether there are diseases and disorders. Over the last few years, the outbreaks of several diseases such as SARS, Ebola, and COVID-19 have increased the attention of many researchers to the study of epidemiology. In this regard, the construction of infectious diseases mathematical models and discussion of their dynamics are very significant, since these models are used to aid prevent and control diseases. Numerous studies on mathematical epidemic models have been examined for a variety of diseases. These models may be classified as either continuous-time [1–5] or discrete-time models. Differential equations are used to create continuous-time models, while difference equations are used to formulate discrete-time models. Recently, discrete models have been extensively employed to evaluate and study infectious diseases as opposed to continuous models since epidemic data are available during discrete time periods. Furthermore, the discrete-time epidemic models display more complicated dynamics than the continuous-time epidemic models. Numerous significant types of discrete models may be found in [6–9].

Vaccination is a significant component of disease prevention strategies across the globe. A vaccine's effectiveness is based on its ability to reduce the number of susceptible individuals in order to prevent infectious diseases from spreading through the population [10]. The effect of vaccination on the dynamics of an epidemic model has been studied by adding a component for the vaccinated individuals to the epidemic model and obtaining the epidemic model with vaccination. For example, in [11] the authors studied the dynamical behavior of a discrete SIR epidemic model with a constant vaccination strategy. The stability and bifurcations of a discrete susceptible-infected-susceptible (SIS) epidemic model with vaccination were investigated in [12], while Xiang et al. in [13] developed a discrete $SIRS$ model that included vaccination and examined its dynamical behavior.

Fractional calculus is a subject in mathematical analysis, where it can be considered as a generalization of integer calculus [14–28]. Despite this, only in the past decades has it been extensively examined, owing to its broad range of use in many areas. Fractional-order derivatives are relatively more accurate than integer-order derivatives because they serve as an effective tool for describing the memory effect throughout all types of processing and materials [29]. Memory effects refer to the fact that the states of systems with a fractional order are determined by all previous states. Over the last several years, numerous academics have concentrated their efforts on discrete fractional calculus [30–32]. Following the proposal of that special field, researchers have become more interested in its applications in neural networks, physics, biology, etc. Meanwhile, since Wu et al. [33] suggested the first chaotic fractional logistic map, the dynamics and control of various fractional-order chaotic discrete-time systems have been intensively studied [34–38].

Fractional-order derivatives have been extensively used in epidemiology, with the majority of these studies focusing on SIR-type models [39–44]. In [45], Selvam et al. investigated the stability of the fractional-order SIR epidemic model using a discretization process, whereas Naik in [46] studied the global dynamics of a fractional-order SIR epidemic model with memory. The analysis and numerical solution of a novel fractional-order SIR dengue model was investigated in [47]. In [48], the stability analysis and bifurcation control for a delayed fractional-order SIR epidemic model with incommensurate order was considered. It is worth noting that all of these studies are based on continuous-time systems. The motivation behind this work, particularly with the current spread of the COVID-19 pandemic, is to assess the benefits of the fractional discrete epidemic model in the field of epidemiology. So far, to the best of our knowledge, the dynamic analysis of a fractional-order discrete-time epidemic model with vaccination based on a Caputo-like difference operator has not yet been investigated. This piqued our interest and inspired us to study the phenomenon and investigate the behavior of a fractional SIR epidemic model with vaccination when the fractional orders are commensurate and incommensurate.

The goal of this article is to contribute to the field of epidemiology by introducing a novel discrete-time SIR epidemic model with vaccination with both commensurate and incommensurate fractional orders. The work is organized as follows. In Section 2, we present the integer-order form of the discrete SIR model and we recast the mathematical model to take the form of the fractional-order discrete SIR model. In Section 3, we look at how the range and type of chaotic behaviors are affected by commensurate and incommensurate fractional orders. Furthermore, we use the $0-1$ test method to distinguish between regular and chaotic behavior. In Section 4, we use the C_0 complexity and approximate entropy $(ApEn)$ to analyze the complexity of the proposed discrete SIR epidemic model. Throughout the paper, numerical simulations are used to demonstrate and verify the results.

2. Mathematical Model

In this section, we introduce the discrete SIR model as an integer-order discrete system, and then we reformulate it as a fractional-order discrete system by employing the Caputo-left difference operator.

2.1. Integer-Order Discrete Model

Consider the following description of the SIR epidemic model described in [49], in which the population densities S_m, I_m, and R_m represent the number of susceptible, infected, and recovered individuals at time m, respectively:

$$\begin{cases} S_{m+1} = S_M - \frac{\alpha}{N} I_m S_m + \beta(R_m + I_m), \\ I_{m+1} = \frac{\alpha}{N} I_m S_m + (1 - \beta - \sigma) I_m, \\ R_{m+1} = (1 - \beta) R_m + \sigma I_m, \end{cases} \quad (1)$$

where $\alpha > 0$ is the contact number, $0 < \beta < 1$ is is the probability of birth, and $0 < \sigma < 1$ is the probability of recovery. The flow diagram for model (3) is seen in Figure 1.

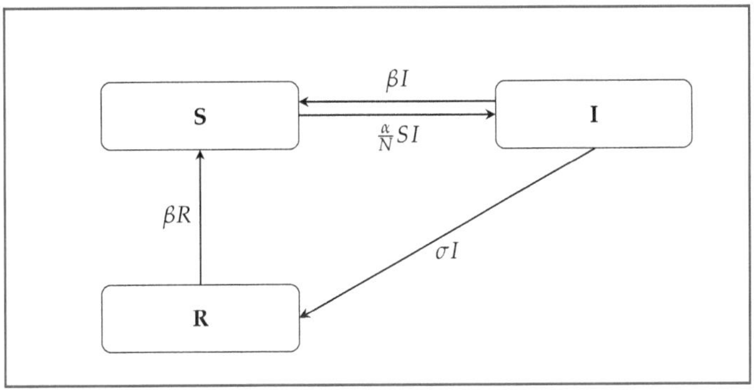

Figure 1. Flow chart of the SIR epidemic model (1).

In [50], the authors focused on the dynamics of an SIR epidemic model by incorporating vaccination in the model (1). The proportion of persons who had been vaccinated was represented by the parameter p, where $0 < p < 1$; and the remainder had a risk of susceptibility to infection in the proportion $1 - p$. The following is the general SIR epidemic model with vaccination, which takes the following form:

$$\begin{cases} S_{m+1} = (1 - p) S_m - \frac{\alpha}{N} I_m S_m + \beta(R_m + I_m), \\ I_{m+1} = \frac{\alpha}{N} I_m S_m + (1 - \beta - \sigma) I_m, \\ R_{m+1} = (1 - \beta) R_m + \sigma I_m + p S_m, \end{cases} \quad (2)$$

(S_0, I_0, R_0) are initial conditions that are positive real values, with $S_0 + I_0 + R_0 = N$. N is the total population size. Furthermore, by employing the relation $S_m + I_m + R_m = N$, we replace R_m by $N - S_m - I_m$ and we get the following system:

$$\begin{cases} S_{m+1} = (1 - p) S_m - \frac{\alpha}{N} I_m S_m + \beta(N - S_m), \\ I_{m+1} = \frac{\alpha}{N} I_m S_m + (1 - \beta - \sigma) I_m, \\ R_{m+1} = (1 - \beta) R_m + \sigma I_m + p S_m, \end{cases} \quad (3)$$

Observing that the two equations concerning (S, I) in the SIR epidemic model (3) do not contain R, they are thus independent of the third one. As a result, the three-dimensional

SIR epidemic model (3) can be reduced to a two-dimensional model, which results in the following equivalent system:

$$\begin{cases} S_{m+1} = (1-p)S_m - \frac{\alpha}{N} I_m S_m + \beta(N - S_m), \\ I_{m+1} = \frac{\alpha}{N} I_m S_m + (1 - \beta - \sigma) I_m, \end{cases} \tag{4}$$

when $\alpha = 4$, $\beta = 0.8$, $\sigma = 0.1$, $p = 0.005$, and $N = 100$, the discrete SIR epidemic model with vaccination (4) exhibits a chaotic attractor as shown in Figure 2.

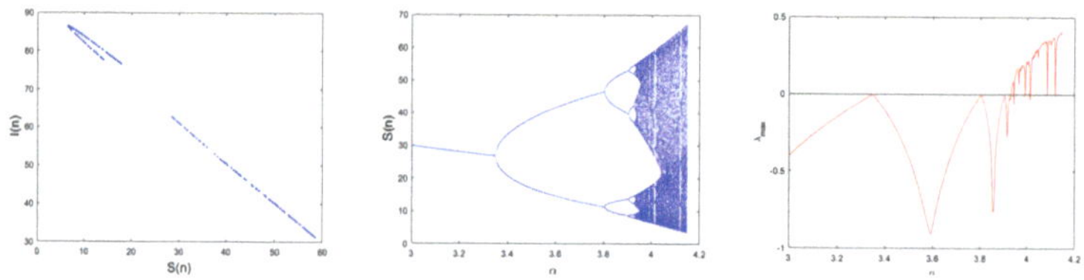

(a) (b) (c)

Figure 2. (**a**) Phase space of the discrete SIR epidemic model with vaccination (4) for $\alpha = 4$, $\beta = 0.8$, $\sigma = 0.1$, $p = 0.005$, $N = 100$, and initial conditions $(S(0), I(0)) = (70, 30)$. (**b**) Bifurcation diagram of the SIR epidemic model (4) versus α. (**c**) The maximum Lyapunov exponents.

2.2. Fractional-Order Discrete Model

In this study, a discrete-time SIR epidemic model with vaccination with both commensurate and incommensurate orders is considered. Herein, the first-order difference of the SIR model (4) is formulated as:

$$\begin{cases} \Delta S(m) = -pS(m) - \frac{\alpha}{N} I(m)S(m) + \beta(N - S(m)), \\ \Delta I(s) = \frac{\alpha}{N} I(m)S(m) - (\beta + \sigma) I(m), \end{cases} \tag{5}$$

As mentioned above, we use the Caputo-left difference operator to create the fractional version of the discrete-time SIR epidemic model with vaccination. The Caputo-left difference operator is defined as follows:

Definition 1. *[51] The γ-Caputo fractional difference operator for a function $h(s)$, is defined by*

$$^{C}\Delta_a^{\gamma} h(s) = \Delta_a^{-(m-\gamma)} \Delta^m h(s) = \frac{1}{\Gamma(m-\gamma)} \sum_{\tau=a}^{s-(m-\gamma)} (s - \tau - 1)^{(m-\gamma-1)} \Delta^m h(\tau), \tag{6}$$

where $s \in \mathbb{N}_{a+m-\gamma}$, $m = \lceil \gamma \rceil + 1$, and $\gamma \notin \mathbb{N}$. $(s - 1 - \tau)^{(m-\gamma-1)}$ and $\Delta^m h(\tau)$ are the falling factorial function and the m-th integer difference operator, respectively, which are defined as

$$(s - \tau - 1)^{(m-\gamma-1)} = \frac{\Gamma(s-\tau)}{\Gamma(s - \tau - m + \gamma + 1)}, \tag{7}$$

and

$$\Delta^m h(s) = \Delta(\Delta^{m-1} h(s)) = \sum_{k=0}^{m} \binom{m}{k} (-1)^{m-k} h(s+k), \qquad s \in \mathbb{N}_a. \tag{8}$$

Remark 1. *If we consider that $m = 1$, we can define the γ-Caputo fractional difference operator by*

$$^{C}\Delta_a^{\gamma} h(s) = \Delta_a^{-(1-\gamma)} \Delta h(s) = \frac{1}{\Gamma(1-\gamma)} \sum_{\tau=a}^{s-(1-\gamma)} (s - \tau - 1)^{(-\gamma)} \Delta h(\tau), \quad s \in \mathbb{N}_{a+1-\gamma} \tag{9}$$

Definition 2. *[30] The γ-th fractional sum for a function h is defined as*

$$\Delta_a^{-\gamma} h(s) = \frac{1}{\Gamma(\gamma)} \sum_{\tau=a}^{s-\gamma} (s-1-\tau)^{(\gamma-1)} h(\tau), \tag{10}$$

with $s \in \mathbb{N}_{a+\gamma}$, $\gamma > 0$. $N_a = \{a, a+1, a+2, \dots\}$ is a time scale, $a \in \mathbb{R}$.

Now, the discrete SIR epidemic model with vaccination (4) may be expressed in fractional order as follows:

$$\begin{cases} {}^C\Delta_a^{\gamma_1} S(s) = -pS(s-1+\gamma_1) - \frac{\alpha}{N} I(s-1+\gamma_1) S(s-1+\gamma_1) + \beta(N - S(s-1+\gamma_2)), \\ {}^C\Delta_a^{\gamma_2} I(s) = \frac{\alpha}{N} I(s-1+\gamma_2) S(s-1+\gamma_2) - (\beta+\sigma) I(s-1+\gamma_2), \end{cases} \tag{11}$$

where $\gamma_{i(i=1,2,3)}$ are the fractional order such that $0 < \gamma_i < 1$. Note that if $\gamma_1 = \gamma_2$, then system (11) is referred to as a commensurate fractional-order system, whereas it is referred to as an incommensurate fractional-order system if $\gamma_1 \neq \gamma_2$.

3. Dynamical Analysis and Numerical Simulations

In this section, we investigate whether the previously suggested fractional-order discrete-time SIR epidemic model with vaccination (11) exhibits chaotic behavior in both the commensurate and incommensurate fractional orders. This study is conducted utilizing many numerical techniques, including bifurcation diagrams, Lyapunov exponent computations, plotting of phase portraits in S–I projection and applying the $0-1$ method.

In order to show these, we first present a theorem that enables us to obtain the numerical formula for the discrete fractional model that is subsequently discussed.

Theorem 1. *[52] For the fractional difference equation*

$$\begin{cases} {}^C\Delta_a^{\gamma_i} y(s) = g(s+\gamma_i-1, y(s+\gamma_i-1)) \\ \Delta^k y(s) = y_k, \quad m = \lceil \gamma_i \rceil + 1, \quad k = 0, 1, \dots, m-1, \end{cases} \tag{12}$$

the unique solution of this initial value problem (1) is given by

$$y(s) = y_0(s) + \frac{1}{\Gamma(\gamma_i)} \sum_{\tau=a+m-\gamma_i}^{s-\gamma_i} (s-\sigma(\tau))^{(\gamma_i-1)} g(\tau+\gamma_i-1, y(\tau+\gamma_i-1)), \quad s \in N_{a+m}, \tag{13}$$

where

$$y_0(s) = \sum_{k=0}^{m-1} \frac{(s-a)^k}{\Gamma(k+1)} \Delta_k y(a). \tag{14}$$

According to Theorem 1, the numerical formula of the fractional discrete-time SIR epidemic model (11) is designed as:

$$\begin{cases} S(s) = S(a) + \frac{1}{\Gamma(\gamma_1)} \sum_{\tau=a+m-\gamma_1}^{s-\gamma_1} (s-\sigma(\tau))^{(\gamma_1-1)} \Big(-p\, S(\tau+\gamma_1-1) - \frac{\alpha}{N}\, I(\tau+\gamma_1-1) \\ \qquad S(\tau+\gamma_1-1) \\ \qquad +\beta\,(N - S(\tau+\gamma_1-1)) \Big), \\ I(s) = I(a) + \frac{1}{\Gamma(\gamma_3)} \sum_{\tau=a+m-\gamma_3}^{s-\gamma_3} (s-\sigma(\tau))^{(\gamma_3-1)} \Big(\frac{\alpha}{N}\, I(\tau+\gamma_2-1)\, S(\tau+\gamma_2-1) \\ \qquad -(\beta+\sigma)\, I(\tau+\gamma_2-1) \Big) \end{cases} . \tag{15}$$

Take $a = 0$ and since $(s-\tau+1)^{(\gamma-1)}$ is equal to $\Gamma(s-\tau)/\Gamma(s-\tau-\gamma+1)$, it follows from Theorem 1 that the numerical formula for (11) is obtained as:

$$\begin{cases} S(n) = S(0) + \frac{1}{\Gamma(\gamma_1)} \sum_{j=0}^{n-1} \frac{\Gamma(n-1-j+\gamma_1)}{\Gamma(n-1-j+1)} \Big(& -p\, S(j) - \frac{\alpha}{N}\, I(j)\, S(\tau+\gamma_1-1) \\ & +\beta\,(N - S(j))\Big), \\ I(n) = I(0) + \frac{1}{\Gamma(\gamma_2)} \sum_{j=0}^{n-1} \frac{\Gamma(n-1-j+\gamma_2)}{\Gamma(n-1-j+1)} \Big(& \frac{\alpha}{N}\, I(j)\, S(j) - (\beta+\sigma)\, I(j)\Big), \end{cases} \tag{16}$$

where $S(0)$ and $I(0)$ are the initial conditions. This is a novel type of fractional-order discrete-time SIR epidemic model with vaccination that possesses "memory effects". As seen in Equation (16), the states $S(n)$ and $I(n)$ are dependent on all previous variables $S(0), S(1), \ldots, S(n-1)$, and $I(0), I(1), \ldots, I(n-1)$.

3.1. Commensurate Fractional Order

3.1.1. Bifurcation Diagram and Maximum LEs

To investigate the dynamic of the commensurate fractional-order discrete-time SIR epidemic model with vaccination given in (11) regarding the fractional order $\gamma = \gamma_1 = \gamma_2$, we set the parameters $N = 100$, $\beta = 0.8$, $\sigma = 0.1$, $p = 0.005$, and initial conditions $(S(0), I(0)) = (70, 30)$. Figure 3 displays the phase space of model (11) for $\gamma = 1$. Figure 4 shows the states of the system for 150 points. Take note that when $\gamma = 1$, the fractional discrete model (11) refers to the integer-order model. Now, using the same initial condition and the same system parameters, the fractional discrete model for different fractional orders γ is presented in Figure 5. We see that as γ decreases, the phase portrait changes its shape between chaotic and periodic trajectories, whereas if we continue to decrease γ, the states of the fractional discrete model diverge to infinity.

The bifurcation diagrams of the fractional discrete model (11) where α varies in the interval $[3, 4.5]$ is presented in Figure 6. Obviously, decreasing the fractional order has an effect on the interval in which chaos occurs, and the bifurcation diagram progressively shifts to the right. For example, when $\gamma = 0.98$, chaos begins to appear at $\alpha = 3.915$ and the maximum chaotic range is reached at $\alpha = 4.128$, whereas when $\gamma = 0.91$, chaos begins to appear at $\alpha = 3.849$ and the maximum chaotic range is reached at $\alpha = 4.047$.

To further study the influence of the commensurate fractional order on the dynamics of the fractional discrete-time SIR epidemic model (11), we plotted the bifurcation diagram for γ as a critical parameter and the findings are presented in Figure 6a. We notice that the fractional discrete SIR epidemic model (11) contains chaos and that the fractional order γ has an influence on the system's dynamics. When $\gamma < 0.8712$, the system (11) exhibits a transient state, which means that until a minimum number of iterations is reached, the proposed system's states approach a limited attractor before diverging to infinity.

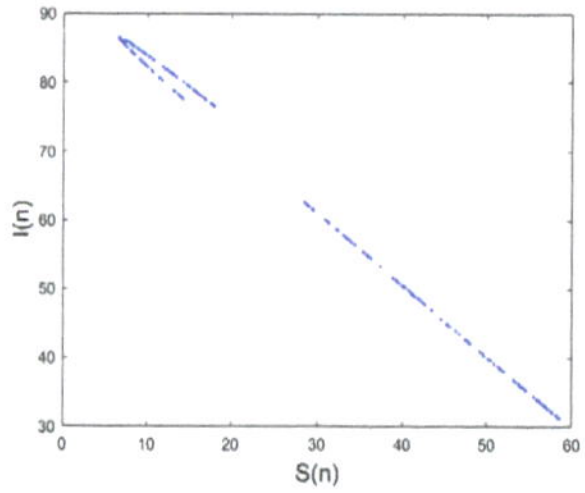

Figure 3. Phase portrait of the fractional SIR epidemic model in (11) with $\gamma = 1$ for $N = 100$, $\beta = 0.8$, $\sigma = 0.1, p = 0.0005$, and initial conditions $(S(0), I(0)) = (70, 30)$.

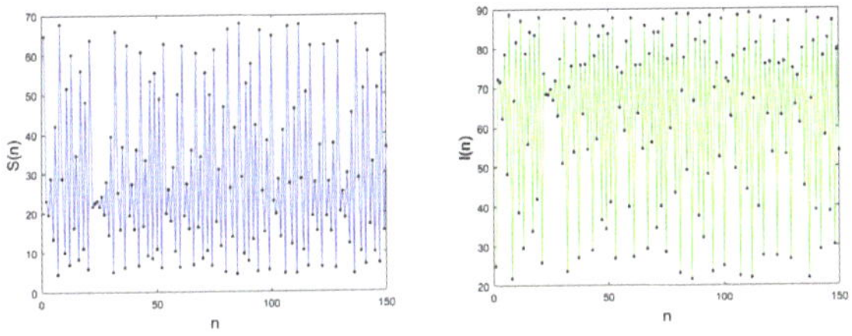

(a)

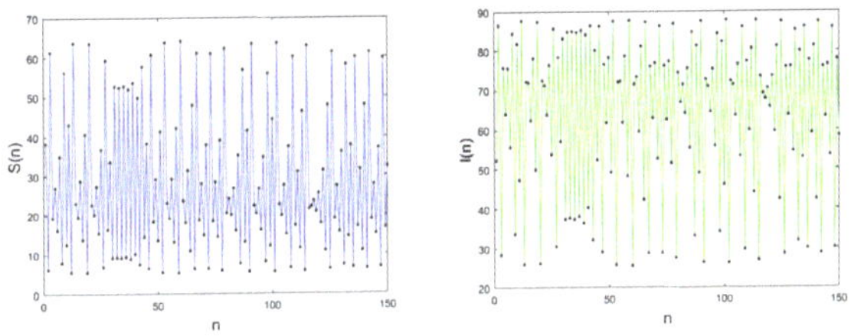

(b)

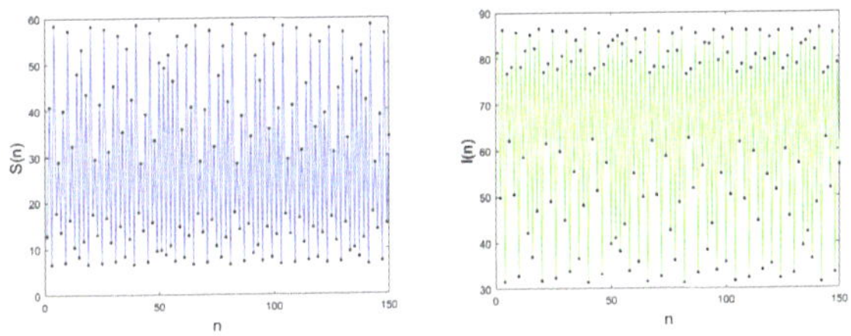

(c)

Figure 4. Time evolution of states for (**a**) $\gamma = 0.89$, (**b**) $\gamma = 0.93$, and (**c**) $\gamma = 1$.

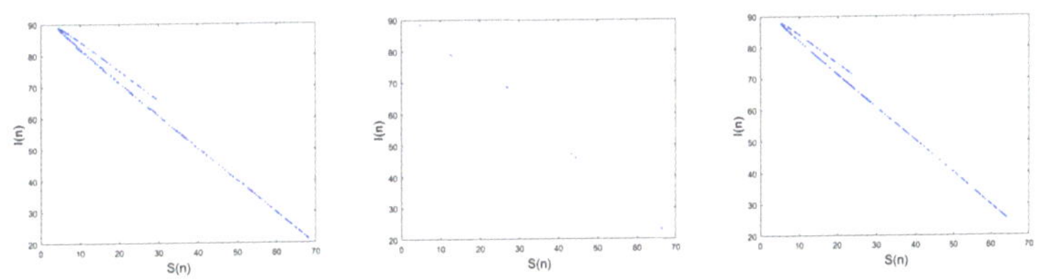

(**a**) $\gamma = 0.89$ (**b**) $\gamma = 0.91$ (**c**) $\gamma = 0.93$

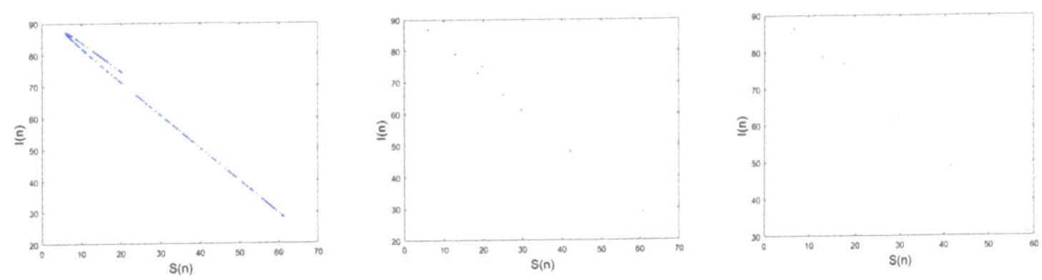

(**d**) $\gamma = 0.96$ (**e**) $\gamma = 0.966$ (**f**) $\gamma = 0.99$

Figure 5. Phase portrait of the fractional SIR epidemic model (11) with commensurate fractional-order values for $N = 100, \beta = 0.8, \sigma = 0.1, p = 0.005$, and initial conditions $(S(0), I(0)) = (70, 30)$.

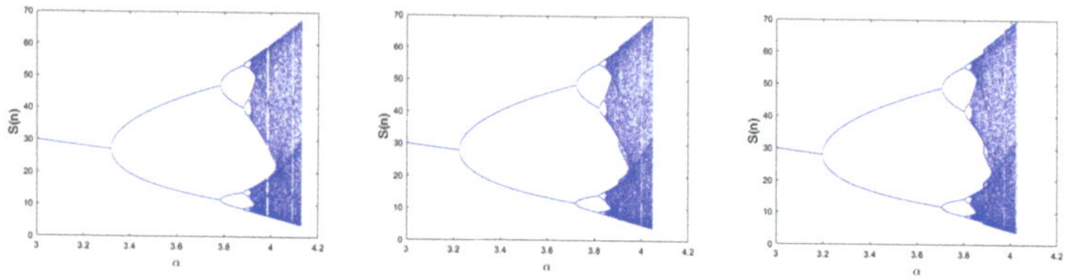

(a) $\gamma = 0.98$ (b) $\gamma = 0.91$ (c) $\gamma = 0.89$

Figure 6. Bifurcation diagrams of the fractional discrete SIR model (11) for different commensurate fractional-order values.

An important tool used in fractional discrete systems is the Lyapunov exponents, which are used in conjunction with bifurcation diagrams to show chaos. As shown in [52], the Jacobian matrix algorithm is used to compute or estimate the maximum Lyapunov exponent, which is calculated in a similar way to the states in the fractional-order discrete model (11). The tangent map J_i is defined as follows:

$$J_i = \begin{pmatrix} e_1 & e_2 \\ f_1 & f_2 \end{pmatrix} \tag{17}$$

where

$$\begin{aligned}
e_1(n) &= e_1(0) + \frac{1}{\Gamma(\gamma_1)} \sum_{j=0}^{n-1} \frac{\Gamma(n-1-j+\gamma_1)}{\Gamma(n-1-j+1)} \Big(-p\, e_1(j) - \frac{\alpha}{N}\, (e_1(j)\, I(j) + f_1(j)\, S(j)) \\
&\qquad - \beta\, e_1(j) \Big), \\
e_2(n) &= e_2(0) + \frac{1}{\Gamma(\gamma_1)} \sum_{j=0}^{n-1} \frac{\Gamma(n-1-j+\gamma_1)}{\Gamma(n-1-j+1)} \Big(-p\, e_2(j) - \frac{\alpha}{N}\, (e_2(j)\, I(j) + f_2(j)\, S(j)) \\
&\qquad - \beta\, e_2(j) \Big), \\
f_1(n) &= f_1(0) + \frac{1}{\Gamma(\gamma_2)} \sum_{j=0}^{n-1} \frac{\Gamma(n-1-j+\gamma_2)}{\Gamma(n-1-j+1)} \Big(\frac{\alpha}{N}\, (e_1(j)\, I(j) + f_1(j)\, S(j)) - (\beta+\sigma)\, f_1(j) \Big), \\
f_2(n) &= f_2(0) + \frac{1}{\Gamma(\gamma_2)} \sum_{j=0}^{n-1} \frac{\Gamma(n-1-j+\gamma_2)}{\Gamma(n-1-j+1)} \Big(\frac{\alpha}{N}\, (e_2(j)\, I(j) + f_2(j)\, S(j)) - (\beta+\sigma)\, f_2(j) \Big).
\end{aligned} \tag{18}$$

Then, the Lyapunov exponents can be given by :

$$\lambda_k(x_0) = \lim_{i \to \infty} \frac{1}{i} \ln |\lambda_k^{(i)}| \quad \text{for} \quad k = 1, 2. \tag{19}$$

where λ_k are the eigenvalues of the Jacobian matrix J_i.

The method described above can be turned into a MATLAB script to calculate the largest LE of the fractional-order discrete-time SIR epidemic model with vaccination (11). Because it is difficult to predict the behavior of the system using analytical techniques, we must rely on approximate numerical approaches, which can be achieved by using MATLAB software. For further information on how to use numerical simulation of some chaotic systems using MATLAB, see the link in [53]. The results of the MLE for system with the same parameters and with initial conditions $(S(0), I(0)) = (70, 30)$ are shown in Figure 7b. As can be seen, the system has positive LE, which shows that the MLE is compatible with the bifurcation diagram depicted in Figure 7a. We can also see that some values of Lyapunov exponents are negative, meaning that the system has periodic attractors, and this is illustrated further by the phase portraits plotted in Figure 5.

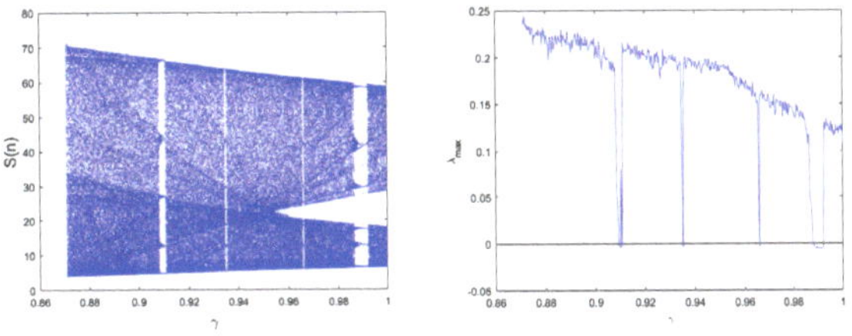

(a) (b)

Figure 7. (**a**) Bifurcation diagram versus γ of the fractional discrete SIR model (11) with $\alpha = 4$. (**b**) The maximum Lyapunov exponent.

3.1.2. The 0−1 Test

To test chaotic behavior in nonlinear systems, one can use a 0−1 test method [54]. Unlike the Lyapunov exponent approach, the 0−1 test acts directly on the time sequence, therefore eliminating the necessity for phase reconstruction. In particular, this technique operates on finite points $(S(n))_{n=1,\dots,N}$ and an arbitrary number $c \in (0, \pi)$. According to the time series $(S(n))$, two terms referred to as the translation variables can be defined for $n = \overline{1, N}$ as $p_c(n) = \sum_{k=1}^{n} S(k)\cos(kc)$ and $q_c(n) = \sum_{k=1}^{n} S(k)\sin(kc)$. We represent the mean square displacement by

$$M_c(n) = \lim_{N\to\infty} \frac{1}{N} \sum_{k=1}^{N} \left[\left(p_c(n+k) - p_c(k)\right)^2 + \left(q_c(n+k) - q_c(k)\right)^2 \right], \quad n \le \frac{N}{10}. \tag{20}$$

The asymptotic growth rate K_c is given by the definition

$$K_c = \lim_{n\to\infty} \frac{\log M_c(n)}{\log n}. \tag{21}$$

The asymptotic growth rate $K = median(K_c)$ or the plotting of p_c and q_c on the p–q plane may be used to assess if chaos occurs on the fractional discrete SIR epidemic model. This means that the dynamics of the proposed fractional discrete SIR epidemic model are nonchaotic when K is close to 0 and the behavior of trajectory in the $(p_c$–$q_c)$ plane is bounded; on the other hand, when K is close to 1 and the trajectory in the $(p_c$–$q_c)$ plane exhibits Brownian-like behavior, the dynamics of the fractional discrete SIR epidemic model are chaotic.

The 0−1 test of the fractional discrete-time SIR epidemic model with vaccination (11) is applied directly to the state $S(n)$. Figure 8 shows the $(p_c$–$q_c)$ plots of the fractional discrete SIR epidemic model (11) with commensurate fractional order. Observing the p_c and q_c trajectories, it is clear that the p_c and q_c trajectories display Brownian-like behavior when $\gamma = 1$ and $\gamma = 0.94$, indicating that the fractional SIR model is chaotic. Otherwise, when $\gamma_3 = 0.99$ and $\gamma_3 = 0.966$, the trajectories in the $(p_c$–$q_c)$ plane exhibit a bounded behavior, so the SIR model with commensurate order yields regular dynamics.

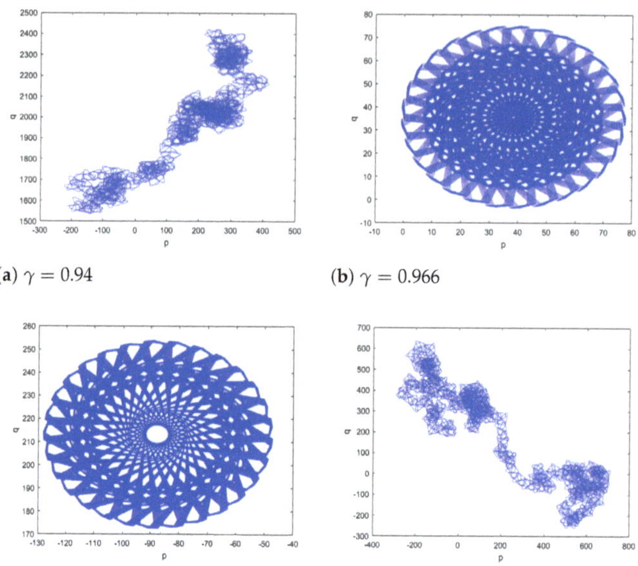

(**a**) $\gamma = 0.94$ (**b**) $\gamma = 0.966$

(**c**) $\gamma = 0.99$ (**d**) $\gamma = 1$

Figure 8. The p–q trajectories of the $0-1$ test of the fractional discrete SIR epidemic model (11) with commensurate orders ($\gamma_1 = \gamma_2$).

3.2. *Incommensurate Fractional Order*

In the following, the chaotic behavior of the proposed fractional SIR model with incommensurate fractional order (11) are carefully analyzed via the computation of bifurcation diagrams, Lyapunov exponents, phase portraits and $0-1$ test method. Similarly, the effects of system parameter α and the fractional-order values on the dynamics of the model are illustrated in detail. We examine the case of incommensurate orders because it is more representative of reality than the case of commensurate orders, as each variable changes and moves independently of the others, implying that the rank of the influencer varies from one equation to another.

3.2.1. Bifurcation Diagram and Maximum LEs

Here, we discuss the dynamics of the fractional discrete SIR model (11) by varying α from 3 to 4.5 with the step size $\Delta\alpha = 0.001$. Figure 9 displays the bifurcation diagrams for the fractional order values $(\gamma_1, \gamma_2) = (0.94, 0.97)$, $(\gamma_1, \gamma_2) = (0.94, 0.99)$, and $(\gamma_1, \gamma_2) = (0.99, 0.94)$, respectively. From Figure 9, we see that the system exhibits periodic doubling or flip bifurcation. When α increases from 3, stability begins with a one-period orbit, then periodic orbits of periods 2, 4, and 8 are seen, which eventually evolve into chaos. In addition, we observe that the states of the system (11) are influenced by the fractional-order values γ_i and the system parameter α, as shown in Figure 9. As can be observed, when the incommensurate fractional orders (γ_1, γ_2) are changed, the interval in which the chaos may be found shifts. Namely, when $(\gamma_1, \gamma_2) = (0.94, 0.97)$ the system is chaotic at $\alpha \in]3.822, 3.907[\cup]3.915, 4.008[\cup]4.014, 4.055[$, while when $(\gamma_1, \gamma_2) = (0.99, 0.94)$ the system is chaotic at $\alpha \in]4.023, 3.136[\cup]4.139, 4.1754[$. As a consequence of these findings, it can be concluded that the suggested fractional discrete SIR epidemic model (11) with incommensurate fractional orders shows a greater variety of chaotic attractors than the fractional discrete SIR epidemic model with commensurate fractional orders. To further investigate the effect of the incommensurate fractional orders on the dynamics behavior of system (11), Figure 10 displays the bifurcation diagram and the maximum Lyapunov exponents with γ_1 as a bifurcation parameter where the initial conditions are $(S(0), I(0)) =$

$(70, 30)$, the system parameter is $\alpha = 4$, and the fractional order is $\gamma_2 = 0.94$. We can see that the states of the system diverge towards infinity as soon as γ_1 goes below 0.8383 and when γ_1 increases, chaos occurs with certain periodic orbits. We can also see that when γ_1 approaches 1, the states become totally periodic. In Figure 10b, the maximum, Lyapunov exponents calculated for the same parameters and initial conditions as in Figure 10a are shown. Obviously, the maximum Lyapunov exponents have a positive number indicating that chaos exists, which agrees with the corresponding bifurcation diagram in Figure 10a.

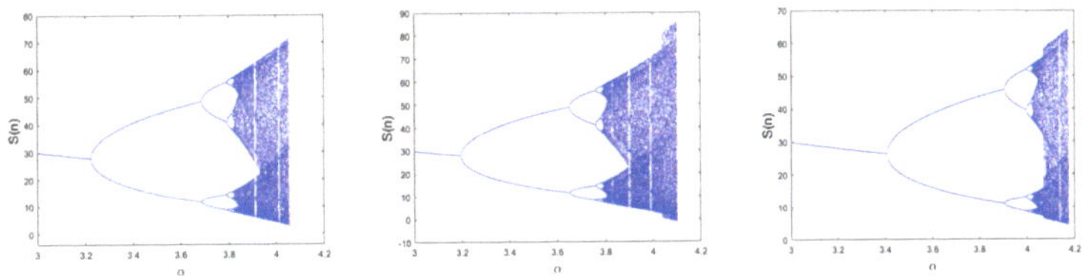

(**a**) $\gamma_1 = 0.94,\ \gamma_2 = 0.97$ (**b**) $\gamma_1 = 0.94,\ \gamma_2 = 0.99$ (**c**) $\gamma_1 = 0.99,\ \gamma_2 = 0.94$

Figure 9. Bifurcation diagrams of the fractional discrete SIR model (11) for different incommensurate fractional-order values.

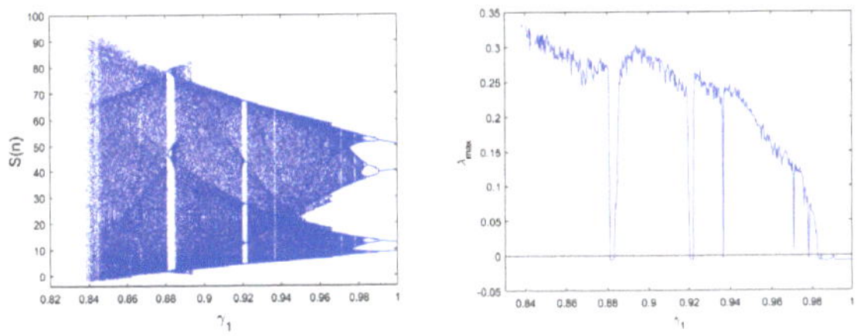

(**a**) (**b**)

Figure 10. (**a**) Bifurcation diagram versus γ_1 of the fractional discrete SIR model (11) for $\alpha = 4$ and $\gamma_2 = 0.94$. (**b**) The maximum Lyapunov exponents.

Now, the dynamic behavior with the variation of the fractional order γ_2 is studied for $\gamma_1 = 0.94$. The bifurcation diagram and its corresponding maximum LE are illustrated in Figure 11. We can see that the dynamical behavior of the fractional discrete SIR epidemic model (11) evolves from periodic to chaos as γ_2 increases. In particular, the proposed SIR model is chaotic when $\gamma_2 \in]0.888, 0.9494[\cup]0.951, 0.979[\cup]0.9836, 1]$; where the maximum LE is positive. The results illustrate that the dynamics behavior of the SIR model is affected by the order γ_2. To illustrate the dynamics of the fractional discrete SIR epidemic model (11) better, phase portraits with different values of (γ_1, γ_2) are presented in Figure 12. From Figure 12, we notice that the proposed SIR model shows different dynamic behaviors for these corresponding different fractional-order values.

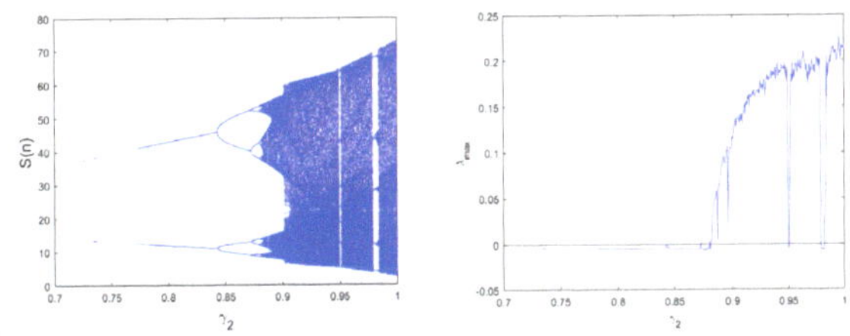

(**a**) (**b**)

Figure 11. (**a**) Bifurcation diagram versus γ_2 of the fractional discrete SIR model (11) for $\alpha = 4$ and $\gamma_1 = 0.94$. (**b**) The maximum Lyapunov exponents.

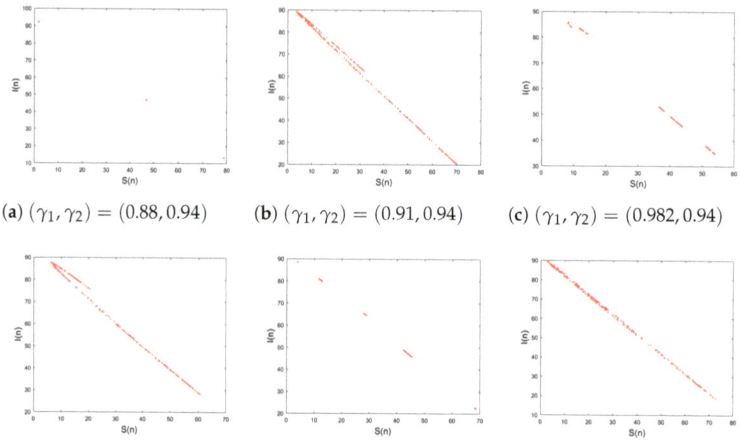

(**a**) $(\gamma_1, \gamma_2) = (0.88, 0.94)$ (**b**) $(\gamma_1, \gamma_2) = (0.91, 0.94)$ (**c**) $(\gamma_1, \gamma_2) = (0.982, 0.94)$

(**d**) $(\gamma_1, \gamma_2) = (0.94, 0.91)$ (**e**) $(\gamma_1, \gamma_2) = (0.94, 0.983)$ (**f**) $(\gamma_1, \gamma_2) = (0.94, 1)$

Figure 12. Phase portrait of the fractional discrete SIR epidemic model with vaccination (11) with incommensurate fractional-order values for $N = 100$, $\beta = 0.8$, $\sigma = 0.1$, $p = 0.005$, and initial conditions $(S(0), I(0)) = (70, 30)$.

To get a better understanding of the influence of the fractional order on the SIR epidemic model, and in light of previous numerical findings, we compare the recent results obtained from the integer-order SIR model with the results obtained from the fractional-order SIR model, which are presented in Table 1. It can be observed that the maximum number of susceptible cases obtained from the fractional-order SIR model is identical to the maximum number obtained from the integer-order SIR model, which is 70, whereas the minimum number of susceptible cases obtained from the fractional-order SIR model is less than the minimum number obtained from the integer-order SIR model. On the other hand, we see that the maximum and minimum numbers of infected cases predicted in the fractional-order SIR model are not identical to those expected in the integer-order SIR model. This confirms the effect of the fractional order on the SIR epidemic model in predicting the number of susceptible individuals and infected individuals.

Table 1. The minimum and maximum numbers of expected cases for the susceptible class and infected class.

Classes	Integer-Order Model (min, max)	Fractional-Order Model (min, max)
Susceptible individuals (S)	$(7, 70)$	$(4, 70)$
Infected individuals (I)	$(30, 86)$	$(20, 90)$

3.2.2. The 0−1 Test

Similarly to the commensurate fractional orders, the 0−1 test was used to evaluate the fractional discrete SIR epidemic model (11) with incommensurate fractional order. The translation components p_c and q_c in the $(p_c–q_c)$ plan are illustrated in Figure 13. As can be observed, for $(\gamma_1, \gamma_2) = (0.91, 0.94)$ and $(\gamma_1, \gamma_2) = (0.94, 0.91)$, the trajectories p_c and q_c display Brownian-like behavior, and a bounded behavior for $(\gamma_1, \gamma_2) = (0.94, 0.983)$ and $(\gamma_1, \gamma_2) = (0.982, 0.94)$, which confirms that the fractional discrete SIR epidemic model (11) has chaotic attractors for $(\gamma_1, \gamma_2) = (0.97, 0.94)$ and $(\gamma_1, \gamma_2) = (0.97, 1)$, and has a regular behavior for $(\gamma_1, \gamma_2) = (0.99, 0.94)$ and $(\gamma_1, \gamma_2) = (0.97, 0.9)$. On the other side, Figure 14 depicts the asymptotic growth rate K with fractional orders (γ_1, γ_2), in which $\alpha = 4$ and $(S(0), I(0)) = (70, 30)$. As we can see, for the majority of values of γ_1 and γ_2, the asymptotic growth rate K approaches 1, implying that the fractional-order discrete SIR model with vaccination exhibits a chaotic behavior. These findings support very well the results of the bifurcation diagrams and the maximum Lyapunov exponents obtained before.

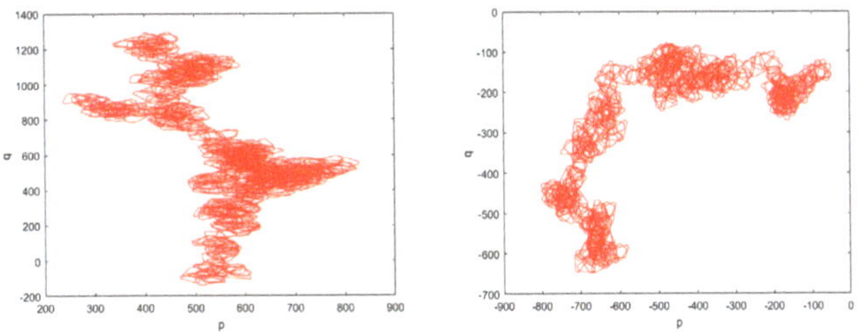

(**a**) $(\gamma_1, \gamma_2) = (0.91, 0.94)$ (**b**) $(\gamma_1, \gamma_2) = (0.94, 0.91)$

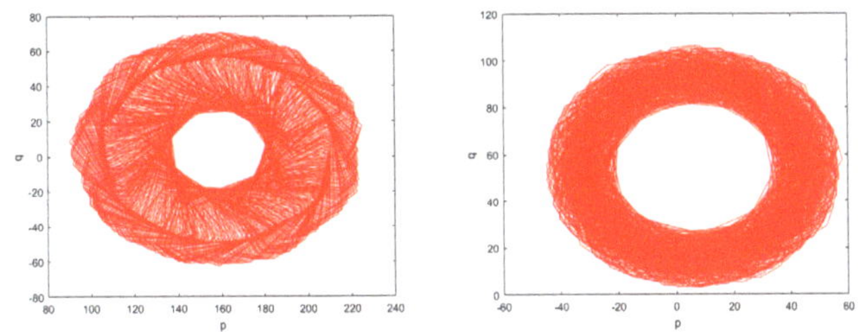

(**c**) $(\gamma_1, \gamma_2) = (0.94, 0.983)$ (**d**) $(\gamma_1, \gamma_2) = (0.982, 0.94)$

Figure 13. The p–q trajectories of the $0-1$ test of the fractional discrete SIR epidemic model (11) with incommensurate orders (γ_1, γ_2).

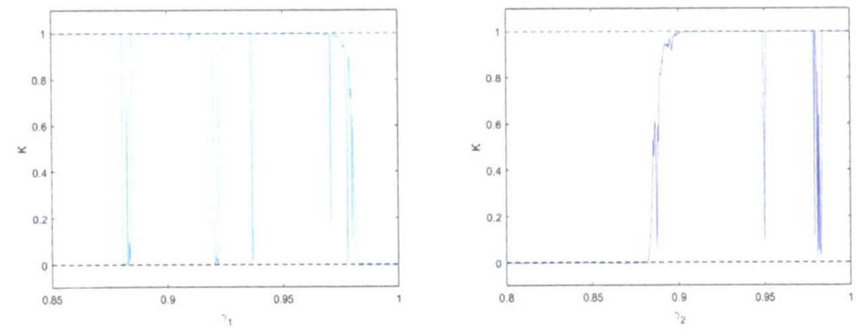

(**a**) (**b**)

Figure 14. The asymptotic growth rate K of the fractional discrete SIR epidemic model (11) with (γ_1, γ_2) as critical parameters for (**a**) $\gamma_2 = 0.94$, (**b**) $\gamma_1 = 0.94$.

4. Complexity Analysis of the Fractional Discrete SIR Epidemic Model with Commensurate and Incommensurate Fractional Orders

For assessing the dynamic characteristics of chaotic systems, one technique is to consider the complexity of the chaotic characteristic. The model becomes more chaotic as the level of complexity increases. In this section, the approximate entropy and the C_0 complexity algorithm are used to assess the complexity of the fractional-order discrete-time SIR epidemic model with vaccination.

4.1. C_0 Complexity

Based on the inverse Fourier transform, the analysis of the complexity of chaotic systems was carried out using the C_0 algorithm. It divides the time series of the system into two components, a series of regular and a series of irregular parts, where the series of irregular parts is what we need. For a sequence $[S(0), S(1), \ldots, S(N-1)]$ with a length of N, and a control parameter r, the algorithm process is defined as follows [55]. Firstly, the discrete Fourier transform of $\{S_n\}$ is determined by

$$X_N(k) = \frac{1}{N}\sum_{j=0}^{N-1} S(j)\exp^{-2\pi i(\frac{kj}{N})}, \quad k = 0,1,..,N-1, \tag{22}$$

and the mean square value is calculated as $G_N = \frac{1}{N}\sum_{k=0}^{N-1}|X_N(k)|^2$. We let

$$\bar{X}_N(k) = \begin{cases} X_N(k) & \text{if} \quad |X_N(k)|^2 > rG_N, \\ 0 & \text{if} \quad X_N(k)|^2 \le rG_N. \end{cases} \tag{23}$$

The inverse Fourier transformation of $\bar{X}_N$ is defined as

$$\bar{x}(j) = \frac{1}{N}\sum_{k=0}^{N-1} \bar{X}_N(k)\exp^{2\pi i(\frac{kj}{N})}, \quad j = 0,1,..,N-1. \tag{24}$$

Then, the C_0 complexity is given by:

$$C_0 = \frac{\sum_{j=0}^{N-1}|S(j) - \bar{x}(j)|^2}{\sum_{j=0}^{N-1}|S(J)|^2}. \tag{25}$$

The C_0 complexity of the fractional-order discrete-time SIR epidemic model with vaccination (11) with varying commensurate fractional-order value γ and incommensurate fractional-order values γ_1, γ_2 are calculated and the result is shown in Figure 15. Interestingly, in the case of commensurate fractional-order values γ, as with the MLE, the C_0 complexity value of the fractional SIR model increases rapidly when γ_i decreases. On the other hand, the fractional discrete SIR epidemic model (11) with incommensurate fractional-order values, in contrast to the case when the model has commensurate fractional-order values, has more complexity when γ_2 approaches 1. Thus, we can see that the C_0 algorithm can measure the complexity effectively. Figure 15a illustrates that model (11) has a higher complexity when $\gamma \in (0.9712, 0.9082]$. When $\gamma_1 = 0.94$, the high complexity region in Figure 15c also exists in the range of $\gamma_2 \in [0.9836, 1]$, and the C_0 complexity increases with the increase of fractional order γ_2.

4.2. Approximate Entropy ($ApEn$)

The approximate entropy is a measure of regularity that quantifies the level of complexity within systems generated by a time series. Generally, time series with larger values of $ApEn$ are considered to be more complex [56]. Note that, the approximate entropy value is dependent on two essential parameters: the similarity tolerance r and the embedding dimension m. In this article, we took $m = 2$ and $r = 0.2std(S)$ where $std(S)$ represents the standard deviation of the data S. Theoretically, the $ApEn$ is calculated as follows:

$$ApEn = -\Phi^{m+1}(r) + \Phi^m(r), \tag{26}$$

where $\Phi^m(r)$ is denoted by

$$\Phi^m(r) = \frac{1}{n-m-1}\sum_{i=1}^{n-m+1}\log C_i^m(r) \tag{27}$$

where $C_i^m(r) = \frac{K}{n-m+1}, i \in [1, n-m+1]$, K is the number of $X(j)$ such that $d(X(j), X(i)) \le r$, and $X(i) = [x(i), x(i+1), ..., x(i+m-1)]$.

The $ApEn$ complexity of the fractional-order discrete-time SIR epidemic model with vaccination (11), with both commensurate and incommensurate orders (γ_1, γ_2) was analyzed and the findings are reported in Table 2. It can be seen that the complexity of the fractional discrete SIR epidemic model (11) varies when γ_i $(i = 1,2)$ varies, and the highest $ApEn$ is found when the model is chaotic, which agrees very well with the maximum LE results. As a result, in order to obtain a relatively high structural complexity, we must be cautious when choosing the values of γ_i in system (11).

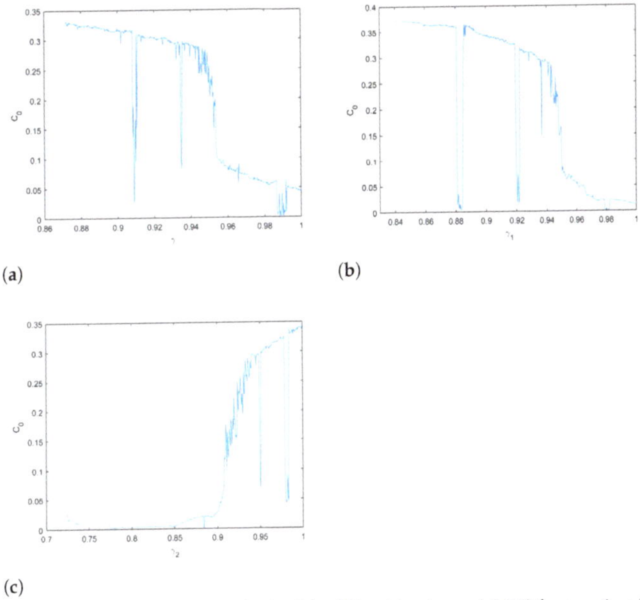

Figure 15. C_0 complexity analysis of the SIR epidemic model (11) for $\alpha = 4$ and $(S(0), I(0)) = (70, 30)$ and with fractional-order values: (**a**) $\gamma_1 = \gamma_2$, (**b**) $\gamma_1 = 0.94$, and (**c**) $\gamma_2 = 0.94$.

Table 2. Approximate entropy test of the SIR model (11) with different fractional-order values γ_i, $i = 1,2$.

γ_1	γ_2	**ApEn**
1	1	0.2552
0.99	0.99	0.0265
0.94	0.94	0.3718
0.94	0.983	0.0503
0.982	0.94	0.1336
0.91	0.94	0.4056

5. Conclusions

In this paper, we dealt with the dynamics of a new fractional-order discrete-time SIR epidemic model with vaccination with commensurate and incommensurate orders. Through phase portraits, bifurcation diagrams, and maximum LEs, the complex dynamics of the proposed system were discussed. In addition, we also calculated the 0−1 test, approximate entropy ($ApEn$), and C_0 complexity of the fractional-order discrete SIR epidemic model for different fractional order values, all of which intended to demonstrate and quantify the complex dynamics of the system. Results indicated that the fractional-order model is more complex than the integer-order model. Furthermore, we showed that the incommensurate fractional orders have a greater effect on the behavior of the system than when the fractional orders are commensurate. The reasonable range of commensurate fractional orders is between $\gamma = 0.8712$ and $\gamma = 1$ while the reasonable range of incommensurate fractional orders is between $\gamma_2 = 0.77$ and $\gamma_2 = 1$. Throughout this work, numerical simulations were used to explain all the findings. Due to the current spread of the COVID-19 epidemic and the onset of the implementation of various vaccination strategies in various countries, we will attempt to use the fractal properties of most of the COVID-19 series in order to establish a connection between the proposed fractional model and the stochastic self-affine characteristics of the COVID-19 times series in future research. In addition, the determination of the coupled time-fractional differential equations of the proposed fractional model using the Caputo fractional derivative, as well as the study of some complex epidemiological models will be considered in our future research.

Author Contributions: Conceptualization, Z.Y.-H., A.A.; H.J., N.D.A. and Y.W.; methodology, Z.Y.-H., A.A., H.J., N.D.A. and Y.W.; software, Z.Y.-H., A.A., H.J., N.D.A. and Y.W.; validation, Z.Y.-H., A.A., H.J., N.D.A. and Y.W.; formal analysis, Z.Y.-H., A.A., H.J., N.D.A. and Y.W.; investigation, Z.Y.-H., A.A., H.J., N.D.A. and Y.W.; writing—original draft preparation, Z.Y.-H., A.A., H.J., N.D.A. and Y.W.; writing—review and editing, Z.Y.-H., A.A., H.J., N.D.A. and Y.W. All authors have read and agreed to the published version of the manuscript.

Funding: The Deanship of Scientific Research (DSR) at King Abdulaziz University, Jeddah, Saudi Arabia has funded this project, under grant no. (FP-035-43).

Institutional Review Board Statement: Not applicable.

Informed Consent Statement: Not applicable.

Conflicts of Interest: The authors declare no conflict of interest.

References

1. Lv, W.; Ke, Q.; Li, K. Dynamical analysis and control strategies of an SIVS epidemic model with imperfect vaccination on scale-free networks. *Nonlinear Dyn.* **2020**, *99*, 1507–1523. [CrossRef]
2. Din, A.; Khan, A.; Baleanu, D. Stationary distribution and extinction of stochastic coronavirus (COVID-19) epidemic model. *Chaos Solitons Fractals* **2020**, *139*, 110036. [CrossRef] [PubMed]
3. Pandey, P.; Chu, Y.M.; Gómez-Aguilar, J.F.; Jahanshahi, H.; Aly, A.A. A novel fractional mathematical model of COVID-19 epidemic considering quarantine and latent time. *Results Phys.* **2021**, *26*, 104286. [CrossRef] [PubMed]
4. Jahanshahi, H.; Munoz-Pacheco, J.M.; Bekiros, S.; Alotaibi, N.D. A fractional-order SIRD model with time-dependent memory indexes for encompassing the multi-fractional characteristics of the COVID-19. *Chaos Solitons Fractals* **2021**, *143*, 110632. [CrossRef] [PubMed]
5. Zha, T.H.; Castillo, O.; Jahanshahi, H.; Yusuf, A.; Alassafi, M.O.; Alsaadi, F.E.; Chu, Y.M. A fuzzy-based strategy to suppress the novel coronavirus (2019-NCOV) massive outbreak. *Appl. Comput. Math.* **2021**, *20*, 160–176.
6. Allen, L.J. Some discrete-time SI, SIR, and SIS epidemic models. *Math. Biosci.* **1994**, *124*, 83–105. [CrossRef]
7. Cao, H.; Zhou, Y.; Ma, Z. Bifurcation analysis of a discrete SIS model with bilinear incidence depending on new infection. *Math. Biosci. Eng.* **2013**, *10*, 1399.
8. Parsamanesh, M.; Mehrshad, S. Stability of the equilibria in a discrete-time sivs epidemic model with standard incidence. *Filomat* **2019**, *33*, 2393–2408. [CrossRef]
9. Parsamanesh, M.; Erfanian, M. Stability and bifurcations in a discrete-time SIVS model with saturated incidence rate. *Chaos Solitons Fractals* **2021**, *150*, 111178. [CrossRef]
10. Keeling, M.J.; Danon, L. Mathematical modelling of infectious diseases. *Br. Med. Bull.* **2009**, *92*, 1. [CrossRef] [PubMed]
11. Rashidinia, J.; Sajjadian, M.; Duarte, J.; Januário, C.; Martins, N. On the dynamical complexity of a seasonally forced discrete SIR epidemic model with a constant vaccination strategy. *Complexity* **2018**, *2018*, 7191487. [CrossRef]
12. Parsamanesh, M.; Erfanian, M.; Mehrshad, S. Stability and bifurcations in a discrete-time epidemic model with vaccination and vital dynamics. *BMC Bioinform.* **2020**, *21*, 1–5. [CrossRef] [PubMed]
13. Xiang, L.; Zhang, Y.; Huang, J. Stability analysis of a discrete SIRS epidemic model with vaccination. *J. Differ. Equations Appl.* **2020**, *26*, 309–327. [CrossRef]
14. Wang, B.; Liu, J.; Alassafi, M.O.; Alsaadi, F.E.; Jahanshahi, H.; Bekiros, S. Intelligent parameter identification and prediction of variable time fractional derivative and application in a symmetric chaotic financial system. *Chaos Solitons Fractals* **2021**, *28*, 111590. [CrossRef]
15. Wang, B.; Jahanshahi, H.; Dutta, H.; Zambrano-Serrano, E.; Grebenyuk, V.; Bekiros, S.; Aly, A.A. Incorporating fast and intelligent control technique into ecology: A Chebyshev neural network-based terminal sliding mode approach for fractional chaotic ecological systems. *Ecol. Complex.* **2021**, *47*, 100943. [CrossRef]
16. Liu, Z.; Jahanshahi, H.; Gómez-Aguilar, J.F.; Fernandez-Anaya, G.; Torres-Jiménez, J.; Aly, A.A.; Aljuaid, A.M. Fuzzy adaptive control technique for a new fractional-order supply chain system. *Phys. Scr.* **2021**, *96*, 124017. [CrossRef]

17. Wang, H.; Jahanshahi, H.; Wang, M.K.; Bekiros, S.; Liu, J.; Aly, A.A. A Caputo–Fabrizio Fractional-Order Model of HIV/AIDS with a Treatment Compartment: Sensitivity Analysis and Optimal Control Strategies. *Entropy* **2021**, *23*, 610. [CrossRef]
18. Wang, Y.L.; Jahanshahi, H.; Bekiros, S.; Bezzina, F.; Chu, Y.M.; Aly, A.A. Deep recurrent neural networks with finite-time terminal sliding mode control for a chaotic fractional-order financial system with market confidence. *Chaos Solitons Fractals* **2021**, *146*, 110881. [CrossRef]
19. Li, J.F.; Jahanshahi, H.; Kacar, S.; Chu, Y.M.; Gómez-Aguilar, J.F.; Alotaibi, N.D.; Alharbi, K.H. On the variable-order fractional memristor oscillator: Data security applications and synchronization using a type-2 fuzzy disturbance observer-based robust control. *Chaos Solitons Fractals* **2021**, *145*, 110681. [CrossRef]
20. Jahanshahi, H.; Sajjadi, S.S.; Bekiros, S.; Aly, A.A. On the development of variable-order fractional hyperchaotic economic system with a nonlinear model predictive controller. *Chaos Solitons Fractals* **2021**, *144*, 110698. [CrossRef]
21. Xiong, P.Y.; Jahanshahi, H.; Alcaraz, R.; Chu, Y.M.; Gómez-Aguilar, J.F.; Alsaadi, F.E. Spectral entropy analysis and synchronization of a multi-stable fractional-order chaotic system using a novel neural network-based chattering-free sliding mode technique. *Chaos Solitons Fractals* **2021**, *144*, 110576. [CrossRef]
22. Wang, B.; Jahanshahi, H.; Volos, C.; Bekiros, S.; Khan, M.A.; Agarwal, P.; Aly, A.A. A New RBF Neural Network-Based Fault-Tolerant Active Control for Fractional Time-Delayed Systems. *Electronics* **2021**, *10*, 1501. [CrossRef]
23. Wang, B.; Jahanshahi, H.; Bekiros, S.; Chu, Y.M.; Gomez-Aguilar, J.F.; Alsaadi, F.E.; Alassafi, M.O. Tracking control and stabilization of a fractional financial risk system using novel active finite-time fault-tolerant controls. *Fractals* **2021**, *29*, 2150155–2150177. [CrossRef]
24. Chen, S.B.; Soradi-Zeid, S.; Alipour, M.; Chu, Y.M.; Gomez-Aguilar, J.F.; Jahanshahi, H. Optimal control of nonlinear time-delay fractional differential equations with Dickson polynomials. *Fractals* **2021**, *29*, 2150079–2150170. [CrossRef]
25. Chen, S.B.; Soradi-Zeid, S.; Jahanshahi, H.; Alcaraz, R.; Gómez-Aguilar, J.F.; Bekiros, S.; Chu, Y.M. Optimal control of time-delay fractional equations via a joint application of radial basis functions and collocation method. *Entropy* **2020**, *22*, 1213. [CrossRef] [PubMed]
26. Jahanshahi, H.; Yousefpour, A.; Munoz-Pacheco, J.M.; Kacar, S.; Pham, V.T.; Alsaadi, F.E. A new fractional-order hyperchaotic memristor oscillator: Dynamic analysis, robust adaptive synchronization, and its application to voice encryption. *Appl. Math. Comput.* **2020**, *383*, 125310. [CrossRef]
27. Soradi-Zeid, S.; Jahanshahi, H.; Yousefpour, A.; Bekiros, S. King algorithm: A novel optimization approach based on variable-order fractional calculus with application in chaotic financial systems. *Chaos Solitons Fractals* **2020**, *132*, 109569. [CrossRef]
28. Jahanshahi, H.; Yousefpour, A.; Munoz-Pacheco, J.M.; Moroz, I.; Wei, Z.; Castillo, O. A new multi-stable fractional-order four-dimensional system with self-excited and hidden chaotic attractors: Dynamic analysis and adaptive synchronization using a novel fuzzy adaptive sliding mode control method. *Appl. Soft Comput.* **2020**, *87*, 105943. [CrossRef]
29. Podlubny, I. *Fractional Differential Equations: An Introduction to Fractional Derivatives, Fractional Differential Equations, to Methods of Their Solution and Some of Their Applications*; Elsevier: Amsterdam, The Netherlands, 1998.
30. Atici, F.M.; Eloe, P.W. Discrete fractional calculus with the nabla operator. *Electron. J. Qual. Theory Differ. Equ.* **2009**, *1*, 1–99. [CrossRef]
31. Diaz, J.B.; Osler, T.J. Differences of fractional order. *Math. Comput.* **1974**, *28*, 185–202. [CrossRef]
32. Anastassiou, G.A. Principles of delta fractional calculus on time scales and inequalities. *Math. Comput. Model.* **2010**, *52*, 556–566. [CrossRef]
33. Wu, G.C.; Baleanu, D. Discrete fractional logistic map and its chaos. *Nonlinear Dyn.* **2014**, *75*, 283–287. [CrossRef]
34. Khennaoui, A.A.; Ouannas, A.; Bendoukha, S.; Grassi, G.; Wang, X.; Pham, V.T.; Alsaadi, F.E. Chaos, control, and synchronization in some fractional-order difference equations. *Adv. Differ. Equ.* **2019**, *2019*, 1–23. [CrossRef]
35. Peng, Y.; He, S.; Sun, K. Chaos in the discrete memristor-based system with fractional-order difference. *Results Phys.* **2021**, *24*, 104106. [CrossRef]
36. Khennaoui, A.A.; Ouannas, A.; Bendoukha, S.; Grassi, G.; Lozi, R.P.; Pham, V.T. On fractional–order discrete–time systems: Chaos, stabilization and synchronization. *Chaos Solitons Fractals* **2019**, *119*, 150–162. [CrossRef]
37. Ouannas, A.; Khennaoui, A.A.; Odibat, Z.; Pham, V.T.; Grassi, G. On the dynamics, control and synchronization of fractional-order Ikeda map. *Chaos Solitons Fractals* **2019**, 123, 108–115. [CrossRef]
38. Hadjabi, F.; Ouannas, A.; Shawagfeh, N.; Khennaoui, A.A.; Grassi, G. On two-dimensional fractional chaotic maps with symmetries. *Symmetry* **2020**, *12*, 756. [CrossRef]
39. El-Saka, H.A. The fractional-order SIR and SIRS epidemic models with variable population size. *Math. Sci. Lett.* **2013**, *2*, 195. [CrossRef]
40. Mouaouine, A.; Boukhouima, A.; Hattaf, K.; Yousfi, N. A fractional order SIR epidemic model with nonlinear incidence rate. *Adv. Differ. Equations* **2018**, *2018*, 1–9. [CrossRef]
41. Momani, S.; Kumar, R.; Srivastava, H.M.; Kumar, S.; Hadid, S. A chaos study of fractional SIR epidemic model of childhood diseases. *Results Phys.* **2021**, *27*, 104422. [CrossRef]
42. Owoyemi, A.E.; Sulaiman, I.M.; Mamat, M.; Olowo, S.E. Stability and Bifurcation Analysis in a Fractional-order Epidemic Model with Sub-optimal Immunity, Nonlinear Incidence and Saturated Recovery Rate. *IAENG Int. J. Appl. Math.* **2021**, *51*, 3.

43. Alaoui, A.L.; Tilioua, M.; Sidi Ammi, M.R.; Agarwal, P. Dynamical Analysis of a Caputo Fractional Order SIR Epidemic Model with a General Treatment Function. In *Analysis of Infectious Disease Problems (COVID-19) and Their Global Impact*; Springer: Singapore, 2021; pp. 17–33.
44. Javeed, S.; Anjum, S.; Alimgeer, K.S.; Atif, M.; Khan, M.S.; Farooq, W.A.; Hanif, A.; Ahmed, H.; Yao, S.W. A Novel Mathematical Model for COVID-19 with Remedial Strategies. *Results Phys.* **2021**, *8*, 104248. [CrossRef]
45. Selvam, A.G.; Vianny, D.A. Discrete fractional order SIR epidemic model and it's stability. In *Journal of Physics: Conference Series*; IOP Publishing: Bristol, UK, 2018; Volume 1139, p. 012008.
46. Naik, P.A. Global dynamics of a fractional-order SIR epidemic model with memory. *Int. J. Biomath.* **2020**, *13*, 2050071. [CrossRef]
47. Ahmad, S.; Javeed, S.; Ahmad, H.; Khushi, J.; Elagan, S.K.; Khames, A. Analysis and numerical solution of novel fractional model for dengue. *Results Phys.* **2021**, *28*, 104669. [CrossRef]
48. Liu, F.; Huang, S.; Zheng, S.; Wang, H.O. Stability Analysis and Bifurcation Control for a Fractional Order SIR Epidemic Model with Delay. In Proceedings of the 2020 39th Chinese Control Conference (CCC), Shenyang, China, 27–29 July 2020; pp. 724–729.
49. Din, Q. Qualitative behavior of a discrete SIR epidemic model. *Int. J. Biomath.* **2016**, *9*, 1650092. [CrossRef]
50. GümüÅŸ, Ã.-z.A.; Selvam, A.G.; Vianny, D.A. Bifurcation and stability analysis of a discrete time SIR epidemic model with vaccination. *Int. J. Anal. Appl.* **2019**, *17*, 809–820.
51. Abdeljawad, T. On Riemann and Caputo fractional differences. *Comput. Math. Appl.* **2011**, *62*, 1602–1611. [CrossRef]
52. Wu, G.C.; Baleanu, D. Jacobian matrix algorithm for Lyapunov exponents of the discrete fractional maps. *Commun. Nonlinear Sci. Numer. Simul.* **2015**, *22*, 95–100. [CrossRef]
53. Petras, I. Fractional Order Chaotic Systems. MATLAB Central File Exchange. Available online: https://www.mathworks.com/matlabcentral/fileexchange/27336-fractional-order-chaotic-systems (accessed on 27 December 2021).
54. Gottwald, G.A.; Melbourne, I. The 0-1 test for chaos: A review. *Chaos Detect. Predict.* **2016**, *915*, 221–247.
55. Ran, J. Discrete chaos in a novel two-dimensional fractional chaotic map. *Adv. Differ. Equ.* **2018**, *2018*, 1–2. [CrossRef]
56. Pincus, S.M. Approximate entropy as a measure of system complexity. *Proc. Natl. Acad. Sci. USA* **1991**, *88*, 2297–2301. [CrossRef]

mathematics

Article

Probabilistic Interpretations of Fractional Operators and Fractional Behaviours: Extensions, Applications and Tribute to Prof. José Tenreiro Machado's Ideas

Jocelyn Sabatier

IMS Laboratory, Bordeaux University, UMR 5218 CNRS, 351 Cours de la Libération, 33405 Talence, France; jocelyn.sabatier@u-bordeaux.fr

Abstract: This paper extends and illustrates a probabilistic interpretation of the fractional derivative operator proposed by Pr. José Tenreiro Machado. While his interpretation concerned the probability of finding samples of the derivate signal in the expression of the fractional derivative, the present paper proposes interpretations for other fractional models and more generally fractional behaviours (without using a model). It also proposes probabilistic interpretations in terms of time constants and time delay distributions. It shows that these probabilistic interpretations in terms of time delay distributions can be connected to the physical behaviour of real systems governed by adsorption or diffusion phenomena.

Keywords: fractional behaviours; probabilistic interpretations; adsorption; diffusion

MSC: 26A33; 34K37; 60E05

Citation: Sabatier, J. Probabilistic Interpretations of Fractional Operators and Fractional Behaviours: Extensions, Applications and Tribute to Prof. José Tenreiro Machado's Ideas. *Mathematics* **2022**, *10*, 4184. https://doi.org/10.3390/math10224184

Academic Editors: Alexandra M.S.F. Galhano, António Lopes and Carlo Cattani

Received: 27 September 2022
Accepted: 4 November 2022
Published: 9 November 2022

1. Introduction

As part of this special issue in tribute to our colleague José Tenreiro Machado, this paper is an analysis of one of his works and serves as a basis for a deeper reflection on the origins of fractional behaviors and their modeling. The work by Professor José Tenreiro Machado analysed here is an interpretation of the fractional derivative operator [1].

Fractional operators remained abstract mathematical objects for a long time before applications in modelling [2] and control [3], but the latter did not give them a physical meaning. Finding interpretations and giving meaning, physical or not, to fractional operators has subsequently been the concern of many researchers. However, most of the interpretations proposed in the literature were not obtained from the observation of a given phenomenon but resulted from purely mathematical discussions [4–9]. In the case of non-commensurate fractional orders, certain interpretations even invalidated the model obtained [10]. But none of these approaches really help to understand the fractional behaviour observed in practice (fractional models and fractional behaviours are two distinct concepts), and they do not inform on the advantages and disadvantages of the fractional operators usually used to model these fractional behaviours (and we must not forget that other models exist [11–13]).

To try to obtain an interpretation which is the reflection of what occurs within a system having a fractional behaviour, one should not lose sight of the fact that these behaviours for the most part result from stochastic phenomena or result from geometries which are stochastic constructions (natural fractals). It is therefore interesting to seek an interpretation of these phenomena, and therefore of the fractional models very often used for their modelling, which is also stochastic in nature. A first attempt was made by Prof Tenreiro Machado in [1] with a probabilistic interpretation of the fractional differentiation operator based on the Grünwald-Letnikov definition of a derivative of fractional order.

In this paper, the idea of probabilistic interpretations for fractional models and behaviours is also used. Compared to [1], it proposes other types of interpretations and a confrontation of these interpretations with the physics of certain systems producing fractional behaviors. As defined in [11], we will say that a system has a fractional behaviours or more accurately a power law—like behaviour if its impulse response or if its frequency response exhibits a power law behaviour in a given time or frequency range, which can be evaluated from the system output autocorrelation function or system output power spectral density if the system input is a white noise. Note that fractional behaviours and fractional models are two distinct concepts. One designates a property of a physical system, the other designates a model class, among a set of model classes that capture fractional behaviours [13]. Thus, after recalling the results published in [1], this paper proposes a first probabilistic interpretation in terms of time constants distribution, resulting from the integral representation of fractional models' impulse response. But this interpretation is only a macro representation of what happens in a system producing a fractional behaviour, and very often, it does not correspond to what happens internally, which is in most cases linked to the sequential movement of a multitude of agents in a constrained environment (the case of diffusion, aggregation and adsorption for example). Thus, another probabilistic interpretation in terms of time-delay distribution is proposed. As shown by this paper, this interpretation describes quite well what happens internally for adsorption or diffusion phenomena. For adsorption, this delay distribution it the time needed for particles to find their places on the adsorbing surface. In the case of diffusion, it describes the time needed for particles to cross a material.

2. Professor Tenreiro Machado's Probabilistic Interpretation of the Fractional Derivative Operator

In [1], our colleague José Tenreiro Machado proposed a probabilistic interpretation of fractional differentiation based on the Grünwald-Letnikov definition of a derivative of fractional order ν of a signal $x(t)$. This derivative is given by:

$$D^{\nu}[x(t)] = \lim_{h \to 0}\left[\frac{1}{h^{\nu}}\sum_{k=0}^{\infty}\gamma(\nu,k)x(t-kh)\right] \quad (1)$$

with

$$\gamma(\nu,k) = (-1)^k \frac{\Gamma(\nu+1)}{k!\Gamma(\nu-k+1)} \quad (2)$$

where Γ is the gamma function and h is the sampling interval.

According to Professor Tenreiro Machado, and from a probability theory point of view, in relation (1):

- the "present", namely the sample $x(0)$, is seen with probability one,
- each sample of the past, namely the samples $x(-kh)$ with $k \in [1, \infty[$ is weighted with the probability $\gamma(\nu,k)$ and the expression $-\sum\limits_{k=0}^{\infty}\gamma(\nu,k)x(t-kh)$ can be viewed as the expected value of the random variable X, $E(X)$, such that $P(X = x(kh)) = |\gamma(\nu,k)|$, $k \in [1, \infty[$.

This probabilistic interpretation is based on time information (samples) of the derivative signal. This idea can be extended to other operators with fractional behaviours and to the spectral content of these operators as demonstrated in the following paragraph.

3. Probabilistic Interpretation Based on the Spectral Content of Operators with Fractional Behaviours

The other probabilistic interpretation proposed can be revealed from the impulse response of the fractional operators considered. It is assumed that $H(s)$ denotes the transfer function of these operators. Using inverse Laplace transform, the corresponding impulse response defined by:

$$h(t) = \mathcal{L}^{-1}\{H(s)\} = \frac{1}{2j\pi}\int_{c-j\infty}^{c+j\infty} H(s)e^{st}ds. \quad (3)$$

A Bromwich-Wagner path Γ is used to compute the integral in relation (3). The value of c is then taken greater than the abscissa of the singular point of $H(s)$. For instance, for the transfer function $H(s) = s^{-\nu}$, $\nu \in [0,\ 1]$, and for $t > 0$, the considered path Γ is represented by Figure 1. As the function $s^{\nu} = 0$ is not defined on $]-\infty, 0]$, the path Γ avoid the negative axis and goes around the point $s = 0$. It is shown by Figure 1.

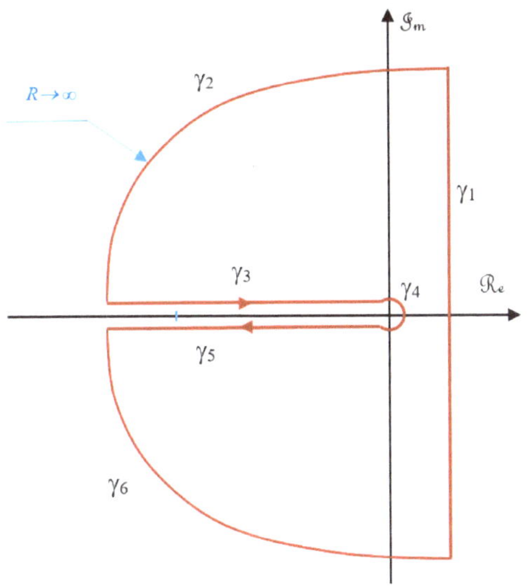

Figure 1. The Γ path used for the computation of the impulse response $h(t)$.

Using the residue theorem, relation (3) can be rewritten as:

$$h(t) = -\frac{1}{2\pi j}\int_{\Gamma-\gamma_1} H(s)e^{ts}ds + \sum_{\substack{poles \\ in\ \Gamma}} \text{Res}\left[H(s)e^{ts}\right], \quad (4)$$

with, in the case of a simple pole p for $H(s)e^{ts}$

$$Res_p\left[H(s)e^{ts}\right] = \lim_{\xi \to p}(\xi - p)H(\xi)e^{t\xi}. \quad (5)$$

The computation of the poles of $H(s)$ is required by relation (4), relation that also shows that that the impulse (3) is made of two parts:

$$h(t) = h_p(t) + h_d(t). \quad (6)$$

The function $h_p(t)$ is computed from the poles of $H(s)$ (residues of the Cauchy method). The function $h_d(t)$ is defined by [14]:

$$h_d(t) = \int_0^{\infty} \mu(z)e^{-tz}dz. \quad (7)$$

In relation (7), the function $\mu(x)$ is defined by [15]:

$$\mu(z) = \frac{1}{2i\pi}\left[H\left((-z)^{-}\right) - H\left((-z)^{+}\right)\right] \text{ with} \mu(z) \in \mathbb{R}. \tag{8}$$

Table 1 provides several impulse responses of fractional transfer functions (with no pole and thus $h_p(t) = 0$ for most) and computed using this method (demonstrations can be found in [16]).

Table 1. Table of impulse responses (inverse Laplace transforms) of some transfer functions with fractional behaviours.

$H_i(s)$	$h_i(t)$
$\frac{\sum_{l=0}^{L} b_l s^{\beta_l}}{\sum_{k=0}^{K} a_k s^{\alpha_k}}$	$\sum_{i=1}^{r}\sum_{j=1}^{n_i} r_{ij} Y_j(t) e^{s_i t} + \frac{1}{\pi}\int_0^{+\infty} \frac{\sum_{k=0}^{K}\sum_{l=0}^{L} a_k b_l \sin((\alpha_k-\beta_l)\pi) x^{\alpha_k+\beta_l}}{\sum_{k=0}^{K} a_k^2 x^{2\alpha_k} + \sum_{0\le k\le l\le K} a_k a_l \cos((\alpha_k-\alpha_l)\pi) x^{\alpha_k+\alpha_l}} e^{-xt} dx$ From [15] with demonstration in [17]. s_i are the poles.
$\frac{1}{s^{\nu}}$	$\frac{\sin(\nu\pi)}{\pi}\int_0^{+\infty} \frac{1}{x^{\nu}} e^{-xt} dx$
$\frac{1}{\left(\frac{s}{\omega_l}+1\right)^{\nu}}$	$\frac{\sin(\nu\pi)}{\pi}\int_{\omega_l}^{+\infty} \frac{{\omega_l}^{\nu}}{(x-\omega_l)^{\nu}} e^{-xt} dx$
$\frac{\left(\frac{s}{\omega_h}+1\right)^{\nu-1}}{\left(\frac{s}{\omega_l}+1\right)^{\nu}}$	$\frac{\sin(\nu\pi)}{\pi}\frac{{\omega_l}^{\nu}}{{\omega_h}^{\nu-1}}\int_{\omega_l}^{\omega_h} \frac{(\omega_h-x)^{\nu-1}}{(x-\omega_l)^{\nu}} e^{-xt} dx$
$\frac{\left(\frac{s}{\omega_h}+1\right)^{\nu}}{\left(\frac{s}{\omega_l}+1\right)^{\nu}}$	$\left(\frac{\omega_l}{\omega_h}\right)^{\nu}\left(\delta(t) + \frac{\sin(\nu\pi)}{\pi}\int_{\omega_b}^{\omega_h} \frac{(\omega_h-x)^{\nu}}{(x-\omega_l)^{\nu}(s+x)} dx\right)$ $\delta(t)$: Dirac impulse
$\frac{1}{s}\frac{\left(\frac{s}{\omega_l}+1\right)^{1-\nu}}{\left(\frac{s}{\omega_h}+1\right)^{1-\nu}}$	$\left(\frac{\omega_h}{\omega_l}\right)^{1-\nu}\left(H_e(t) + \frac{\sin((1-\nu)\pi)}{\pi}\int_{\omega_l}^{\omega_h} \frac{(x-\omega_l)^{1-\nu}}{x(\omega_h-x)^{1-\nu}} e^{-xt} dx\right)$ $H_e(t)$: Heaviside step function
$\frac{e^{-z\sqrt{as+b}}}{\sqrt{as+b}}$ $a > 0, \quad b > 0, \quad z > 0$	$\frac{1}{\pi}\int_{\frac{b}{a}}^{+\infty} \frac{\cos\left(z\sqrt{ax-b}\right)}{\sqrt{ax-b}} e^{-xt} dx$

The Laplace transform of function $h_d(t)$ is given by:

$$h_d(s) = \int_0^{\infty} \frac{\mu(z)}{s+z} dz. \tag{9}$$

For a practical implementation, relation (9) must be truncated and discretized. But this implementation method is not very efficient. A large number of terms are needed in the sum resulting from the integral discretization. A solution consists in applying a change of variable to (9) before discretization [18]. Let $z = e^x$ and thus $dz = e^x dx$ this change of variable. Relation (9) becomes:

$$h_d(s) = \int_{-\infty}^{\infty} \frac{\mu(e^x)}{\frac{s}{e^x}+1} dx. \tag{10}$$

Relation (10) can thus be viewed as the expectation of a random variable X, such that $P\left(X = \frac{1}{\frac{s}{e^x}+1}\right) = \mu(e^x)$, $x \in \mathbb{R}$. Physically, if a real system is characterized by a fractional model such as those in Table 1 (or others), relation (10) implicitly provides information on the probability of encountering the time constant $\frac{1}{e^x}$ in the model.

To illustrate this probabilistic interpretation, it is now proposed to use relation (10) to design a filter of the form

$$F_i(s) = \frac{1}{M}\sum_{k=1}^{M}\frac{1}{\frac{s}{e^x}+1} \tag{11}$$

in which the corner frequencies e^{x_k} (or the time constant $\frac{1}{e^{x_k}}$) are randomly selected with a probability density $\mu(e^x)$ where

$$\mu(e^x) = \frac{\sin(\nu\pi)}{\pi}\frac{\frac{e^{\nu x}}{a^\nu}}{1+2\frac{e^{\nu x}}{a^\nu}\cos(\nu\pi)+\frac{e^{2\nu x}}{a^{2\nu}}} \tag{12}$$

i.e., the function $\mu(e^x)$ involved in the impulse response of the transfer function (see line 1 of Table 1)

$$F_f(s) = \frac{1}{\left(\frac{s}{a}\right)^\nu+1} \quad a>0 \;\; 0<\nu<1. \tag{13}$$

The function $\mu(e^x)$ given by relation (12) is represented by Figure 2 for various values of ν and $a = 10$. It is interesting to note that all these curves resemble a normal law centred on the value $log(a) = log(10) \approx 2.3$.

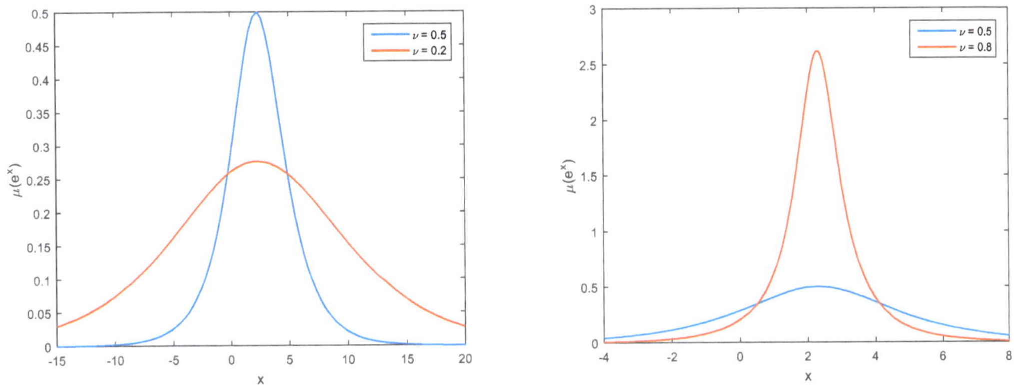

Figure 2. Function $\mu(e^x)$ of relation (13) for various values of ν.

The cumulative distribution function

$$F_c(y) = \int_{-\infty}^{y}\mu(e^x)dx \tag{14}$$

(evaluated numerically) is represented by Figure 3 for several values of ν and $a = 10$.

To obtain the corner frequencies e^{x_k} of relation (11), the following algorithm is used:

- *generate a random number u_k from the standard uniform distribution in the interval [0, 1];*
- *compute $x_k = F_c^{-1}(u_k)$.*

In the case of $\nu = 0.4$ (in this case $h_p(t) = 0$) and $a = 10$, the cumulative distribution function provided by relation (14) and the histogram of x_k generated by the algorithm above are represented by Figure 4. These figures are obtained after 50×10^6 random numbers u_k have been generated. A comparison of the gain and phase diagrams of the transfer functions $F_i(s)$ and $F_f(s)$ is proposed in Figure 5. This comparaison validates the probabilistic interpretation done relating to time constants.

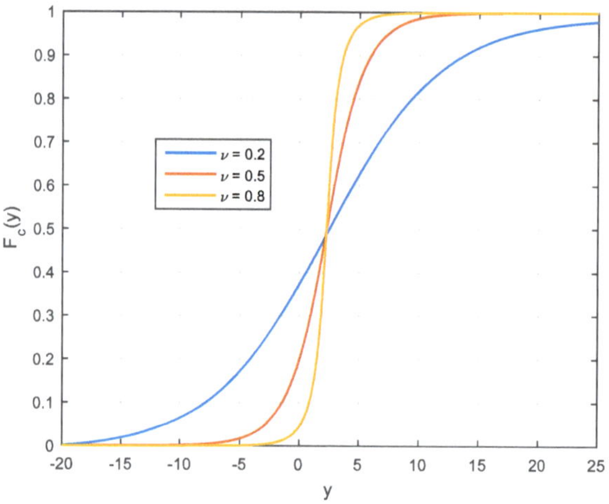

Figure 3. Function $F_c(y)$ of relation (14) for various values of ν and $a = 10$.

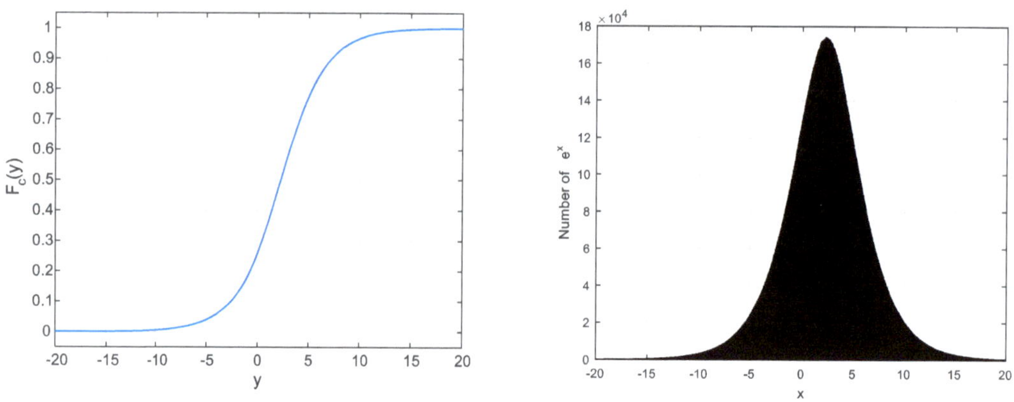

Figure 4. Cumulative distribution function provided by relation (14) (**left**) and the histogram of x_k obtained (**right**) in the case of $\nu = 0.4$ and $a = 10$.

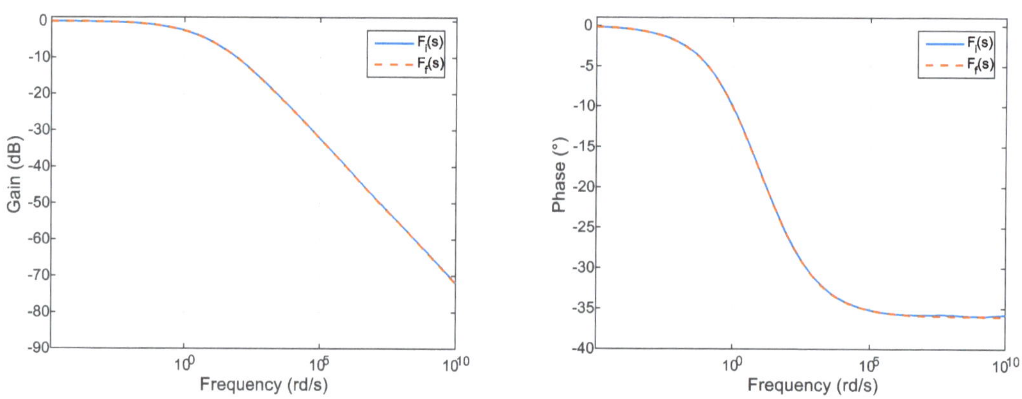

Figure 5. Comparison of the gain (**left**) and phase (**right**) diagrams of the transfer functions $F_i(s)$ and $F_f(s)$.

The probabilistic interpretation of fractional behaviors proposed in this section leads to the following remarks.

Modeling a real system by a fractional model amounts to considering that this system has a distribution of time constants (or corner frequencies) whose probability of occurrence is governed by the function $\mu(e^x)$.

This interpretation highlights that some fractional models, in particular those which, in the Laplace domain, involve fractional powers of the Laplace variable (s^ν), induce the probability, admittedly low but not zero, of the presence in the model of infinitely large and infinitely small time constants, which is physically not realistic [19].

With a view to proposing more physically realistic models than the fractional models of the type mentioned above, it would be possible to work directly on a model having a substantially equal number of parameters and of the type

$$h_d(s) = \int_{x_l}^{x_h} \frac{\mu(e^x)}{\frac{s}{e^x}+1} dx. \tag{15}$$

The characterization of a real physical phenomenon using such a model would then consist in identifying the probability density $\mu(.)$ after having described it by an appropriate function and the parameters x_l and x_h. These two parameters can be fixed a priori from information on the sampling frequency of the measured signals and on the bandwidth (response time) of the physical phenomenon modeled. The author will describe this identification method in more detail in another paper.

This interpretation in terms of time constants distribution is interesting but does not necessarily reflect the physics of the modelled system. This is particularly the case of phenomena such as diffusion, adsorption, and aggregation, in which entities (atoms, molecules, people, etc.) evolve in a more or less constrained spatial domain.

In order to obtain a description that is even closer to the physical reality of systems with fractional behaviours, the following interpretation in terms of delays is proposed.

4. Probabilistic Interpretation Based on the Delay Approximation of Operators with Fractional Behaviours

The previous probabilistic interpretation based on time constants distribution does not well describe what happens in phenomena such as adsorption, which is now described.

4.1. Description of Adsorption Phenomena

Adsorption is the phenomenon which consists of the accumulation of a substance at the interface between two phases (gas-solid, gas-liquid, liquid-solid, liquid-liquid, solid-solid). It has its origin in the intermolecular forces of attraction, of varied nature and intensity, which are responsible for the cohesion of the condensed, liquid or solid phases [20,21]. Adsorption on solids is frequently used for the separation and purification of gases or the separation of solutes in liquids [22]. Adsorption is used in many industrial and academic applications [23,24] and in particular as water purification [22] and sensors [25,26].

Random Sequential Adsorption of RSA, is an idealized stochastic process often encountered in the literature to study chemical adsorption phenomenon. In the 1D case, it is first Flory [27] and Rényi [28] (car-parking problem) who studied RSA. The 2D case was also investigated in [29–31]. These latest studies focus on the final value of particles concentration and also on the kinetic behaviour of RSA.

To define RSA process, let consider a square plane substrate of edge length L (thus leading to a surface L^2). Disk particles are supposed by this substrate and disk radius is supposed small in relation to substrate size $(R \ll L)$. During RSA process, particles fall sequentially onto the substrate. At each process iteration, the position of the fall is fixed randomly with a uniform distribution. A falling particle remains attached to the surface only if its surface covers a still free surface of the substrate, otherwise, the adsorption attempt fails. At the beginning of the process (time t_0), the substrate surface is supposed empty.

In the case $R = 0.5$ and $L = 100R$, the density $\theta(t)$ of the covered surface is shown in Figure 6 as a function of trials (number of discs that fell on the surface) which is denoted t in the sequel. The fluctuations of the value of $\theta(t)$ between several simulations are represented by error bars. These fluctuations result in the randomness of the process.

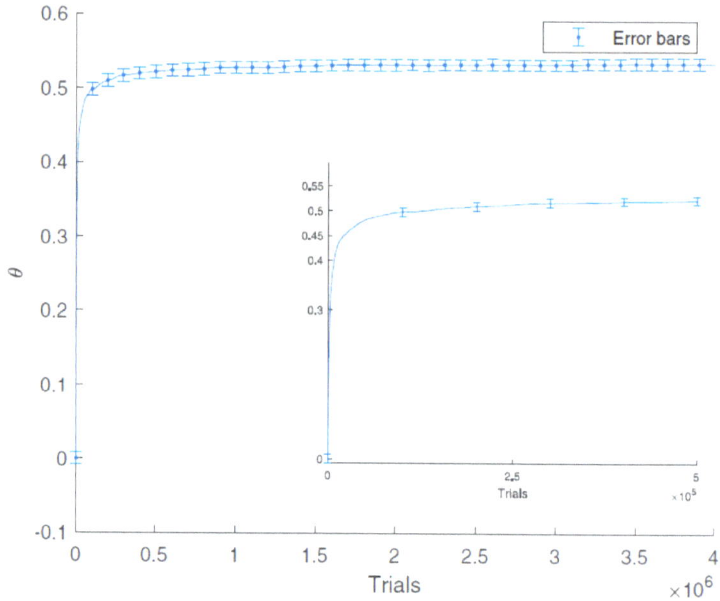

Figure 6. Density of the occupied area as a function of trials.

For high coverage regimes, it is suggested in the literature [31–35] that the covered surface can be described by a power law:

$$\theta_\infty - \theta(t) \sim t^{1/2}, \tag{16}$$

in which θ_∞ is the value of $\theta(t)$ when t goes to infinity.

The RSA process and adsorption thus result in the stochastic behaviour of agents (atoms, molecules, people, etc.) in a constrained geometry, and the distribution of time constants struggles to explain the overall behaviour of all these agents. A time constant is indeed linked to a continuous time process whereas the placement of a disk in the RSA process is sequential.

4.2. Time Delay Distribution for Fitting Fractional Behaviours

José Tenreiro Machado's probabilistic interpretation published in [1] was based on the sample distribution given in the Grünwald-Letnikov definition of a fractional derivative. But one can also see in this definition an interpretation which relates to a distribution of delay induced by the operator s^ν and which can be generalized to other fractional operators and behaviors. Indeed, Laplace transform applied to relation (1) leads to:

$$\mathcal{L}\{D^\nu[x(t)]\} = s^\nu x(s) = x(s)\lim_{h\to 0}\left[\frac{1}{h^\nu}\sum_{k=0}^{\infty}\gamma(\nu,k)e^{-khs}\right] \tag{17}$$

and thus

$$s^\nu = \lim_{h\to 0}\left[\frac{1}{h^\nu}\sum_{k=0}^{\infty}\gamma(\nu,k)e^{-khs}\right]. \tag{18}$$

Using relation (17) and regarding the definition of the operator s^{ν}, from a probability theory point of view it can be said that the delay operators e^{-khs} with $k \in [1,\infty[$ are weighted with the probability $\gamma(\nu,k)$ and the expression $-\sum_{k=0}^{\infty}\gamma(\nu,k)e^{-khs}$ can be viewed as the expected value of the random variable X, $E(X)$, such that $P\left(X = e^{-khs}\right) = |\gamma(\nu,k)|$, $k \in [1,\infty[$.

This idea can of course be extended to many other fractional operators. This is now highlighted graphically with the fractional integration operator $s^{-\nu}$. To explain how such a behaviour can be approximated by a delay distribution, the step response of the transfer function

$$H(s) = s^{-\nu} \text{ with } \nu = \frac{1}{2} \tag{19}$$

is analysed. It is defined by

$$y(t) = \frac{t^{1/2}}{\Gamma(3/2)} \tag{20}$$

and is represented by Figure 7. This figure shows that this time response can be approximated by a distribution of delayed steps with the same magnitude M, thus permitting the following approximation in the Laplace domain:

$$Y(s) = \mathcal{L}\{y(t)\} \approx \frac{M}{s}\sum_{k=1}^{N} e^{-t_k s} \tag{21}$$

and thus

$$s^{-\nu} \approx M\sum_{k=0}^{N} e^{-t_k s} \text{ with } (t_k)^{1/2} = \Gamma(3/2)kM. \tag{22}$$

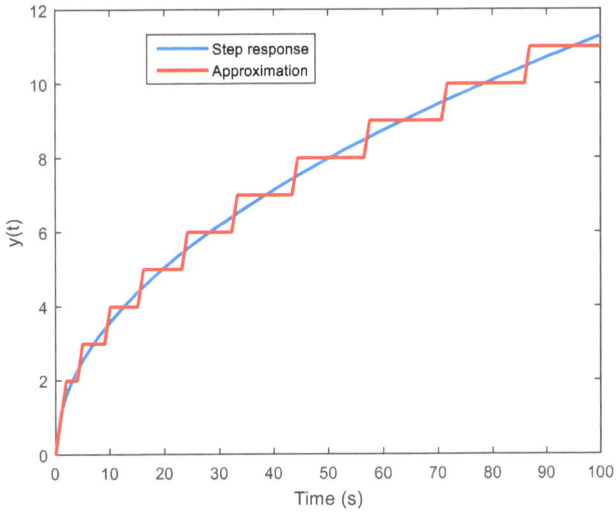

Figure 7. Approximation of a fractional integrator step response by a distribution of delayed steps.

The frequency response of the approximation given by relation (21) is represented by Figure 8. This figure shows that the gain diagram decreases with a slope equal to -20ν dB per decade and a constant phase equal to -90ν degrees, like the frequency response of a fractional integrator.

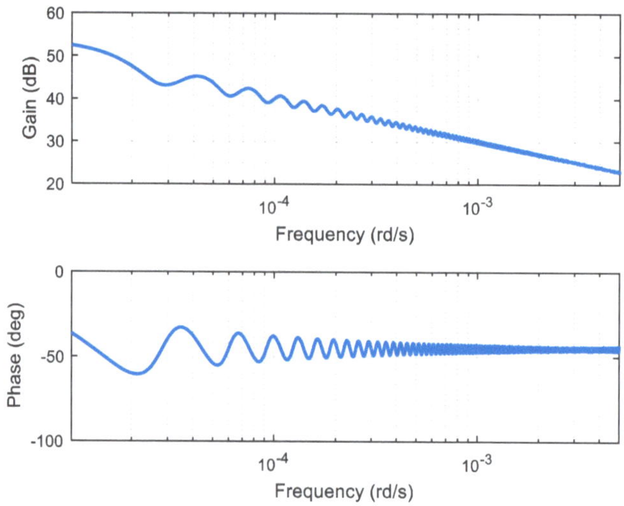

Figure 8. Frequency response of the approximation of a fractional integrator time response by a distribution of delayed steps.

This approximation is of course not unique and an interesting one can be obtained by considering the logarithm of the fractional integrator step response as a function of the logarithm of time, namely a straight line whose slope is ν. As shown by Figure 9, this straight line can be approximated by a distribution of steps of magnitude A that occurs at time t_k that are linked by the recurrence equation involving a constant factor r:

$$log(t_{k+1}) - log(t_k) = r \text{ and thus } log(t_k) = log(t_0) + kr \tag{23}$$

or

$$t_k = t_0 e^{kr}. \tag{24}$$

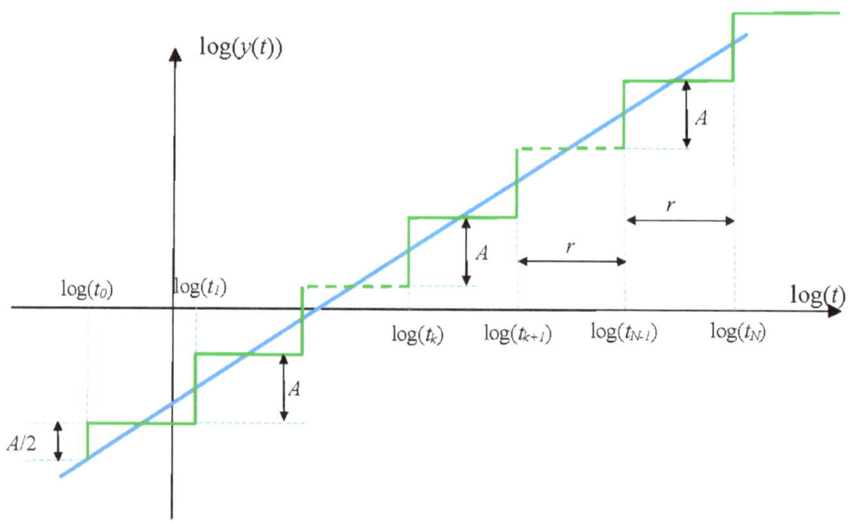

Figure 9. Approximation of a fractional integrator time response by a recursive distribution of delayed steps.

In such a situation, the fractional integrator step response admits in the Laplace domain the approximation

$$Y(s) = \mathcal{L}\{y(t)\} \approx \frac{1}{s}\sum_{k=0}^{N} M_k e^{-t_k s} \tag{25}$$

where the magnitudes M_k are defined by:

$$M_0 = \frac{1}{\Gamma(3/2)} t_0^{\nu} \left(\frac{t_1}{t_0}\right)^{\nu/2} \tag{26}$$

$$M_k = M_{k-1}\left(\frac{t_k}{t_{k-1}}\right)^{\nu} = M_{k-1}e^{\nu r} = M_0 e^{k\nu r}. \tag{27}$$

An approximation of a fractional integrator time response can thus be obtained using a recursive distribution of time delays:

$$s^{-\nu} \approx \sum_{k=1}^{N} M_k e^{-t_k s}. \tag{28}$$

The resulting step response approximation in the time domain is shown by Figure 10.

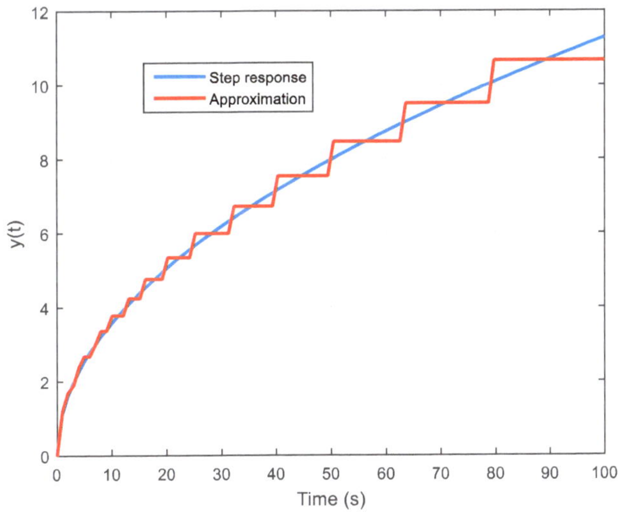

Figure 10. Approximation of a fractional integrator step response by a recursive distribution of delayed steps (relation (28)).

However, in the previous two approximations, the lag between two delays is not constant as in relation (17). To satisfy this constraint the following approximation can be used

$$s^{-\nu} \approx \sum_{k=1}^{N} M_k e^{-kT_0 s} \tag{29}$$

with

$$M_k = \frac{y((k+1)T_0) + y(kT_0)}{2}. \tag{30}$$

The accuracy of this approximation with a constant gap T_0 between two consecutive delays is represented by Figure 11 in the time domain (with $T_0 = 4$ s, $\nu = 0.5$) and in the frequency domain (with $T_0 = 0.02$ s, $\nu = 0.5$).

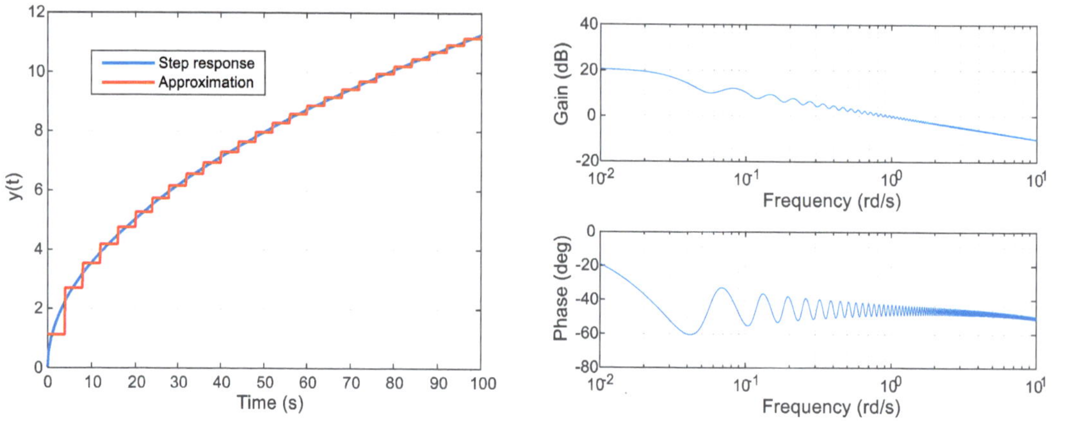

Figure 11. Comparison of the fractional integrator step response and the approximation (29) (**left**) and frequency response of the approximation (29) (**right**).

This approximation method and thus the associated probabilistic interpretation can be extended to numerous fractional transfer functions and behaviours. We have for instance

$$\frac{K_f}{1+\frac{s^\nu}{\omega_l}} \approx \sum_{k=1}^{N} M_k e^{-kT_0 s}. \tag{31}$$

A comparison of the filter step response and its approximation is done in Figure 12 with $N = 10000$, $T_0 = 0.5$ s $K_f = 1$, $\omega_l = 0.1$ rd/s and $\nu = 0.6$. Note that the coefficients M_k can be computed from the filter step response using relation (30) in which $y(t)$ denotes the filter step response. Thus, this approximation method can be applied directly to measures resulting from the step response of a real system exhibiting a fractional behaviour. It is interesting to see that this kind of approximation is close to the IIR filter-based approximations proposed in the literature for fractional transfer functions [36–38].

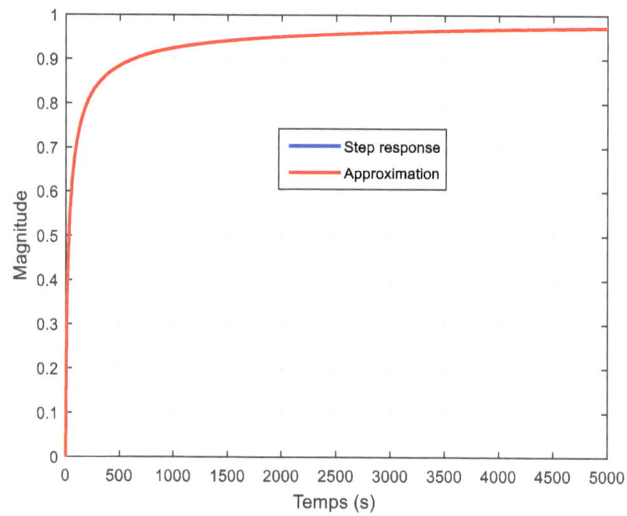

Figure 12. Comparison of the step response of the fractional filter in relation (14) and its approximation.

In relation (31), coefficients M_k/K_f can be viewed as the probability to have a delay of duration kT_0 in the fractional behaviour studied as

$$\sum_{k=0}^{N} \frac{M_k}{K_f} = 1. \tag{32}$$

This probability distribution of M_k/K_f as a function of k ($k \in [1, 100]$) for the transfer function of relation (31) is represented by Figure 13, again with $N = 10{,}000$, $T_0 = 0.5$ s, $K_f = 1$, $\omega_l = 0.1$ rd/s and $\nu = 0.6$.

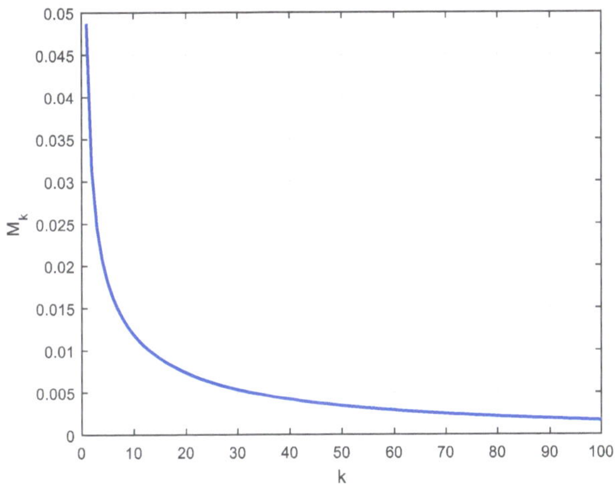

Figure 13. Distribution of coefficients M_k/K_f in relation (31) for $k \in [1, 100]$.

It must be noted that the idea of modeling fractional behaviors by means of a delay distribution was also used in another form in [39].

4.3. Adsorption Phenomena to Illustrate the Interest of this Probabilistic Delay Interpretation

The adsorption phenomenon and the RSA process are again considered. The algorithm described in Section 4.1 for the RSA process is used to generate data with disk particles of radius $R = 0.5$ that fall on a square with an edge length $L = 50$. The response obtained in terms of surface coverage θ as a function of trials is shown by Figure 14, and will be denoted $\theta(t)$ in the sequel. This figure also shows the placement of the disks at the end of the process.

The response $\theta(t)$ can be approximated using a distribution of delay as in relation (31). The delay distribution is similar to the one presented by Figure 13. On the other hand, from the response $\theta(t)$, it is possible to compute the number of disks N_k that find a place on the square in the time interval $[(k-1)T_0, kT_0[$. The numbers obtained are divided by ε_∞, the value of $\theta(t)$ when t goes to infinity, which is close to 0.547 according to [31–33]. The resulting N_k/ε_∞ are then represented by Figure 15 as a function of k for $k \in [1, 100]$.

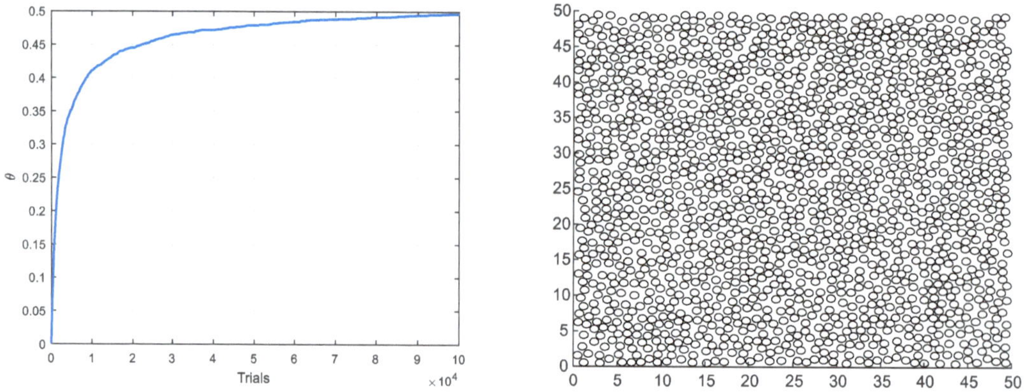

Figure 14. Surface covered as a function of trials during the RSA process (**left**) and disk placement at the end of the process (**right**).

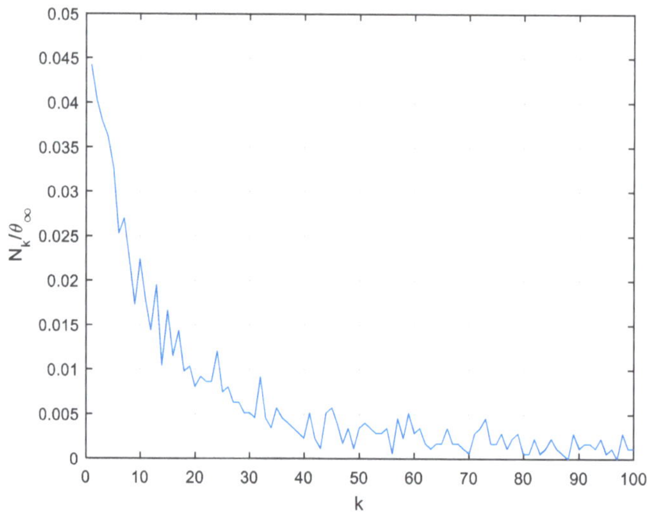

Figure 15. Representation of N_k/ε_∞ for $k \in [1, 100]$.

One can note a very large similarity between Figures 13 and 15, and thus between the values of M_k/K_f and N_k/ε_∞. Consequently, the third probabilistic interpretation in terms of delay distribution leading to the expansion (31) has a physical meaning. The probability to find a time delay with duration kT_0 in the fractional behaviour produced by the RSA process is also the probability that N_k/ε_∞ disks find a place in the time interval $[(k-1)T_0, kT_0[$.

4.4. Another Example of Possible Physical Interpretation

The third statistical interpretation detailed in Section 4.2 which describes the probability that a delay is induced by a system that produces a given fractional behaviour may have other physical interpretations. This is the case for diffusion. It is well known that diffusion can be physically described through particle random walk [40]. It is also well known that diffusion produces fractional behaviours of order $\frac{1}{2}$ or different from $\frac{1}{2}$ in complex media such as fractal media [41]. But an interpretation based on delay distribution is also possible.

This interpretation is represented by Figure 16. The medium is assumed to be constituted of channels of various lengths. Before the water hits the medium, the channels are assumed to be empty of fluid. When the fluid pressure $P(t)$ appears at the input side, it creates a flow inside the channels such that the total flow at the output side denoted $Q(t)$ is the sum of the flow produced by each channel, denoted $Q_k(t)$:

$$Q(t) = \sum_k Q_k(t). \tag{33}$$

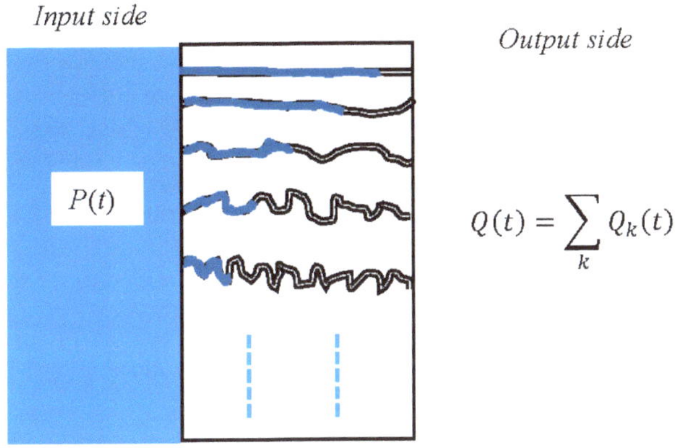

Figure 16. Interpretation in terms of time delay distribution of the fractional behaviour produced by diffusion phenomena.

The flow $Q_k(t)$ depends on the pressure $P(t)$ and on the inverse of the hydraulic resistance R_k of each channel. Moreover, due to the difference in the channel length, each flow $Q_k(t)$ reaches the output side with a time delay T_k depending on the channel length, thus leading for each $Q_k(t)$ to the relation:

$$Q_k(t) = \frac{1}{R_k} P(t - T_k) \tag{34}$$

and thus for the total flux:

$$Q(t) = \sum_k \frac{1}{R_k} P(t - T_k). \tag{35}$$

If Q_∞ denotes the steady state value of the flow $Q(t)$ when $P(t)$ is a unit step, then with such a modelling approach, it can be said that the coefficients $1/(Q_\infty R_k)$ can be viewed as the probability to have a delay of duration T_k in the fractional behaviour produced by the diffusion phenomena. More physically, $1/(Q_\infty R_k)$ can be connected to a distribution of channel length and to the probability to find the corresponding channel in the studied system.

5. Conclusions

This paper proposes extensions of a probabilistic interpretation of fractional derivative operator that can be found in the literature [1]. The proposed interpretations are extensions because they concern other fractional operators and more generally fractional behaviours, and also by the nature of the random variables involved in these interpretations.

A first interpretation is derived from the impulse response of various fractional order transfer functions. These impulse responses are fitted with a distribution of time constants weighted by a function that can be interpreted as the probability to find these time constants in the system. However, this interpretation only gives a macroscopic view of what happens in a real system producing a fractional behaviour. Many systems that exhibit fractional behaviours are stochastic systems in that they are based on random and sequential kinetics of a multitude of agents in a constrained space. This is the case of adsorption or diffusion for instance.

A second interpretation is thus proposed. It is based on the approximation of a fractional behaviour using a distribution of time delays. The resulting probabilistic interpretation provides information on the probability of a given time delay to be present in a system. This interpretation is particularly interesting because it also allows a physical interpretation of the phenomena that take place in the system having a fractional behaviour. This is highlighted with the adsorption phenomenon that can be approximated by the RSA (Random Sequential Adsorption) process. RSA is a stochastic process in which particles sequentially incide a substrate at uniformly randomly chosen surface positions. A particle remains on the surface only if the target site is empty. As the substrate fills up, one has to wait longer and longer for a new disk to find its place. It is this notion of delay that is found in the second probabilistic interpretation proposed. This interpretation is also used to propose a physical interpretation of diffusion.

But this work is also a tribute to our colleague Professor Tenreiro Machado who was the first to propose in the literature a probabilistic interpretation of fractional derivative operator. Professor Tenreiro Machado, Dear José, we wish you were here, with us, to discuss again these probabilistic interpretations.

Funding: This research received no external funding.

Conflicts of Interest: The author declares no conflict of interest.

References

1. Tenreiro Machado, J.A. A probabilistic Interpretation of the Fractional-Order differentiation. *J. Fract. Calc. Appl. Anal.* **2003**, *6*, 73–80.
2. Cole, K.V.S.; Cole, R.H. Dispersion and Absorption in Dielectrics-I Alternating Current Characteristics. *J. Chem. Phys.* **1941**, *9*, 341–352. [CrossRef]
3. Manabe, S. The non-integer Integral and its Application to control systems. *J. Inst. Electr. Eng. Jpn.* **1961**, *80*, 589–597.
4. Nigmatullin, R.R. A fractional integral and its physical interpretation. *Theor. Math. Phys.* **1992**, *90*, 242–251. [CrossRef]
5. Rutman, R.S. On physical interpretations of fractional integration and differentiation. *Theor. Math. Phys.* **1995**, *105*, 393–404. [CrossRef]
6. Ben Adda, F. Geometric interpretation of the fractional derivative. *J. Fract. Calc.* **1997**, *11*, 21–52.
7. Gorenflo, R. Afterthoughts on interpretation of fractional derivatives and integrals. In *Transform Methods and Special Functions, Varna'96*; Rusev, P., Dimovski, I., Kiryakova, V., Eds.; Institute of Mathematics and Informatics, Bulgarian Academy of Sciences: Sofia, Bulgaria, 1998.
8. Mainardi, F. Considerations on fractional calculus: Interpretations and applications. In *Transform Methods and Special Functions, Varna'96*; Rusev, P., Dimovski, I., Kiryakova, V., Eds.; Institute of Mathematics and Informatics, Bulgarian Academy of Sciences: Sofia, Bulgaria, 1998.
9. Podlubny, I. Geometric and physical interpretation of fractional integration and fractional differentiation. *J. Fract. Calc. Appl. Anal.* **2002**, *5*, 357–366.
10. Dokoumetzidis, A.; Magin, R.; Macheras, P. A commentary on fractionalization of multi-compartmental models. *Pharm. Pharm.* **2010**, *37*, 203–207. [CrossRef]
11. Sabatier, J.; Farges, C.; Tartaglione, V. Some alternative solutions to fractional models for modelling long memory behaviors. *Mathematics* **2020**, *8*, 196. [CrossRef]
12. Sabatier, J. Modelling Fractional Behaviours Without Fractional Models. *Front. Control. Eng.* **2021**, *2*, 716110. [CrossRef]
13. Sabatier, J.; Farges, C.; Tartaglione, V. Fractional Behaviours Modelling: Analysis and Application of Several Unusual Tools. In *Intelligent Systems, Control and Automation: Science and Engineering (ISCA, Volume 101) Series*; Springer: Berlin/Heidelberg, Germany, 2022.
14. Montseny, G. Diffusive representation of pseudo-differential time-operators. *ESAIM Fract. Differ. Sys-Tems Models Methods Appl.* **1998**, *5*, 159–175. [CrossRef]
15. Matignon, D. Stability properties for generalized fractional differential systems. *ESAIM Proc.* **1998**, *5*, 145–158. [CrossRef]
16. Sabatier, J. Introduction of new kernels and new models to solve the drawbacks of fractional integration/differentiation operators and classical fractional-order models. In *Emerging Methodologies and Applications in Modelling Fractional Order Systems, Chapter Fourteen*; Ahmed, G.R., Farooq, A.K., Lobna, A.S., Eds.; Academic Press: Cambridge, MA, USA, 2022; pp. 551–586.
17. Sabatier, J.; Farges, C.; Merveillaut, M.; Fenetau, L. On observability and pseudo state estimation of fractional order systems. *Eur. J. Control* **2012**, *3*, 260–271. [CrossRef]

18. Sabatier, J. Beyond the particular case of circuits with geometrically distributed components for approximation of fractional order models: Application to a new class of model for power law type long memory behaviour modelling. *J. Adv. Res.* **2020**, *25*, 243–255. [CrossRef]
19. Sabatier, J. Fractional Order Models Are Doubly Infinite Dimensional Models and thus of Infinite Memory: Consequences on Initialization and Some Solutions. *Symmetry* **2021**, *13*, 1099. [CrossRef]
20. Brunauer, S.; Deming, L.S.; Deming, W.E.; Teller, E. On a Theory of the van der Waals adsorption of gases. *J. Am. Chem. Sciety* **1940**, *62*, 1723–1732. [CrossRef]
21. Rouquerol, F.; Rouquerol, J.; Sing, K.; Llewellyn, P.; Maurin, G. *Adsorption by Powders and Porous Solids: Principles, Methodology and Applications*, 2nd ed.; Academic: Oxford, UK, 2014.
22. Bonilla-Petriciolet, A.; Mendoza-Castillo, D.I.; Reynel-Ávila, H.E. *Adsorption Processes for Water Treatment and Purification*; Springer: Berlin, Germany, 2017.
23. Czelej, K.; Cwieka, K.; Colmenares, J.C.; Kurzydłowski, K. Insight on the interaction of methanol-selective oxidation intermediates with Au- or/and Pd-containing monometallic and bimetallic. *Langmuir* **2016**, *32*, 7493–7502. [CrossRef]
24. Czelej, K.; Cwieka, K.; Kurzydłowski, K. CO2 stability on the Ni low-index surfaces: Van der Waals corrected DFT analysis. *Catal. Commun.* **2016**, *80*, 33–38. [CrossRef]
25. Halil, H.; Menini, P.; Aubert, H. Novel microwave gas sensor using dielectric resonator with SnO2 sensitive layer. *Procedia Chem.* **2009**, *1*, 935–938. [CrossRef]
26. Nikolaou, I.; Hallil, H.; Conédéra, V.; Deligeorgis, G.; Dejous, C.; Rebiere, D. Inkjet-printed graphene oxide thin layers on love wave devices for humidity and vapor detection. *IEEE Sens. J.* **2016**, *16*, 7620–7627. [CrossRef]
27. Flory, P.J. Intramolecular reaction betweenNeighboring Substituents of Vinyl Polymers. *J. Am. Chem. Soc.* **1939**, *61*, 1518–1521. [CrossRef]
28. Rényi, A. *On a One-Dimensional Problem Concerning Random Space Filling*; Publication of the Mathematical Institute of the Hungarian Academia of Sciences: Budapest, Hungary, 1958; Volume 3, p. 109.
29. Viot, P.; Tarjus, G.; Ricci, S.; Talbot, J. Random sequential adsorption of anisotropic particles. I. jamming limit and asymptotic behavior. *J. Chem. Phys.* **1992**, *97*, 5212–5218. [CrossRef]
30. Hinrichsen, E.; Feder, J.; Jøssang, T. Geometry of random sequential adsorption. *J. Stat. Phys.* **1986**, *44*, 793–827. [CrossRef]
31. Feder, J.; Giaever, I. Adsorption of ferritin. *J. Colloid Interface Sci.* **1980**, *78*, 144–154. [CrossRef]
32. Ciesla, M.; Ziff, R.M. Boundary conditions in random sequential adsorption. *J. Stat. Mech. Theory Exp.* **2018**, *2018*, 043302. [CrossRef]
33. Zhang, G.; Torquato, S. Precise algorithm to generate random sequential addition of hard hyperspheres at saturation. *Phys. Rev. E* **2013**, *88*, 053312. [CrossRef]
34. Tartaglione, V.; Farges, C.; Sabatier, J. Nonlinear dynamical modeling of adsorption and desorption processes with power-law kinetics: Application to CO2 capture. *Phys. Rev. E* **2020**, *102*, 052102. [CrossRef]
35. Tartaglione, V.; Sabatier, J.; Farges, C. Adsorption on Fractal Surfaces: A Non Linear Modeling Approach of a Fractional Behavior. *Fractal Anf Fract.* **2021**, *5*, 65. [CrossRef]
36. Jiang, C.X.; Adams, J.L.; Carletta, J.E.; Hartley, T.T. Hardware implementation of fractional-order systems as infinite impulse response filters. *IFAC Proc. Vol.* **2006**, *39*, 408–413. [CrossRef]
37. Barbosa, R.S.; Machado, J.A.T.; Jesus, I.S. A General Discretization Scheme for the Design of IIR Fractional Filters. In Proceedings of the Seventh International Conference on Intelligent Systems Design and Applications (ISDA 2007), Rio de Janeiro, Brazil, 20–24 October 2007; pp. 665–670.
38. Romero, M.; de Madrid, A.P.; Mañoso, C.; Vinagre, B.M. IIR approximations to the fractional differentiator/integrator using Chebyshev polynomials theory. *ISA Trans.* **2013**, *52*, 461–468. [CrossRef]
39. Sabatier, J. Power Law Type Long Memory Behaviors Modeled with Distributed Time Delay Systems. *Fractal Fract.* **2020**, *4*, 1. [CrossRef]
40. Masuda, N.; Porter, M.A.; Lambiotte, R. Random walks and diffusion on networks. *Phys. Rep.* **2017**, *716–717*, 1–58. [CrossRef]
41. Zhang, B.; Yu, B.; Wang, H.; Yun, M. A fractal analysis of permeability for power-law fluids in porous media. *Fractals* **2006**, *14*, 171–177. [CrossRef]

MDPI AG
Grosspeteranlage 5
4052 Basel
Switzerland
Tel.: +41 61 683 77 34

Mathematics Editorial Office
E-mail: mathematics@mdpi.com
www.mdpi.com/journal/mathematics

www.ingramcontent.com/pod-product-compliance
Lightning Source LLC
Chambersburg PA
CBHW041359210926
51850CB00013B/3122